# Women Filmmakers
# in Early Hollywood

STUDIES IN INDUSTRY AND SOCIETY
Philip B. Scranton, Series Editor

Published with the assistance of the
Hagley Museum and Library

RELATED TITLES IN THE SERIES

Elspeth H. Brown, *The Corporate Eye:
Photography and the Rationalization of American Commercial Culture,
1884–1929*

Clark Davis, *Company Men:
White-Collar Life and Corporate Cultures in Los Angeles, 1892–1941*

Pamela Walker Laird, *Advertising Progress:
American Business and the Rise of Consumer Marketing*

JoAnne Yates, *Structuring the Information Age:
Life Insurance and Technology in the Twentieth Century*

# WOMEN FILMMAKERS
## *in*
## EARLY
## HOLLYWOOD

Karen Ward Mahar

The Johns Hopkins University Press
*Baltimore*

The Johns Hopkins University Press
2715 North Charles Street
Baltimore, Maryland 21218-4363
www.press.jhu.edu

Library of Congress Cataloging-in-Publication Data

Mahar, Karen Ward, 1960
Women filmmakers in early Hollywood / Karen Ward Mahar.
p. cm.
Includes bibliographical references and index.
ISBN 0-8018-8436-5 (hardcover : alk. paper)
1. Women in the motion picture industry—United States. 2. Women in motion pictures.
3. Motion pictures and women—United States. I. Title.
PN1995.9.W6M32 2006
791.43′028092273—dc22
2006002413

A catalog record for this book is available from the British Library.

To my parents,

J. Frances Ward & Ralph V. Ward

# CONTENTS

Illustrations follow pages 76 and 178.

# PREFACE

A few years into my Ph.D. program I was lucky enough to find work doing background research for an oral history project sponsored by the Women in Film Foundation. The Legacy Series creates and preserves oral histories of important women in television and film, and it was just getting started. Amazingly, my first assignment was Lillian Gish. Within a few days I was surprised to discover that the ethereal actress directed a film entitled *Remodeling Her Husband* (1921), starring her sister, Dorothy. This led me to Anthony Slide's *Early Women Directors* (1977), which described a then nearly forgotten era in which women directed films by the dozen. Unfortunately, my interview with Lillian Gish never took place, as her health rapidly declined (she passed away in 1993), but my historical encounter with her inspired this project. As this book goes into production, scholarship on women in the early film industry has become so large that there are international conferences on the topic. The fascinating work produced by film studies scholars over the last decade has enriched this study, which is, like me, historically grounded. It is my fond hope that my work will converse with that of other scholars who have posed different questions and employed different skills.

There is no possible way to thank all of those who assisted in this project from its origin to its completion. I sincerely apologize to anyone whom I may have overlooked. First, I thank the friends I made in graduate school ( and before): Jo Westbrook, Tyler Anbinder, Laurie Pintar, George Potamianos, Jill Fields, Wendy Holliday, Myrna Donahoe, and especially Kathy Fuller-Seeley, whose support and scholarly generosity knows no bounds. With sadness I remember Clark Davis, a graduate school colleague, friend, and author of *Company Men*, a title in this series, who tragically passed away at the beginning of his scholarly career. I could not have completed this project without the institutional support I received from the University of Southern California History Department, the Harold Hastings McVicar Fellowship, and the Haynes Foundation Dissertation Fellowship. I also thank the American Historical Association and the Library of Congress, for awarding me the J. Franklin Jameson Fellowship for 1996–1997, and the Organization of American Historians, for co-awarding the dissertation form of this book with the Lerner-Scott Prize in 1997. I am grateful to the powers that be at Texas A&M University at Corpus Christi for granting me a semester's leave to work on the book and to my former colleagues there, particularly Pat Carroll, Anthony Quiroz, and Robert Wooster; my current colleagues in the history department at Siena College; Angel Kwolek-Folland, for her encouragement and help; the extremely generous Kevin Brownlow, for providing the single most enjoyable day of research by meeting me in London, showing me Weber's *Where Are My Children?* (the Dutch version), giving me stills right out of his files, sharing his amazing index of names and dates, and even providing lunch and tea; Mrs. Jan Zilliacus, for sharing her memories of her father, film director Larry Trimble, who coproduced films with Florence Turner and Jane Murfin; archivists at the British Film Institute, particularly Margaret Hennessey, in the Stills, Posters, and Designs department, who helped me find many of the stills in this book; Cari Beauchamp, generous biographer of screenwriter Frances Marion; the librarians of the University of Southern California Cinema-Television Library, especially Ned Comstock; the librarians at the Margaret Herrick Library; and especially the amazing Madeline Matz of the Motion Picture division of the Library of Congress, who not only guided my research while I was on a postdoctoral fellowship but took stills of trade journal advertising out of pure generosity. Toward the end of this project I received a much needed boost of scholarly camaraderie from Jane Gaines of the Duke University Women Pioneers Film Project.

Finally, I offer heartfelt gratitude to those who exercised Herculean patience while waiting for the final version of this manuscript, particularly my parents, Ralph V. Ward and J. Frances Ward; former dissertation advisers Lois W. Banner, Steven J. Ross, and Edwin J. Perkins; and especially Robert J. Brugger of the Johns Hopkins University Press. Joe Abbott's careful editing improved the manuscript on its way to press. Of course, any errors are my own. I must finally acknowledge the patience, fortitude, support, and good humor of John F. Mahar, whom I met at the very dawn of my research and who has lived with this project as long as I have. But no one will be happier to see this book reach its completion than our daughter, Sean Marie.

# WOMEN FILMMAKERS
# IN EARLY HOLLYWOOD

# MAKING MOVIES AND
# INCORPORATING GENDER

*In no line of endeavor has woman made so emphatic an impression than
in the amazing film industry . . . [O]ne may not name a single vocation
in either the artistic or business side of its progress in which
women are not conspicuously engaged.*
—Robert Grau, "Woman's Conquest in Filmdom"

*[A]ny position which a man has occupied in the new industry has been,
and is now, occupied by a woman.*
—Mlle. Chic, "The Dual Personality of Cleo Madison"

In the 1910s and early 1920s the American film industry offered women opportunities that existed in no other workplace. Female stars like Mary Pickford, Mabel Normand, and Gloria Swanson earned some of the highest salaries in the world, and many more women worked in creative roles behind the camera. In any given production the screenplay was likely to have been penned by a woman, as was the continuity script, the step-by-step guide outlining all production activities. A female director may have guided the female star, who quite often worked for her own production company. Some women did it all. Lois Weber, the most famous female filmmaker of this period, was a screenwriter, actress, director, and producer, often on the same project.[1] After the shooting ended, a woman may have edited the film, a female censor may have re-edited it, a female exchange owner may have distributed it, and a female manager might have exhibited it in her theater.

When in 1920 the *Ladies' Home Journal* predicted that within five years "the feminine influence will be fully 'fifty-fifty' in 'Studio Land,'" it was more than just wishful thinking.[2]

Women filmmakers (broadly defined to include a range of production activities) were present in most facets of the American film industry by 1909. At the height of their activity, between 1918 and 1922, women directed forty-four feature-length films, headed more than twenty production companies, wrote hundreds of produced screenplays, became the first agents, and held positions as editors and heads of scenario and publicity departments.[3] Women outside of the United States entered filmmaking at about the same time: Elvira Notari in Italy; Olga Wohlbrück, Lotte Reiniger, and Thea von Harbou in Germany; Olga Preobrazhenskaya in Russia; Ester Shub in the Ukraine; Lottie Lytell in Australia; Germaine Dulac in France; and Adriana and Dolores Ehlers in Mexico. No country, however, witnessed the number of women filmmakers that the United States did, nor did any other country experience the swift expulsion of women from filmmaking in the years to follow.[4] Although women retained their positions as screenwriters, it was a different story elsewhere on the lot. By the mid-1920s, female directors and producers, many of whom were critically and commercially successful, found themselves defined as unfit. Girls who wanted to become editor's apprentices were discouraged. Female stars no longer started their own production companies. By 1927, on the eve of the sound era, director Lois Weber advised young women to avoid filmmaking careers. "Don't try it," she cautioned. "You'll never get away with it."[5] Only one female film director, Dorothy Arzner, sustained a successful career in mainstream Hollywood during the so-called golden age of the 1930s and 1940s. Not until the 1970s would the numbers of women directing and producing feature films begin to increase.[6]

Why did the early American film industry offer such a range of opportunities to women at this time? Just as important, why did these opportunities disappear? This study is not the first to recognize the work of these early filmmakers or to ask these questions. In fact, scholarship on women in the early American cinema has blossomed over the past decade. Scholars and archivists are making heroic efforts to discover extant films by early women filmmakers and are producing fascinating analyses of the films themselves. Recent biographies of filmmakers Dorothy Arzner, Alice Guy Blaché, Nell Shipman, and Lois Weber have significantly advanced our knowledge of the contribution of women to the early American cinema and are symptomatic of a turn within film studies scholarship toward history, albeit with the insights of feminist and film theory.[7] This study contributes to this emergent body of scholarship by offering a historical analysis of the gendering of filmmaking. It uses concepts from the sociology of gender and work, methodology and context from women's and business history, and insights from feminist film studies to create an overview of gender and film production from 1896, when the Vitascope first premiered, to the rise of the studio system in the mid-1920s. By stepping back and taking a broad view, by including the experiences of women in all facets of the film industry, and by contextualizing the experiences of women in the film industry with those of women in other industries, we can expand our understanding not only of women in the early film industry but also of women in American business.

⌒‿⌒

FOR MORE THAN A DECADE it appeared that creative and powerful women were going to be the norm in the American film industry, not the exception. Women filmmakers were said to be at the beginning of their stories—that the exceptional would soon be unexceptional and that workaday parity with men was in the near future. Yet it was not to be. In the largest sense the rise and fall of the woman filmmaker was related to changing strategies employed by the expanding industry. When longer "story" films encouraged the adoption of a theatrical model of production, actresses found work both in front of and behind the camera. When conservative middle-class reformers condemned the nickelodeons as too dangerous for "impressionables," assumptions of female moral superiority encouraged the industry to embrace and promote women in their midst. When the image of the New Woman serial heroine appeared to draw in young female fans, serial queens graced the covers of the trade magazines. When highbrow social problem films attracted middle-class patrons, uplifter Lois Weber became the most celebrated director in the industry. But when executive control and financial legitimacy became paramount, women disappeared.

This argument does not seem to leave much room for female agency, but that is not the case. Despite claims that women were naturally suited to melodrama, for example, women made all kinds of films, from cliffhanger serials to comedies, as well as what would later be termed "women's films." During the period between the nickelodeon boom and the end of World War I, industry leaders had to consider what their often largely female audiences wanted, what would draw a well-heeled audience, and what would keep censorship advocates at bay, and they could do little more than guess at what sorts of films might help them achieve their goals. As a result, in this era before the consolidation of the classical Hollywood style, studio heads were open to a variety of film forms. It was true that some women, such as those directing at Universal, were expected to make films for female and juvenile audiences, but those who were displeased with the established studios could and did leave for the freedom of independent production.

Does this mean that female screenwriters, directors, producers, and editors were able to realize their own vision? Did women filmmakers make a difference on the screen? Any study of filmmakers, even one aimed at uncovering gender hidden in industrial processes, presumes that the question of who gets to make movies makes a difference. The gender of the filmmaker undoubtedly influences the final product, and movies were arguably the most influential cultural medium in the world during the silent era. In the 1910s and early 1920s, when female filmmakers were most active, motion pictures had yet to compete with radio or television for the attention of the masses. Even after radios became widely available in the 1920s, the public kept its "movie habit." In 1928, at the close of the silent era, sixty-five million Americans visited movie theaters every week, a number representing more than half of the country's population.[8]

If we accept the idea that women filmmakers made a difference on the screen, then we are talking about authorship. Authorship, however, is a contentious concept, even though many feminist film scholars reject the theoretical "death of the author."[9] As Jane Gaines recently pointed out, filmmaking was, and still is, a collaborative process, but even beyond this fact is the context within which these women worked. How much of Lois Weber's celebrated philosophy of "uplift" was sincere? Did she frame her work and persona in recognition of the industry's needs? Was Helen Holmes really a daredevil daughter of a railroad man, and is that why she wrote and starred in hair-raising railroad serials? Did Ruth Roland really personally supervise every scene in *The Adventures of Ruth*, as her advertisements claimed?[10] Did Clara Kimball Young really make the films she desired when she became an independent producer, or was she railroaded by her partner Harry Garson? We can rarely know the answers to such questions. Even when we have memoirs and autobiographies to guide us, we are still troubled by the multivalent reasons why film industry insiders say what they do, or if they really said it at all, given the existence of publicity departments since the 1910s. For this reason film scholars must approach even oral histories with caution. How many anecdotes, originally written by the studio, were repeated so often that even the subject, often advanced in age, could no

 4

longer distinguish between reality and publicity? And what of motivations? As Amelie Hastie argues, as much as we may wish to believe everything filmmaker Alice Guy Blaché claimed in her memoir, we must realize that the forgotten Guy Blaché wanted to write herself back into film history, a history that was by then peopled with "great men" such as D. W. Griffith and Georges Méliès. Under these circumstances she may well have claimed more authorial control than she had in fact enjoyed.[11] While approaching the material with caution, this study will take the leap of faith that we can know enough from the primary sources to piece together a history of gender in the early American film industry.

I BEGIN BY REJECTING THE "EMPTY FIELD" THESIS, which assumes that because the field of filmmaking was new, it was ungendered. All work emerges from some previously gendered context, and the film industry is no exception. The question is how, when, and why the American film industry became supportive of women in powerful and creative roles, and what caused that support to disappear? For the purposes of understanding gender, I find three periods to correspond to the primary strategies employed by the film industry to secure growth and profits, which in turn influenced the industry's perception of the woman filmmaker: (1) the "technological" decade, from roughly 1896 to 1908; (2) the period of "uplift," from about 1908 to 1916; and (3) the period of "big business," from 1916 to the end of the silent era, in 1928. This trajectory ultimately resulted in the rise of the studio system and the classical Hollywood style. Outside of this trajectory women created two-reel serial thrillers and slapstick comedies from 1912 to 1922, which, although subject to many of the same changes in production as longer feature films, came to a somewhat different end and are thus treated separately.

The prologue discusses the first phase, from roughly 1896 to 1908, when the American film industry emerged within a masculinized context. The first moving picture devices were understood as electrical novelties, and the inventors, mechanics, and entrepreneurs who became the first filmmakers made profits not from the moving pictures but from the ownership and control of cameras, projectors, and patent rights. Even the first films were often male-oriented products, demonstrating the wonders of the equipment with images of daring stunts or peep-show themes associated with masculine amusements. Women did contribute but primarily in the film laboratories, where they labored at sex-segregated jobs.

Part I begins with the second phase, from roughly 1908 to 1916, when women became visible in the film industry. The popularity of longer films implied that viewers and buyers were paying attention to the dramatic qualities of the films themselves; thus manufacturers turned to the stage to find experienced thespians. These actors and actresses brought with them an egalitarian work culture, which expected both men and women to assist with production. It was here, on the shop floor as it were, that the first male and female professional film directors and producers emerged, as the entrepreneurs who owned the studios largely left those who created the films to their own devices. When stardom struck a handful of these actresses, they were well placed to start their own production companies.

At the same time, the film industry began organizing and creating a fairly unified voice, represented by trade organizations and trade journals, to answer charges from middle-class reformers that the movies, now shown in cheap theaters known as nickelodeons, were harming children and other "impressionables." A censorship crisis in 1908 and 1909 inspired the uplift movement, which aimed to secure cultural legitimacy for the movies. During this phase the industry advocated self-censorship, took the so-called legitimate stage as its model, and lauded the female filmmakers and exhibitors in its midst. Drawing from the traditional belief that women were morally superior to men, the industry assumed that women would create cleaner films (and no doubt depended on other Americans to make the same assumption). The woman who most demonstrated the positive impact of the female filmmaker was Lois Weber, whose cycle of highbrow social problem films drew as much critical and popular acclaim as the films of D. W. Griffith.

The middle section of the book, Interlude, consists of a single chapter detailing women who created the short-film genres of serials and comedies. Whereas Lois Weber demonstrated the allegedly natural uplifting qualities of women, these female filmmakers embraced the New Woman by creating hair-raising thrillers and anarchistic slapstick comedies starring themselves. Despite the fact that the New Woman filmmaker contradicted the "True Woman" filmmaker (Weber), the industry, searching for what audiences desired (as well as trying to uplift its image), embraced these filmmakers, too. These short one- and two-reelers, which arrived in 1912 and disappeared by 1922, were among the most popular films of the 1910s. Although also subject to the increasing control of the central producer, their demise came at the hands of censorship advocates rather than the studio system.

Part II concerns the third phase, the period of "big business," which began in 1916, just after a second censorship crisis ended the uplift movement. In one sense the uplift movement succeeded, because better-heeled patrons were now going to the movies. These patrons were not only viewing better fare but were also paying more to see their favorite stars. Once again, a number of female stars were able to use their leverage to start their own production companies. In another sense, however, the uplift movement failed. The serious social problem films embraced by reformers within the industry, such as Lois Weber, drew the ire of conservatives by exploring controversial themes such as birth control. Industry leaders turned away from the kinds of social problem films associated with women and began to advocate the idea that films should be primarily for entertainment. Women filmmakers became marginalized when the film industry's new strategy aimed not at cultural legitimacy but financial legitimacy. As studios looked to vertically integrate by buying chains of theaters, they needed the kind of funds only Wall Street could provide. Wall Street, in turn, would not accept "unbusinesslike" methods and sent its representatives to show Hollywood studios how to become worthy of the stock exchange. As studios adopted new (masculinized) business attitudes, and as the number of independent theaters shrank, there was no room for the female filmmaker. Even women who had proven records of filmmaking success found themselves standing outside the studio gates as the vertically integrated studio system emerged. There was one exception: Dorothy Arzner, the subject of the epilogue.

THERE WAS NO MEMO CIRCULATED TO STUDIO HEADS asking them to eliminate women filmmakers in the 1920s. Rather, a shift in ideology brought gender roles within the film industry in line with those of other industries. These processes were largely invisible to the women and men who were making movies and even to the studio executives, efficiency experts, and investors who initiated such changes. Understanding the rise and decline of the woman filmmaker in the silent era requires sensitivity to changing discourse regarding sexual difference, and it requires understanding the changing parameters of a volatile industry. However, these women also existed within the larger parameters of early twentieth-century society, when, as Nancy Cott observes, "more than one generation now collided, those who had been brought up in women's sphere (of varying cultural traditions) and those whose experience was just as much shaped by factory or office, coeducational schooling, urban social life, municipal reform efforts, or political action in clubs, unions, temperance or socialist organizations."[12] Gender ideology in the early twentieth century was in flux; an older domestic and maternalist ideal for women overlapped with the emergent New Woman, creating a tense dialectic that did not easily give way to synthesis. All of these film-makers could be considered New Women by virtue of their employment in the most modern of industries, yet the majority worked with male part-ners, typically a husband or lover. And when those partnerships ended, so did many women's filmmaking careers. Even when a financial disaster de-stroyed a company, it was often the women who "retired," whereas their male partners enjoyed many more years of filmmaking in the employment of others. The fact that the women filmmakers as a whole were largely elimi-nated within the span of a few years suggests that forces beyond the personal were at play. Yet while we search for industry-wide patterns, we must also account for the personal experiences of these filmmakers, who were, after all, also women living in the early twentieth century, with all the conflicted expectations, freedoms, and constraints that implies.

The short answer to the question "what happened to the female film-maker?" is that she became marginalized as the film industry became a Wall Street–defined, vertically integrated big business. But there is much more to this story. The fate of the female filmmaker during the silent era illustrates how industrial growth and change can unexpectedly open as well as close opportunities for women, the way that shifts in gender perception can ac-company shifts in industrial strategy, and the tangled ways that the multiple voices of society, work, family, and self determine the course of women's lives and careers.

# "The Greatest Electrical Novelty in the World"

## Gender and Filmmaking before the Turn of the Century

On an April evening in 1896 Koster and Bial's Music Hall unveiled Edison's "Latest Marvel" to a New York vaudeville audience. The wizard of Menlo Park, already known for his phonograph and Kinetoscope "peephole" moving picture device, did not disappoint. The Vitascope, the first commercially successful American film projector, threw startlingly lifelike images onto a screen. The "umbrella dance," waves crashing at Dover, a comedic Mutt and Jeff–style boxing match, a marching band, and a serpentine dancer were familiar Kinetoscope fare, but their larger-than-life verisimilitude awed the crowd. A writer from the *New York Mail and Express* exclaimed that "every change" of the umbrella dance was "smooth and even," and "there was absolutely no hitch" in the waves at Dover, which was so impressive "it had to be repeated many times."[1] The Vitascope, claimed one early exhibitor, was "the greatest electrical novelty in the world."[2]

The earliest American movies emerged within an already masculinized context. The very first films were not an art form or a replacement for live theater, worlds where women existed as artists, actresses, playwrights, and managers, but commercialized sensations drawn from the highly masculinized setting of the inventor's laboratory. Although the venue for the movies was typically a mixed-sex setting, such as a vaudeville house, a tent show, a church basement, or a visit by an itinerant exhibitor, the trifecta of science, mechanical arts, and commerce in which the Vitascope emerged ensured the film industry was gendered at the start. The first filmmakers did employ women but only to perform routinized film processing tasks deemed appropriate to their sex in largely segregated settings. For male entrepreneurs, however, the film industry's first decade suggested adventure, autonomy, and riches.

THE VITASCOPE WAS PRESENTED AS A POPULAR SCIENCE ATTRACTION. "All the Town Is Talking about Edison's Astonishing Vitascope!" sang out an ad in the *Providence Journal*; it "Puzzles Scientists, Baffles Analysis," and "Creates Round-Eyed Wonder."[3] The name itself, combining the learned languages of Latin (*vita*, "life") and Greek (*scope*, "to see"), was meant to legitimate the projector as a scientific instrument, as did the names of its antecedents, such as the Phenakistoscope (1849), the Kinematascope (1861), and the Phasmatrope (1870).[4] Audiences understood the Vitascope's premiere within the context of related popular science entertainments, particularly the familiar and popular magic lantern. The magic lantern, a forerunner of the projector, threw images on a wall by placing a light source behind colored glass. Although at the turn of the century the magic lantern often projected pleasant photographs or painted slides, science lecturers used it to project actual specimens encased in glass slides. Viewers were made aware of the magic lantern in their midst as the male "professors" who exhibited these devices provided lectures and often specifically reminded the audience of the technical means by which the images before them were created.[5] Such entertainments doubly enhanced masculine associations, for not only was the equipment the focus of attention, but so, too, was the knowledgeable male narrator who explained and demonstrated it. In this vein the most commonly cited scientific antecedent to the moving picture is the step photography of English-born photographer Eadweard Muybridge. In the 1870s Muybridge set up still cameras to capture sequential photographs in order to study animal and human movement, and he delivered illustrated lectures for both education and entertainment, using a dissolve between sequential slides to suggest movement. Muybridge conducted some of his work at the University of Pennsylvania and published the results of his study in 1887 under the title *Animal Locomotion*. He ultimately created a moving picture device called the zoopraxiscope and inspired Thomas Edison to consider the challenge of moving pictures.[6]

Moving pictures themselves emerged as a feat of "applied science," a term used by engineers at the end of the nineteenth century. (The word *technology* would not gain wide currency until well into the next century.) Engineering, like science (and especially medicine), was undergoing a process of professionalization at the end of the nineteenth century that led to the marginalization of women just as the film industry emerged. Female engineering students had been welcomed into the new technically focused land grant colleges that emerged after the Civil War. Often excelling in calculus and geometry, the first women in engineering schools used their math skills to demonstrate ability and often earned multiple degrees. But in a conscious effort to keep the field from becoming feminized, by the end of the century engineering programs began stressing "hands-on experience" and leadership in the field, activities that barred female students. By the end of the century "the ability to 'handle men,'" attests historian Ruth Oldenziel, "remained the true hallmark of the successful engineer."[7]

Female inventors outside academe found similar treatment. When Charlotte Smith and Matilda Joslyn Gage tried to include female inventors in the historical record in the nineteenth century, they discovered that the Patent Office omitted one out of four female applicants and particularly omitted the names of women who invented machinery.[8] By the time that the moving picture industry emerged at the turn of the century, Darwinists argued that women suffered from arrested development and were thus outside the bounds of progress, making it seem extremely unlikely that women could contribute to the field of applied science.[9] Indeed, women were often seen as antithetical to technology altogether, flummoxed by the mysteries of labor-saving domestic appliances and dangerous behind the wheel of the century's new automobiles. Such imagery was contested by many women, such as suffrage advocates who embraced the automobile as the symbol of women's emancipation.[10] But while women (and often advertisers) demonstrated feminine competence over consumer technology, the masculinization of engineering work progressed.[11]

In the first decade of the film industry, from 1896 to roughly 1906, the masculinized arena of applied science joined with the masculinized ethos of the marketplace as Edison and his competitors battled over the sales of cameras and projectors (and the movies that went with them). Edison made his reputation as the "businessman's inventor," thanks to his early work with Jay Gould and other industrialists who desired communications technology. Edison entered the amusement field in the late 1880s, after it became clear that his phonograph was impractical for office use. He sold the device for entertainment purposes as an alternative, and by 1890, coin-operated phonograph parlors proved commercially successful. One year later Edison patented a moving picture camera, the Kinetograph, which made films for the Kinetoscope, a tall wooden box with a peephole through which a standing individual could view a loop of moving picture film. Some argue that Edison could have developed a film projector at this time, but his already successful phonograph parlors directed his thinking toward coin-operated machines for individual viewing.[12]

By the late nineteenth century, the American public was "accustomed to getting new sensations from [Edison] as regularly as they would put a nickel into a slot."[13] When advertised as "Edison's latest marvel," the Vitascope moving picture projector appeared within a familiar and marketable context. In fact, Edison did not actually invent the Vitascope. Edison's early focus on the individualized peephole cabinet proved a mistake. After the Kinetoscope had reaped stunning profits its first year, returns fell precipitously, and its future looked grim. By 1896, as inventors in the United States and abroad perfected projection systems, it was too late for Edison to overtake them. When approached by two entrepreneurs who owned the rights to a working projector, Edison listened to their proposal. Their machine had experienced a lackluster debut at a state fair in Atlanta the previous autumn, and the men thought they might have better luck if the famous Edison played the role of inventor. With his own efforts at projection fizzling, Edison agreed to call the Vitascope his "latest marvel" for a cut of the profits. The Vitascope was a smashing success, spawning dozens of licensees and imitators almost immediately and creating a new, highly competitive, and highly masculinized, industry.

Mere months after the Vitascope's debut, hundreds of moving picture cameras and projectors ticked and hummed across the country, but it was not good news for Edison. The French Lumière brothers devised a light (sixteen-pound) hand-cranked camera-printer unit called the *cinématographe*, which proved more popular than Edison's equipment when it entered the American market in the early summer of 1896. Shortly thereafter, a superior "flickerless" picture from the American Biograph and Mutoscope Company appeared, using a camera and projector designed by Edison's former motion picture engineer, W. K. L. Dickson.[14] Added to this mix were cameras and projectors devised by individual mechanics. Edison immediately revised a patent application dating from the Kinetograph and filed it with the U.S. Patents Office, "claiming that he had created a device for moving pictures long before anyone else, and that any subsequent machines therefore infringed his patent." Edison's numerous lawsuits, meddling detectives, and alleged thugs colored the first decade of the new American film industry but did not prevent entrepreneurs from making and exhibiting moving pictures.[15]

Memoirs and early film histories reflecting on this period present a rough-and-ready business culture that undoubtedly exaggerated manly control and lawless competitiveness. Film historian and producer B. B. Hampton hyperbolically claimed that despite Edison's efforts, "anyone who could rent, buy, or borrow any form of 'box' that would hold a lens and roll of film could become a picture producer." All one had to do was "set up the camera anywhere, 'shoot' almost anything in motion, develop the negative, print the positives, and sell them practically at his own price."[16] With a winking boastfulness, cameramen Fred F. Balshofer and Arthur C. Miller recalled how they and their peers visited "speakeasy" camera supply stores in the 1890s.[17] Eluding Edison "became rather amusing," recalled another early filmmaker, "something like a game of hide-and-seek."[18] Since most films were made outdoors, companies allegedly sent out "decoy groups" to throw off Edison's detectives so that the real filmmakers could work unhindered. The cameramen of the Independent Moving Picture Company (forerunner of Universal) claimed to hide their cameras in an icebox. Certainly the oft-told tale that filmmakers relocated to Los Angeles in order to flee across the border from Edison's thugs cast early film entrepreneurs as outlaws worthy of a western. In these accounts one senses a mischievous and gleeful escape from the law of the father.

The historical accuracy of the details is less important than the participant's sense that filmmaking during this period was a manly adventure. Conservative ideology held that although there was a place for female proprietors in the economy, women were unsuited to expressions of overt commercialism.[19] Female milliners, dressmakers, and purveyors of beauty products, for example, worked within a preindustrial apprentice-craftsman tradition, making the products they sold to a limited group of exclusively female consumers.[20] Although penning many nineteenth-century best-sellers, female authors retained their gentility by working under the paternalistic guidance of the "gentleman publisher," who himself projected an image of literary dignity rather than crass commercialism.[21]

For men filmmaking offered not only a means to make profits but access to a kind of work-related masculinity that had almost disappeared in America by the start of the twentieth century. However apocryphal, memoirs and early film histories highlight luck, pluck, and Horatio Alger—like "rags to respectability" trajectories for the first filmmakers in an era of increasing bureaucratization and class conflict.[22] Neither of the partners who launched the Vitagraph Company, for example, would have inspired much confidence before creating one of the major motion picture firms of the silent era.[23] James Stuart Blackton, a quick-sketch vaudeville artist, worked as a writer and a cartoonist when engagements were slack. His partner, Albert E. Smith, was a magician who worked as a bookbinder when not onstage. Both were barely twenty in 1897 when they raised $800 to buy an Edison projector and a supply of films. They ultimately created one of the most long-lived and successful studios of the silent era. Similar entrepreneurial scenarios took place in other cities. William Selig of Chicago, a magician and owner of a traveling minstrel show, tried to build his own projector when he discovered that his machinist had already built a cinématographe. Using the blueprints, he had a copy made and began making and exhibiting films in 1897, incorporating as the Selig Polyscope Company in 1900. Sigmund Lubin of Philadelphia, a seller of optical and magic lantern goods, asked one of the original inventors of the Vitascope to help him construct a projector. By 1897 Lubin was selling projectors for $150 and was making his own films. A booming success as a film producer, he was the first film entrepreneur to vertically integrate his business when he began acquiring moving picture theaters in 1906.[24]

This first generation of filmmakers enjoyed a level of control over their work that was becoming increasingly rare. Working for others, punching a time clock, and shuffling papers at a desk blurred the time-honored definitions of manhood recognized by generations of self-employed farmers, artisans, and businessmen. The skills required for early moving picture photography, by contrast, came from the machine shop and were honed by hours behind a camera. Early cameras had to be cranked by hand at precisely the correct frames per second to achieve a look of natural movement. If the operator cranked too fast, the image would be seen in slow motion when projected on the screen. If cranking was performed too slowly, the image would appear to leap comically across the frame at superhuman speed.[25] More important, early moving picture cameras were prone to picture-ruining static and were susceptible to extremes of temperature and humidity that often led to breakdowns. The specialized skills required to operate and repair the first moving picture cameras were nearly impossible to acquire outside of the technician's shop. According to Albert E. Smith there were "so many mechanical troubles" with early cameras at Vitagraph that the company replaced the still photographers it originally hired to shoot moving pictures with "mechanics from the machine shop." After having been trained in the rudiments of photography, the mechanics "worked out very satisfactorily."[26] Similarly, D. W. Griffith's cinematographer, G. W. "Billy" Bitzer, was originally hired as an electrician by Biograph in 1896.[27]

The cameraman was the central figure in early film production. Although there is scholarly debate over the existence of a "cameraman system" of production, which argues that the cameraman made all major filmmaking decisions, versus a collaborative partnership system in which someone with theatrical leanings worked with and even changed positions with the cameraman, the operator was the equivalent of the player who brings the bat and ball to a sandlot baseball game—filmmaking could not occur without him.[28] Indeed, perhaps at no other time was the status of the camera operator so elevated. During the industry's first decade, when the competitive focus was on the construction and operation of cameras and projectors rather than the movies themselves, it was the person able to make, repair, improve, and operate these devices who made the entire industry possible. When the industry entered its second decade and competition shifted from the equipment to the films themselves, the cinematographer (as he came to be known) would become subordinate to the director, but at this moment the camera operator was the pivot of production. It was the cameraman who, like the preindustrial artisan, knew the secrets of the trade: how to make the product from start to finish. Given its technological requirements and artisanal integrity, it is no surprise that cinematography remained the most intensely masculinized position in the film industry throughout the twentieth century.[29]

Even the physical demands of early American cinematography defined filmmaking as unsuitable for women. The camera used in Edison's carefully guarded Black Maria studio was so heavy it had to be moved on metal tracks. The camera developed by Dickson for the American Mutoscope and Biograph Company was light enough to travel, but at 350 pounds it still took two men to lift it onto a tripod. It was possible to build a lighter and easier-to-use camera, as was done in France by the Lumière brothers in 1895. The Lumière cinématographe required no electricity, and it was a projector as well, but the operation of all early cameras required a level of mechanical experience that was difficult for women to acquire. The suggestion that gender is built into technology—that, in the words of Michel Callon, "Machines carry the word of those who invented, developed, perfected and produced them"—seems apt here.[30] In fact, the masculinity of cinematography continued even when moving picture cameras became lighter and easier to use.

As a point of reference, the photography trade began in a very similar fashion in the mid-nineteenth century. When commercial photography emerged in the 1850s, still cameras, like early moving picture cameras, were extremely large and very heavy (fifty to seventy pounds), requiring highly skilled operators who were expected to construct their own darkrooms and develop their own glass plates. The first photographers, like the first cinematographers, were entirely male. However, within a few years a handful of women accepted the challenge of photography, and they were joined by more women as still cameras became lighter and easier to use.[31] By the late nineteenth century, as the scientific uses of photography gave way to aesthetic concerns, advice literature argued that photography was an art, requiring the feminine traits of "abnegation and devotion" and a "delicate touch." Not that the transition was effortless. Catherine Weed Barnes, an important female photographer, felt compelled to actively promote photography as "suitable work for ladies." Her article entitled "Why Ladies Should Be Admitted to Membership in Photographic Societies" (1889) suggested that there was resistance to female infiltrators. Relying on traditional gender ideology to make her point (as would the first female filmmakers), Barnes argued, "In this work there are many occasions where the gifts, supposed by the most moss-grown tradition to specially belong to women, serve them well, and make the visible result better even than that of men."[32] By 1900 female professional photographers were no longer an aberration, and thanks to the greatly simplified Kodak camera, first introduced in 1889 and marketed especially to women, female amateur photographers were a common sight.[33]

Although a growing number of women became amateur and professional still photographers, cinematography remained steadfastly male for most of the twentieth century. Perhaps one reason why cinematography retained its early masculine associations was because the purpose of early cinematography was blatantly commercial. Amateur moving picture cameras were marketed throughout the silent period, and advertisements even showed women operating the small cameras. A 1924 Bell and Howell advertisement for the $185.00 "Filmo Automatic CineCamera," a 4.5-pound, handheld device, included a photo of the camera being used by a young woman to denote the light weight and ease of operation. But amateur moving picture cameras remained expensive, and few sold.[34] Moving picture photography was, by definition, professional cinematography, and the moving picture camera held far more profit potential than did the still camera. Professional still photographers might do a brisk local business, but moving picture photography promised national and international distribution.

But neither commercial potential nor technical demands adequately explain the sustained masculinization of cinematography for nearly a century. The missing piece is the cinematographers themselves: they actively gendered the occupation of cinematography. The film industry appeared at a moment when traditional measures of white, native-born manliness shifted and disappeared. New Women entered public life as reformers, college students, workers, and patrons of new forms of leisure; immigrants threatened the well-being of white workers and white supremacy; and factories and offices replaced small businesses and artisan's shops. White men were said to be overcivilized, soft, unable to fulfill their role in a Darwinist survival-of-the-fittest world, and as they failed, so too might the nation. Prescriptions for this malaise began to be implemented into educational settings, as well as leisure pursuits, and included sports, scouting, and a vigorous life of exercise.[35] Theodore Roosevelt, the most famous advocate of an aggressively masculine ethos, outlined such a program in 1899 in a speech before the Hamilton Club of Chicago entitled "The Strenuous Life," in which he celebrated the man "who has those virile qualities necessary to win in the stern strife of actual life" and urged his listeners to "be glad to do a man's work, to dare and endure and to labor."[36] Cinematographers reflected this manly ethos, creating a work culture and identity centered on courage, athleticism, and self-control. The first issue of *American Cinematographer* in 1920 described "The Cameraman":

> He must, first of all, be able to take good pictures, apart from that, he must necessarily be a brave man and ready to attempt anything asked of him. He must be clear-headed, so that he can stand on the edge of a sky-scraper, and lean over the top of a precipice, for that matter. He must perch himself in almost incredible angles, and perhaps stand waist deep in the river or ocean. He must stand steadily by his work when some wild beast comes menacingly close, when the other members of his party can run to shelter, and all the while he must steadily crank, and see that his camera is not injured by fire, animals or water, and it is a matter of record that very valiant deeds are performed by the cameraman, deeds that few actors or directors care to brave.[37]

In their memoirs early cinematographers emphasized mastering the camera and taking personal risks to turn out footage of patriotic appeals, safaris, boxing matches, and other forms of manly derring-do. Mounted horsemen, for example, "charged the camera" in *Roosevelt's Rough Riders.* Indeed, the Spanish American War actualities filmed by Edison and the Biograph companies expressed the conflated virility of soldiers and nation and, by implication, of the cameraman who captured such images, even though battle scenes were staged in military camps.[38]

On a more subtle level the very act of taking moving pictures was inscribed with gendered meanings. In the first years of the American cinema, filmmakers purposefully wandered through public areas searching for locations or for actuality footage.[39] Recent scholarship centering on the gendered meanings of movement, space, and geography suggest that this freedom is a potent source of the masculinity adhering to moving picture photography. The nineteenth-century concept of the flaneur, "the wandering gentleman observer," defined a level of spectatorship that was accessible only to men who could wander and loiter at will in the public sphere with little danger to person or reputation. Until the arrival of circumscribed sites that allowed female spectatorship, such as department stores, fairs, and most important, the cinema, the female equivalent of the flaneur was the streetwalker.[40] And in all cases the cinematographer's role was *to look*, to become the very eyes of the imaginary spectator. Although refined in more recent work, this active look was originally identified by feminist film theory as a male privilege, replicated in the classical Hollywood film, in which the active (male) protagonist gazes at the passive (female) object.[41] The language used to describe photography—*taking* a picture, *capturing* an image, *shooting* a movie—suggests action, possession, and even aggression. Mastery over space itself was implied in the American Society of Cinematographer's first motto: "Give us a place to stand and we will film the universe."[42]

The act of boldly moving through space, taking physical risks, and capturing the male gaze on film made cinematography more transgressive for women than still photography. Many of the professional female photographers who found acceptance within the film trade became specialists in domestic photography—portraits of families, children, and flowers, all of which could be safely taken in the confined space of the indoor studio.[43] In contrast, only three female professional cinematographers appear in the silent era, and their careers were exceedingly brief.[44] We cannot know how many women might have been interested in professional cinematography but found it impossible either to gain the necessary skills or employment. As it organized, cinematography, like most other professionalizing fields, appeared gender neutral but in fact restricted membership to men. The published aim of the American Society of Cinematographers (ASC) when it was established in 1919 appeared objective: "to advance the art of cinematography through artistry and technological progress, to exchange ideas and to cement a closer relationship among cinematographers." The origins of the ASC, however, can be traced back to two men's clubs, both formed in 1913: the Cinema Camera Club of New York City and the "semi-exclusive" Static Club formed in Los Angeles. In 1918, when the clubs evolved into the American

Society of Cinematographers, the society announced that "a strict mode of invitational method of membership entrance" would be enforced.[45] Common gatekeeping methods to elevate a field's status included educational requirements and licensing, but in the absence of both in this early period, the ASC and the clubs that preceded it guarded their membership through exclusivity.[46] Professional cinematographers required "masculine" traits of stamina, bravery, technical expertise, and seriousness of purpose, despite the development of lighter cameras and a redefinition of cinematography as an art (a definition that should have opened up cinematography to women).

WOMEN *DID* PARTICIPATE in the production branch of the film industry during its first decade, and they did so in significant numbers. Once regular film production was underway, the menial work required to reproduce the first movies for exhibitors to rent or buy drew from a different paradigm—not the inventor's laboratory or the mechanic's shop but the nineteenth-century factory. Edison, Biograph, Vitagraph, and other early film companies were not only wrangling over cameras, projectors, and patent rights; they were also competing, often fiercely, in the business of grinding out a sufficient number of films to meet demand. Even the names of some early firms, such as the Edison Manufacturing Company, reflected the openly commodified perception of the first studios, fittingly called "film factories." While a debate eventually raged over whether the movies were an art or a business, the presence of the film factory provided a steady definition of the moving picture industry as just that—an industry. Inside the earliest film factories many of the daily activities centered on film processing, work that remained remarkably stable and sex-typed throughout the silent period.

By 1900, shortly before film factories multiplied, almost half of all single American women earned a wage.[47] Since women worked for less pay than did men, and since film processing consisted of similar tasks performed by women in other industries, it is no surprise to find women employed within the first film factories despite the masculine nature of shooting films. Edison and the American Mutoscope Company built factories as early as 1895 to provide moving pictures for their respective peephole machines: the Kinetoscope and the Mutoscope. In the first five to ten years of filmmaking, these factories were rather simple. Typically, the cameraman developed the film with the aid of a male assistant, who was needed to help maneuver the large slatted drums on which negatives were wound to be developed and dried. The men suspended the drums over large troughs and manually rotated the film through the developing and rinsing fluid. They then reeled the film onto another drum or rack to dry.[48] No illustrations of the Edison Kinetoscope laboratory exist, but such illustrations do exist for Edison's competitor: the American Mutoscope Company, which used a flip-card technique in its peephole cabinet. Etchings published in *Scientific American* show that women performed film processing duties at the Mutoscope by at least 1897, the year that the American Mutoscope Company (soon to be the American Mutoscope and Biograph Company, and then simply the Biograph Company) became the premiere moving picture company in the United States.[49] The etchings also illustrate a workplace segregated by sex. In the darkroom three men tote the large slatted wooden drums carrying negatives to troughs of developing fluids. In a much larger room at least two dozen women dry, cut, retouch, and inspect the films. On the left side of the room three women retouch strips of film frame-by-frame with paintbrushes while the film strips dry on enormous drums. In the rear of the room several women peer through projector-like film-cutting machines.[50]

The photography trade established the precedent of using women to process film. The photography boom began in the early 1840s, after the introduction of the daguerreotype. Edward Anthony, one of the largest wholesalers of photographic apparatus and supplies, hired women to cover and gild the hinged cases that protected and displayed the finished photographs. Men also performed this work, but since it was clean and not too physically demanding, it fit the developing definition of appropriate work for women.[51] Women assumed various other duties as the photographic industry developed. By 1888 Emilie Colston, writing for *Photographic News*, commented that "the duties of the reception room—retouching, printing, mounting, and colouring—are already acknowledged to be within the sphere of women's abilities."[52] By 1890 the female retoucher cut a familiar figure. A writer for *Anthony's Photographic Bulletin* portrayed her in a fictional portrait as somewhat pathetic, emphasizing the monotony and low status of the work. The retoucher was a "nobody" who sat each day in front of a studio window, "her head bent over her easel, a magnifying glass in her eye, a pointed pencil in her hand, busy from dawn to dusk taking out wrinkles and putting in dimples."[53]

Large-scale photographic developing and printing, the kind performed by women in the early film studios, did not appear until the 1880s, when George Eastman introduced the Kodak camera. After using the one-hundred-frame roll of film, already loaded in the camera, the customer mailed the entire camera to the Eastman factory in Rochester, New York, and received developed photographs back via the post office. A photograph of Eastman's photo-finishing laboratory, circa 1889, shows a large, well-lighted room, much like that of the Mutoscope factory, populated mostly by women workers. But whereas only men developed motion picture film for Mutoscope, some women, along with adolescent boys, appear to be developing the Kodak photographs. This difference may be accounted for by the fact that the development of commercial motion pictures was a far more critical task than the development of amateur photographs.[54]

Like most work assigned to women, the tasks performed by women in the photography trade required what were assumed to be feminine skills: dexterity, neatness, and the ability to perform detailed, routine, fairly low-skilled tasks.[55] When the moving picture industry adopted similar processing techniques, it adopted similar employment practices as well. A closer look at gender in the film factory circa 1910 affords us the opportunity to examine the characteristics of movie processing work assigned to men and women at a point in which the moving picture industry was well established. Aside from a few minor differences, the major processing tasks and the gender assigned to each was the same in every studio.

The first major step after "shooting" the film, development of the negative determined the quality of the finished original film from which all copies would be made. The cinematographer controlled this critical process, making developing decisions based on the conditions present when the shot was taken, such as lighting, temperature, and humidity. The cinematographer could correct for over- and underdevelopment or manipulate the process for artistic effect. The laboratory technicians assisting the cinematographer in the negative developing room necessarily worked in close collaboration with the cameraman and were always male. At the large Vitagraph film factory on Long Island, large, waist-high vats of chemicals dominated the developing room, around which men in soiled overalls tended the wide racks of film. After development, workers wound the negative on large drums and set them aside to dry, a process repeated for the positive print and tinted films.[56]

Editing was the next step. The head of the negative department and the film's director decided what should be edited and where the titles should be inserted. Direct cutting of the negative (rather than editing from the positive print, as was done later) was probably preferred because it was a rather simple process. Films before 1910 were quite short, but even when movies reached one thousand feet, the scenario laid out all the shots in a numbered order. With few scenes and rarely any retakes, early films seem simple by modern standards. Nevertheless, editing was skilled work and thus done by men. Women, however, cut the negative at the points determined by the editor and joined the appropriate scenes together. At the Vitagraph plant men and women worked side by side in the negative department, though at different tasks. A photograph from 1912 revealed two men examining a long strip of film held between their fingers, perhaps deciding where to trim the scene in question. Young women seated at desks, each with a wastebasket and a towel nearby, appear to cut the negative according to the dictates of the editors.[57]

The next step sent the negative to the printing room, where women operated printing machines in near-darkness. At the Pathé Frères plant in New Jersey women stood all day in front of the machines, frequently disassembling them to wipe away dirt and dust.[58] At Vitagraph women described as "experienced printers" sat on backless stools and "guided and manipulated" the film through the machines. A superintendent observed their work to ensure required qualities such as density. Although the superintendent at Vitagraph appeared to be male, this may not have been true at all studios. The American Film Manufacturing Company listed Miss Anna Gallaghan as forewoman of this department in 1910.[59]

Once taken from the printing machines, the film needed further development, and once again film development constituted men's work. At Pathé "women wound the film on big wooden frames [that were] turned over to men seated before the tanks"; these men did the actual developing. The developed prints then ran on tracks to "hypo baths" and to the wash room, where running water cleared the film of chemicals. The film then continued on the track to the drying room. At Vitagraph a photograph of the developing room and the washing and cleaning room shows men attending to these activities.[60] After the film dried, it had to be polished, a task delegated to women because of their alleged dexterity. A visitor to the Selig film studio in Chicago in 1909 noted the "nimble fingered girls" who cleaned each section of film.[61] In 1911 future editor Viola Lawrence, a teenager, worked at Vitagraph hand-polishing films.[62] The Pathé Frères factory, always in the vanguard, used a machine by 1910 that polished and cleaned film.[63] In 1912, when the Crystal Film Company in New York mechanized film polishing, it boasted that its machine could perform the work that "formerly required ten girls to do."[64]

Alleged feminine dexterity also made coloring films a woman's job. Again, precedents for tinting as women's work existed in the photography trade and in the magic lantern slide trade. Daniel H. Briggs of Massachusetts, a popular lantern slide manufacturer in the mid-to-late nineteenth century, "had a dozen women working on the slides, which were often passed down the line, with each woman specializing in a different color."[65] Magic lantern slides continued to be used in moving picture theaters to entertain patrons between reels, but as the movies themselves ultimately reduced the demand for colored slides, prices fell drastically. By 1909 an editorial in *Moving Picture World* claimed that manufacturers were "running sweatshops, where they are taking their lost profits out of the girls who color slides."[66]

Some of these colorists found work in the moving picture industry. One Miss E. M. Martine of East Orange, New Jersey, described as an "expert film colorist" of twelve years, specialized in refreshing worn moving picture prints by adding color.[67] Nearly a guaranteed audience draw, colored films were created by tinting (dipping the film strips in vats of dye) or hand painting. Of the two, the latter was far more naturalistic and desirable, but it was extremely costly. Tediously painted frame-by-frame, a hand-colored film cost twenty-six cents per foot at a time when a standard uncolored film cost only ten cents per foot. One contemporary estimated that if a manufacturer wished to release only one colored film a month, and sell one hundred prints of it, he needed a constantly employed staff of one hundred skilled colorists.[68] Few films, then, were colored by hand, and they were typically done by special order. The one exception was the colored films of Pathé Frères. This French company patented a stenciling machine able to color long strips mechanically. Though technology replaced most of the manual work with a machine, it did not affect the sex of the workers in the coloring department. At Pathé's Paris factory four hundred young women attended these machines, and a smaller number of women retouched the frames by hand.[69]

Since the developing racks could handle only two hundred feet of film, every positive print of a film exceeding two hundred feet—nearly every film by 1910—was joined together by hand before being sent out.[70] This simple but laborious procedure, requiring "dexterity but not skill," was quite literally a textbook example of a female-typed job.[71] Indeed, a writer from *Moving Picture World* exclaimed after a 1910 visit to the Vitagraph plant that film joining was "a most congenial occupation for a number of girls and young women." The workplace was tidy, the women seemed happy, and he bragged that "this branch of service had opened up new and clean opportunity for many to earn a good living, free from many of the objectionable features of factory life."[72] Still, the duties of the film joiner were not easy. Film joining required diligent attention and speed, but joiners had few tools to help them. With no guide to align the film strips, and without a clamp to hold the cemented pieces together as they dried, the film joiner searched for the correct frame, snipped the film, brushed on odorous glue out of a bottle, and held the splice together with her fingers.[73]

Understandably, mistakes could be made in developing, editing, tinting, or splicing. Thus, before each film left the factory, it was inspected for errors. In 1910 Pathé Frères employees took the finished films to long, darkened rooms, where female inspectors sat two by two at small tables in front of small white squares on which the films were projected. Each inspector had a button that, when pushed, recorded the frame on which an error occurred.[74] Film importer George Kleine used mostly female film inspectors, who worked for $7 to $12 a week, but the few male inspectors he employed received $2 to $5 more weekly, with the exception of one particularly well-paid female inspector.[75] At the Selig Polyscope Company in 1919, all films were "subjected to the scrutiny of a lady examiner."[76]

The type of work offered to women in the film factories, then, fit within the culturally defined arena of women's work at the turn of the century: it was performed indoors, it did not require great strength or invite danger, and it required "dexterity but not skill." As in other kinds of factory work, women printers, cutters, joiners, and polishers were to be nimble fingered but not creative. Like the new clerical jobs opening up for women, film inspecting was clean and required some education and skill, but it was tedious work. And like clerical work, most of the jobs in the film factories appear to have been limited to white women. Existing etchings and photographs do not reveal any women of color working in early film factories, and there are no references to race or ethnicity in contemporary descriptions. Film historian Charles Musser's conclusion that the early film industry was a "white" world appears to extend to the laboratory as well.[77]

THE EARLY FILM INDUSTRY ENFORCED the two most important factors that kept women's wages low and limited their opportunities: the sexual division of labor and occupational sex-typing. Women worked in tedious, semiskilled jobs within the film factories, whereas men operated the cameras and projectors and made the creative decisions. The films themselves, though improving in quality, were made to demonstrate and advertise the equipment that made them possible. Even journalists writing about the film industry before 1908 tended to credit the machine itself with making the films, erasing the human labor involved.[78] Unlike the photography trade, there was little basis in the early film industry to argue that women's finer sensibilities would make them natural filmmakers. Finer sensibilities were not required. The moving picture was a product, the cameraman a technician, and the scientific, mechanical, and commercial aura that permeated the industry masculine. As long as the battle for the marketplace could be won largely through possession of superior technology—better cameras, projectors, film stock, and the patents to protect them—filmmaking would remain virtually an all-male endeavor. But to become more than just a novelty, the moving pictures themselves needed to attract audiences for their own sake. The industry achieved this, and more, over the next decade, experiencing explosive growth that not only lined the coffers of entrepreneurs but altered the neatly inscribed masculine world inhabited by the first cinematographers and entrepreneurs.

PART ONE

# EXPANSION, STARDOM & UPLIFT

Women Enter the American
Movie Industry,
1908–1916

The period from roughly 1908 to 1916, between the rise
of the nickelodeon and the consolidation of the central-
producer system, was without question the most promising moment for women in the history of the American film industry. First, the spike in demand for moving pictures at this time
encouraged producers to adapt methods and personnel from the
stage, an arena in which women enjoyed unique power. Second, the
confluence of the rise of film stardom and an independent movement gave the first stars enough leverage to enter production on
their own. And finally, the threat of state regulation inspired a gendered approach by the industry to gain middle-class legitimacy.

All of these developments helped to make women welcome in
the American film industry during this period, but it was the uplift movement that secured their position. Since the protection of
public morality and the definition of good taste fell squarely on the
shoulders of the middle-class matron, uplift was an inherently gendered project. Strategies included improving the theaters, cleaning

up the films and live acts, and redefining the movies as akin to the stage. All of these strategies elevated the position of women vis-à-vis the American film industry: as patrons, theater managers, actresses, directors, producers, and even (begrudgingly) as reformers. Thus, during this period women were both present within the industry and poised as a group to shape the outlines of the American cinema as it matured from a "penny amusement" to the most popular and lucrative commercial entertainment in the United States. At the end of this period the American cinema's ideal filmmaker was Lois Weber, the "domestic directress" who embodied uplift as a middle-class matron, a reformer, and an unquestionably talented filmmaker.

# CHAPTER ONE

# A Quiet Invasion

*Nickelodeons, Narratives, and the First Women in Film*

On November 13, 1909, *Moving Picture World* ran a small editorial under a large headline that proclaimed, in capital letters, "A Woman Invades the American Moving Picture Industry." The woman in question was Frida Klug, a representative of an Italian film exchange and "the only lady . . . to our knowledge to grapple with the intricacies of the film importing and renting business." Arriving in Chicago, which had become a national film distribution center, the "vivacious Hungarian brunette" found herself shut out by the Chicago alliance of film distributors. It may have been because she was a woman. But it also may have been because Americans blamed risqué imports for a recent wave of procensorship activity (although her company was known for its high-class product). Or it may have been because her company was not allied with the Motion Picture Patents Company, an Edison-controlled attempt to monopolize the industry. The explanation offered by *Moving Picture World* was that the snub stemmed from a "technicality, namely, that the business house she represents has no office in this country," but of course Klug was in Chicago for just that purpose. Chastising the Chicago men, the editor of *Moving Picture World* tried to make up for their faux pas by "extending her a most hearty welcome into the moving picture field." A few months later, *Moving Picture World* reported that Klug successfully established an office on the East Coast.[1]

In the chaotic expansion of the nickelodeon era, which began in 1905, the American film industry experienced massive industrial change. This change brought with it a gender revolution. Frida Klug was, in fact, something of a latecomer. By the time the "girl" from Turin crossed the Atlantic, women were already working in the exhibition and production branches of the American film industry: as theater owners and managers, actresses and scenario writers, and directors and producers. Though women were rarely mentioned in the trade press, the film industry included them in other positions as well. When a column entitled "With the Film Men" noted Agnes Egan Cobb's resignation from the Itala Film Company in 1913, she was said to have already "served in executive positions with some of the biggest concerns in the country."[2] Although this gender revolution changed the face of the American film industry, the arrival of women into its various branches was rarely as traumatic as it appears to have been in Chicago.[3] Women arrived not as "invaders" but rather as a consequence of industrial reorganization. Between 1903, when a new distribution system created the basis for the explosion of little storefront theaters, and 1907, when a new mode of production was firmly in place, every aspect of the industry changed. The presence of women was but one result of a holistic transformation.

From the motion picture's public debut in 1896 to about 1903, the manner in which films were made, distributed, and exhibited changed little. Movies, made by men, were mere minutes in length and sold by the foot. They were exhibited in vaudeville theaters, traveling exhibits, or in the increasingly popular penny arcades. As we have seen, the basis for competition during these years was not so much the films as superior cameras and projectors, and, hopefully, the patent rights to go with them.[4] In 1903, however, two events occurred that shifted the basis of competition from cameras and projectors to the films themselves: the rise of the narrative film and the creation of the first film exchanges. These events initiated a process that would soon change the way movies were made, exhibited, and distributed, bringing women into an industry where movies had been, in the words of Charles Musser, made "by men, for men."[5]

THE RISE OF THE NARRATIVE, OR "STORY," FILM was a departure from the newsreels, scenics, and travelogues that dominated pre-1903 cinema. By the start of the twentieth century the novelty of seeing everyday events onscreen had worn thin, and the popularity of moving pictures tottered. Some filmmakers attempted to tell stories before 1903, but such efforts were hamstrung by technology. It was difficult to tell a comprehensible story with a beginning, a middle, and an end when films were only a few minutes long.[6] This situation changed just after the turn of the century, when Edison technician Edwin S.

Porter invented a projector able to handle a one-thousand-foot reel of film, making it possible to exhibit a fifteen-minute movie uninterrupted. Recognizing the potential popularity of little fifteen-minute "playlets," Edison built a studio in the heart of the New York theater district, close to props and experienced personnel. At this new Edison studio Porter created the famous nine-shot *Life of an American Fireman* (1902–3), followed by *The Great Train Robbery* (1903). Within a matter of months Edison's competitors were also making or importing narrative films. By 1904 narratives made up a little more than half of all copyrighted film titles in the United States.[7]

After the introduction of the story film the number of moving pictures screened in vaudeville theaters, penny arcades, and traveling exhibitions rose significantly.[8] The increased demand for movies encouraged the production of fiction films. Unlike newsreels and travelogues, where much money was wasted while cameramen and employees waited for unpredictable events or perfect weather, the production of story films could be planned in advance and executed on a schedule.[9] The Biograph Company was the first to fully realize the efficiencies of narrative production; Biograph built a studio with sufficient indoor light to allow filming despite inclement weather or darkness. Other companies (at least those with sufficient capital) soon built their own studios. But although this development represented a major shift in film production, it had little effect on the gender of filmmakers. Men still made the movies.

The arrival of the little narratives increased the popularity of moving pictures, but they were still sold in an inefficient manner. People wanting to exhibit a film—owners of penny arcades, vaudeville impresarios, or traveling lecturers—had to choose between the lesser of two evils. Would-be exhibitors might spend a great deal of money to buy the film outright from the manufacturer, running it over and over for different audiences. Or they could hire an expensive exhibition service, such as that offered by the French Lumière company, consisting of films, projector, and projectionist. Both options were costly, so there was little to recommend movie exhibition as a profitable endeavor.

This situation changed in 1903 with the appearance of the first film exchanges. Rather than buying films directly from the manufacturer, exhibitors across the country could now rent a variety of movies from a local film exchange for reasonable prices. It is now that we begin to observe women outside of the film factory, working in the collaborative partnership configuration that Charles Musser identifies in the industry prior to 1907, only in film exhibition rather than production. Women in collaborative partnerships tended to work with male relatives, continuing the structure of the family firm, a traditional business form that offered women a legitimate if often circumscribed role. Kathryn Fuller-Seeley's research reveals that Fannie Shaw "Harris" Cook worked in partnership with her husband Bert Cook, taking the Cook and Harris High Class Moving Picture to towns in central New York and northern New England from 1903 to 1911. The Cooks were inspired to form their partnership when Bert could not find employment with itinerant exhibition companies if his wife worked and traveled with him. "Ironically," Fuller-Seeley notes, one of the managers who refused to hire them was a woman, Mrs. Alonzo Hatch, who needed help running the Alonzo Hatch Electro Photo Musical Company after her husband was burned in a projector fire. It was Fannie Cook who reassured Bert in 1903 that they would be a success once they became their "own bosses" and that she was "the little Girl that is in for making money." As Fuller-Seeley notes, there were probably many more women working as itinerant exhibitors.[10]

Now that films could be reasonably rented from exchanges, it made sense for the exhibitor to stay in one place, as the local audience could be presented with different films every few days. In 1905 the first five-cent movie theater opened in Pittsburgh. By 1907 these little theaters, now known as nickelodeons, were "multiplying like guinea pigs" all over the nation.[11] In that year *Moving Picture World* boasted that the nickelodeon "has attained that importance where we may no longer snub it as one of the catch-pennies of the street."[12] By 1910 just under 20 percent of the nation's population visited nickelodeons every week, generating gross receipts of $91 million.[13] Scholars disagree over the class origins of the first audiences for the movies, disputing whether the movies appealed only to the working class at this time or whether the well-to-do patronized the nickelodeon. But all note the high proportion of female patrons from the earliest days of the nickel theaters.[14] Like the working-class men who attended urban nickelodeons, women had little spending money or leisure. For working-class women the nickelodeon was a particular boon. In the words of one observer, women previously confined to sitting on the doorstep and watching "the teams go by" could now afford a glimpse of "real life."[15] One writer for the *Manchester (NH) Mirror* reported watching one "matronly" woman, armed with bundles from the day's shopping, convince a similarly burdened woman to take a break at a nickelodeon. When her friend protested, she replied, "It's only a nickel," and that she would be able to get her husband "all the better supper" for having rested before going home.[16] In addition, everyday clothes were quite acceptable, as were small children.[17] Even for more prosperous women, and women outside major cities, the nickelodeon was a welcome novelty. In 1908 a "lady correspondent" from the *Boston Journal* sheepishly admitted that she had contracted "the moving picture habit," and while waiting for an equally embarrassed friend outside a nickelodeon, she described the ease with which women entered the little theaters. After a family of three—a mother, father, and a toddler—came a young woman who "looked as though she might be employed in one of the great department stores." Then arrived three women, evidently "winding up an afternoon's shopping in town" before "returning to their homes to preside over their own supper tables and afterward put the babies to bed."[18]

For a few women the nickel theater offered not only entertainment but also employment. Nickelodeons were often run as family businesses.[19] Even outside of family businesses, however, women found employment in nickelodeons. The typical nickelodeon operation had eight positions: manager, cashier, doorkeeper, usher, projectionist ("operator"), pianist, singer, and janitor. Bert and Fannie Cook illustrate the gender typing of exhibition work that existed even before the nickelodeon: "Bert was the manager of the troupe of four performers and an advance man; he also served as projectionist and occasional soloist. Fannie was musical director, accompanist, ticket seller, and treasurer."[20] Only the positions of projectionist and ticket seller were solidly sex-typed.

Occasionally a female member of a family-owned theater might operate the projector, but this was rare.[21] Running the projector during the nickelodeon period was dangerous and mechanically demanding. The light source was not an encased bulb but an arc light, created by alternating currents surging through two slightly separated carbon rods. Arc lights needed constant care and sputtered dangerously close to the flammable nitrate film. In the summer of 1907 Ohio alone reported two to three projector-related fires a week. To protect patrons, local ordinances required projectionists to work in steel operating booths, which often reached temperatures of 113 degrees Fahrenheit and were filled with dust from the burning arc light. Additionally, projectionists were responsible for keeping both film and projector running smoothly. They were expected to know how to install electrical wire, make wire joints, operate electrical resistance devices like rheostats and transformers, obtain the sharpest focus based on the size of the theater and the type of screen, choose the correct lens, and splice broken films. Keeping both projector and film running required dozens of tools, including three different kinds of pliers, two screwdrivers, a cabinet rasp for sharpening the carbons, two different types of hammers, and a small gasoline torch for soldering wire joints. This was not clean, repetitive, unskilled, or low-risk work, so it fell far outside the definition of appropriate work for women.[22]

Although believed to be too demanding and dangerous for women, projector operation soon became the province of boys. While cinematographers retained an artisanal work culture, projector operation quickly deteriorated into a sweated trade as thousands of new nickelodeons opened after 1906. Efforts were made to form "operator's schools," but most nickelodeon-era projectionists learned on the job in deplorable conditions, laboring up to twelve hours a day, six to seven days a week in hot, cramped, and dust-filled sweatboxes. It was little wonder that operators were the first in the industry to attempt unionization in 1907. But although projector operation became a sweated and low-paid occupation, it retained its aura of technical skill and masculinity.[23] After World War I, when the Red Cross published a textbook aimed at retraining disabled soldiers for jobs as projectionists, the long list of subjects included the use of an arc lamp, transformers, mercury arc rectifiers, rheostats, motor generators, batteries, construction of lenses, the "theory of light," the "construction and care of projecting machines," and the "handling, care, and repairing of films." The list of topics suggests a conscious definition of the work as highly skilled.[24] In fact, projectionists patrolled the borders of their trade as carefully as the cinematographers: by the end of the silent era the International Alliance of Theatrical Stage Employees and Moving Picture Operators declared that if women attempted to "invade" the projection room, it would exclude them by law.[25]

Whereas projectionists were male, their sex-typed counterparts were the young women who sold the tickets. Cashiering was a feminized job, most notably in department stores, where female clerks provided a genteel touch at low pay.[26] The purpose of the "girl in the box office" was not only to sell tickets at the window but to attract potential patrons from the street, a tradition that continued beyond the nickelodeon era and resulted in the familiar glass-enclosed box office located as close to the sidewalk as possible.[27] Nickelodeon cashiers made roughly $4 to $6 a week, the same wage as female department store clerks and semiskilled factory workers. After the cashier sold the tickets, they were taken by the doorkeeper inside, who was usually male. But some female cashiers were required to do more than just sell tickets. At one busy Chicago nickelodeon, for example, the female cashier took "entire charge of the house" during the slow supper hour, acting as cashier, doorkeeper, and usher. At another nickelodeon the cashier also led the between-reel sing-alongs.[28] Nickelodeons needed a pianist to accompany the movies and sing-alongs, and here, too, a woman might find opportunity for work. But as the nickelodeons began to expand and improve after 1908, most exhibitors began hiring professional pianists, singers, and vaudeville talent to entertain patrons between the reels. But even among professionals, female entertainers were common.[29]

The two most important occupations within the nickelodeon were owner and manager. Both were often one and the same, and it appears that partnerships were common. Before 1907, nickelodeons were relatively inexpensive business propositions and appealed to small investors who were often, but not always, from immigrant and working-class communities.[30] They were also frequently female. From the time that *Moving Picture World* first began printing lists of new nickelodeon owners in 1907, the names of women appear consistently. In 1907 the trade journal reported that Mrs. A. R. Lewis, of Salina, Kansas, sold out half an interest in "the Nickelodeon" to E. H. Brown, and was "preparing to open another amusement house here in the near future." In June of 1909 Mrs. Burns of Ottawa, Kansas, bought the half-interest of her partner, Miss Pearl Chalmers, in the Yale nickelodeon. In August Anna Kiel of Chicago secured a permit to build a moving picture theater on Twelfth Street, and Mrs. Stewart leased the airdrome theater of Warrensboro, Missouri, to F. J. Bailey. That same month Mrs. F. F. Fuller "bought all the moving picture theaters" in Hartford City, Indiana.[31] Bert and Fannie Cook settled down to run a nickelodeon in 1911.[32] It is difficult to know which women (or men, for that matter) actively managed their theaters, in comparison to those for whom owning a theater was simply an investment.

There are several possible reasons women were attracted to nickelodeon ownership. First, American women enjoyed a long history of proprietorship. Evidence suggests that 10 to 25 percent of all women in preindustrial America were "engaged in entrepreneurship" and that half the urban retailers in seventeenth-century America were women. Typically women worked in fields considered an extension of their "natural" duties in the home, such as cooking and dressmaking or running boardinghouses, inns, restaurants, or taverns, although exceptions were found among widows who continued their husband's trade.[33] It is interesting that as the separate spheres ideology of the early nineteenth century encoded public space as male and defined the home as the proper place for women, female entrepreneurs continued to run businesses serving an increasingly male clientele. Christine Stansell's study of women in nineteenth-century New York City revealed that when combined with the sale of cheap liquor, female-run businesses such as groceries, inns, and coffeehouses became "bawdy houses," catering to "sexual license, male rowdiness and 'bonhomie.'"[34] This was true for the stage as well. The number of women managing theaters and traveling troupes increased after 1820, when rowdy masculine audiences and balconies of prostitutes characterized the American theater.[35] Women, then, already existed as proprietors of commercial amusements well before the arrival of the nickelodeon.

What put nickelodeons in the reach of female entrepreneurs was the fact that they required little cash up front. As Angel Kwolek-Folland discovered, women were "penny capitalists"; compared to men's, their businesses were "undercapitalized, more ephemeral, and short of credit."[36] In addition, nickelodeon management required little experience, at least early in the boom. A trade journal printed a nickelodeon "recipe" in 1907, beginning with the ingredients: one storefront shop seating two hundred to five hundred persons, chairs, a phonograph, one "young woman cashier," one electric sign, one projector and projectionist, one screen, one piano, one barker, and one manager. After that, add "a few brains and a little tact." Then you "open the doors, start the phonograph, and carry the money to the bank."[37]

Despite the relative ease of entry into film exhibition, this was the most important branch of the industry during the nickelodeon era. It was the exhibitor who created the nickel theater experience, arranging the brief films into coherent programs, choosing the live entertainment, and often interpreting the films for patrons by providing a lecture to accompany the screening. Indeed, exhibitors were encouraged to manipulate the films to suit their audiences. "If the piece grows dull at any point," noted *Moving Picture World* in 1907, "the manager can take a pair of shears and cut out a few yards" to liven it up.[38] Our current perception of movie theater proprietors as mere franchisees from large distribution agencies would not become accurate until after the mid-1920s. The fact that women were a part of this branch of the film industry during the nickelodeon era meant that they helped to shape the experience of the first filmgoers. Yet female exhibitors did not attract any particular attention in the pages of the trade journals until the censorship crisis of 1908 and 1909, after which they were often praised for naturally running cleaner shows than their male counterparts. But before that moment their presence elicited little comment in an industry that was still small and decentralized.

THE POWER OF EXHIBITORS TO CONTROL THEIR OWN SHOWS—and thereby provide a more attractive product than their competitors—was not absolute. Other than on-the-spot editing there was little exhibitors could do to improve the movies, and the high demand meant that there was little incentive among producers to improve the product. According to one frustrated exhibitor, nickelodeon owners were at the mercy of the film exchange operator, who in turn "must take whatever the manufacturer chooses to push on him."[39]

Films did begin to improve, but at first this was merely a by-product of demand. By 1907 most nickelodeons were requiring a fresh program of films every day.[40] As manufacturers worked furiously to produce mostly narrative films, hiring actors by the day, it became clear that they needed to reprise a tradition that was nearly lost in the legitimate theater—the resident stock company, with its fairly permanent group of actors filling the standardized roles of leading lady, leading man, character actors, comedians, and inge-nue.[41] As film producers began hiring cinematic stock companies, women entered the production branch of the industry as actresses. Vitagraph was the first to assemble a stock company in 1907.[42]

As many historians note, the film actor was not the envy of her or his compatriots on the stage. Nearly all early movie actors and actresses were drawn from the mass of unemployed stage actors who gathered in New York each summer. Without work until the touring season began in the fall, many were financially desperate when they first stepped before a camera.[43] The standard $5 a day for movie work was good pay, but for several reasons the work was less than appealing. "Canned drama" suffered from a poor repu-tation, and personal ego, even among unemployed thespians, was an issue. Worse yet, some manufacturers looked on professional stage actors and ac-tresses as a necessary evil. When the writer of the *Cyclopedia of Motion-Pic-ture Work* (1911) took his readers through a mock production, he described one scene as requiring "a bunch of troublesome actresses" and thus "a large expense for wages."[44] Although this point of view might have been extreme, early movie actors were not coddled as *artistes*.

For women the industry's hiring of professional actors proved especially helpful, since previously female roles were sometimes played by men. Mabel Rhea Dennison claimed in 1909 that an actress's "chances of making a liv-ing have been increased by the rise of the biograph machines . . . Every year there has been an increased demand for women to pose," Dennison stated, "and indications are that the demand will go on increasing, for instead of one concern in the field, there are now fifteen at least."[45] Like most actors, the first film actresses hailed from the lower end of the boards: cheap vaude-ville, touring melodramas, and low-priced plays. Future film star Florence Lawrence ended up in the movies after her mother's traveling troupe fell on hard times. The teenaged vaudeville veteran and her mother traveled to New York in 1906 with the hope that Lawrence would land a role in an upcoming play, but the summer ended with no engagement and so, too, the fall. Running out of savings, mother and daughter resorted to loitering at the door of the Edison company, where actors were casually handpicked for the day's work. Allegedly due to her ability to ride a horse, Lawrence was cast in *Daniel Boone; or, Pioneer Days in America.* She found to her surprise that moving picture work was not entirely distasteful.[46] Gene Gauntier, a stage actress specializing in melodramas, tried to save money "against the rainy day when the season would end and [she] must go back to New York." In the summer of 1906, however, her savings were gone by June. She thought of moving pictures, "a new opening for actors," but "looked upon them with scorn." When a fellow stage actor, Sidney Olcott, convinced her to at least give the movies a try, she auditioned for Biograph. On her first day before the camera, the filmmaker asked Gauntier, a nonswimmer, to dive off a cliff. She did it and was rewarded with the valuable respect of Biograph director Frank Marion, who "for several days would not even consider another lead-ing woman."[47] Once they got over their initial distaste, many stage actors and actresses found moviemaking intriguing and the possibility of steady employment attractive.

The hiring of stage actors signaled the beginning of a major shift in the production of moving pictures. As output soared between 1905 and 1908, manufacturers realized that they needed to make movies more like the stage, so they constructed studios that more and more resembled theaters.[48] To make story films, manufacturers required the building of sets, the use of props, an array of costumes, and all the other accoutrements of the theater.[49] Now called "studios," the film factories boasted "all that pertains to the the-ater except the auditorium."[50] All of these theatrical needs required person-nel, and film manufacturers quite logically looked to the cheapest source of trained talent—their own actors and actresses.[51] Within the stock companies both men and women were not only allowed to perform offstage duties but were expected to do whatever tasks necessary to speed production along. Vitagraph cofounder J. Stuart Blackton described this practice as "doubling in brass," an old minstrel term meaning the need to perform double duty.[52] Florence Turner, Vitagraph's first leading lady, not only assisted Vitagraph's wardrobe mistress (her mother) but attended "to all the business affairs of the concern as well as act. I was a sort of handywoman! I paid the staff and the artistes, I kept the books, and was clerk, cashier, accountant, and actress all rolled into one."[53] Blanche Lasky, who coproduced and acted on the variety stage with her brother Jesse, worked with him in his new filmmaking concern as well. At the New York offices of the Jesse Lasky Company, an interviewer noted in 1914 that Blanche "passes the final decision on every important matter that comes up in the multifarious businesses of the company, from the question of which novel or what play is to be made into moving pictures to the way in which the leading lady out on the coast shall do her hair."[54] While men always dominated the numbers of film directors and producers, even after the rise of the feature film, women were able to try their hand at nearly all types of production work. Frances Marion (no relation to director Frank [Francis] Marion) remembered that her first movie job in 1914, as an assistant to Lois Weber, allowed her to try "every kind of job I could find except emptying the garbage pails." She worked in costume, set design, and in the editing room, and ultimately became a filmmaker herself.[55] At about the same time, writer Beulah Marie Dix recalled that in addition to writing scenarios, she worked as an extra, tended the lights, and "spent a good deal of time in the cutting room."[56] At the Griffith studio Lillian Gish helped the laboratory technician develop negatives, and when she wasn't busy in the laboratory, she worked with the head electrician on lighting effects and screen tests.[57]

Many early film actors grumbled about their offscreen duties, which often required manual labor and kept them at continuous work throughout the day in a manner resembling a factory.[58] But not until 1911, when renegade actor Maurice Costello flatly refused to build sets and paint scenery, were actors relieved of such duties if they wished to be. Historians interpret Costello's moment of thespian indignation as progress, indicating the rise of the distinct profession of film acting.[59] But it was the fluidity between the emerging crafts, the theatrical tradition of "doubling in brass," that created opportunities for women, as well as men. It kept each craft roughly equal in status and lowered incipient gender boundaries.

Manufacturers turned to their stock companies not only for mundane administrative tasks but for help in production. Film manufacturers assumed that newly hired stage actors, regardless of their sex and however minor their previous stage careers, brought with them a modicum of dramatic know-how. Although memoirs are fraught with distortions, the reminiscences of several producers and actors suggest a relative lack of hierarchy and abundance of collaboration in the period between roughly 1907 and 1909. At

Carl Laemmle's Independent Moving Picture Company (IMP), "the actors would go out at night and see a [stage] show, and in the morning come back and make a picture."[60] The collaborative work culture that distinguished filmmaking in the nickelodeon era was especially evident at the Kalem production unit headed by Sidney Olcott between 1907 and 1910. Responsible for budgeting and directing the films, as well as for enforcing discipline, Olcott turned the production tasks over to his company of actors.[61] Leading actress Gene Gauntier wrote the scenarios, and on rainy days the men of the company scouted for locations. After the locations were chosen, the "plan of procedure was mapped out." Olcott expected an actor to take "upon himself certain tasks" and "lend a hand in all emergencies." In the evenings all the members gathered to hear the scenario read aloud, discuss casting and makeup, and edit the films. After dinner they would gather in a makeshift projection room, "pencil and notebooks in hands, our number augmented by any outsiders who had worked in the film as well as by friends," and offer a "running comment of criticism and praise" as they watched the day's rushes. Before retiring, "a discussion in Mr. Olcott's room [covered] every little detail."[62]

Under this new theatrically based system of production, the film director supervised the overall performance in front of the camera, leaving the cameraman to focus on the technical demands of moving picture photography.[63] Indeed, it is accurate to dub this new worker the director-producer. Although work culture varied from studio to studio, it was increasingly the role of the director to assume responsibility for the film's subject, action, cast, crew, location, and budget. Ultimately, the director supervised the cinematography and editing.[64]

Directors gained a certain amount of authority and respect within the studio, and those chosen to direct had to exhibit some talent. But in the rush to meet demand, the position of director-producer remained remarkably open. A mere six months after joining Biograph in 1906, an undistinguished actor known as Larry Griffith gained his chance behind the camera and was soon directing under the name D. W. Griffith. Just a year before becoming a director for Kalem in 1907, Sidney Olcott belonged to New York's army of unemployed stage actors.[65] Even gender proved no barrier. Lois Weber, a promising musical actress, emerged from a brief sabbatical from the stage by writing, directing, and playing the lead in her first movie in 1907. Actress Gene Gauntier found herself codirecting, as well as writing and performing, less than a year after joining Kalem in 1906.[66]

In addition to director-producer, the role of scenarist, or screenwriter, emerged during the nickelodeon era as well. While impromptu acting before the camera still occurred, scenarios (a brief outline of the film's story and action) became standard as films grew to one thousand feet. Early film manufacturing companies often expected their new director-producers to write their own scenarios, as "surely no author is better qualified," but scenarios were accepted by anyone with imagination or a good memory.[67] Actors and actresses commonly contributed scenarios; before the first successful copyright infringement case (over Kalem's 1907 *Ben Hur*), actors drew plot ideas from their previous stage productions. But studios also solicited story ideas from freelance writers, stole them from other studios, and even recruited them from the public at large through contests. Given the interest of women in the movies and popular fiction, the winners of such contests were frequently female. In 1909 Evangeline Sicotte of New York City won $150 in the Georges Méliès Scenario Contest for her script "The Red Star Inn," and Florence E. Turner of Brooklyn won third place, receiving $50 for "The Fiend of the Castle."[68] A scenario submitted by Mrs. Clemens to the *St. Louis Times* not only resulted in a cash prize but also reached the screen in 1910 as a film entitled *The Double.*[69]

Unsolicited scenarios were accepted by many studios well into the mid-1910s. The definition of the freelance scenario writer as an "amateur" made it amenable to women, a label that stuck even when writers were paid for their work, mostly because it paid too little to provide a living. Outside of hefty prizes for contest winners, scenario writers generally received $5 to $15 a scenario—a sum considered paltry by a writer for *Moving Picture World* in 1910.[70] The pay was immaterial to dozens and perhaps hundreds of film fans. In 1911, tired of receiving scenarios from hopeful readers, *Moving Picture World* published a list of film manufacturing companies and their addresses under "Attention, Scenario Writers."[71]

As thousands of scenarios arrived at studios on a monthly basis, studios set up scenario departments where outside scenarios could be read and analyzed and where staff writers might work on original plots, draw from already published material, and work on intertitles, which had been almost an afterthought before the nickelodeon era.[72] Women often populated the first scenario rooms. Mrs. Breta Breuil worked as a scenario editor for the Vitagraph Company from 1910 to 1913, when she was replaced by Mrs. Catherine Carr.[73] Historian Wendy Holliday found that screenwriting in the early 1910s created a particularly "modern" heterosocial work culture in which male and female writers, like actors and actresses, were roughly equal, having a hand in all phases of production. "It was all very informal in those early days," recalled writer Beulah Marie Dix. "Anybody on the lot did anything he or she was called upon to do."[74]

Recruited from these early stock companies, the first directors, producers, and writers of the movies enjoyed unusually low craft and gender boundaries. This was not simply due to the high demand for films or the novelty of filmmaking. It was true that film exchanges needed films and that studio executives were primarily concerned that sufficient products flowed out each week. But it was also true that established theatrical practices, the tradition of "doubling in brass," made the fluidity between craft and gender boundaries possible. These men and women were not inventing film craft out of whole cloth; most of the work closely resembled the production of staged drama. The only exception was cinematography, where masculine associations prevailed. Otherwise, men and women moved smoothly between acting, writing, directing, and other sundry duties behind the camera.

This is clearly illustrated by the early career of Gene Gauntier. In 1907 Gauntier considered herself a stage actress temporarily dabbling in the movies for Biograph. That year Frank Marion, the Biograph director for whom she had dived off a cliff, decided to start his own company. Marion convinced Samuel Long, manager of the Biograph film laboratory, to join him, along with film importer George Kleine. When their company, Kalem, was founded in 1907, Marion hired Gauntier as leading lady and Sidney Olcott, heretofore an actor, as director-producer. Marion appointed himself scenarist and, according to Gauntier, hastily scribbled abbreviated stories on the backs of business envelopes for Olcott to decipher.[75]

Before 1907 was over, the increased workload forced Marion to give up scenarios, and since "there were no trained writers" in the new medium of the film narrative, he asked Gauntier to give it a try. She fashioned a scenario from a 1905 stage melodrama, "Why Girls Leave Home," in which she had once played. Although her first effort was, in her own estimation, "hopeless," she soon became "the mainstay of the Kalem Scenario Department." Each week she played the female leads in two pictures while writing two or three scenarios.[76]

Gauntier left the movies to act on the stage in late 1907, but when her tour ended in the spring of 1908, Kalem offered her a summer salary of $20 a week. For this she was to "write, assist Mr. Olcott in the direction, and act when I liked." She immediately accepted, but Biograph also requested an interview, and she stopped by her old studio out of curiosity. Henry Marvin of Biograph made her a similar offer—"scenario editor, studio manager, supervising director, but no acting." Since she was satisfied with the Kalem offer, and was friends with Sidney Olcott besides, she "audaciously" told Biograph that she would work for no less than $40 a week. To her surprise, they agreed. Eventually she accepted the offer. In an indication of the quickly moving division of labor, Gauntier's duties at Kalem were now split between four people—Robert Vignola as assistant director; Marion Leonard as leading lady; Kenean Buel, who looked for locations, props, and costumes; and Frank Marion, who once again took up scenario writing after Gauntier's departure.[77]

Working conditions at Biograph did not suit the restless Gauntier. She supervised last rehearsals, wrote and bought scenarios, and interviewed actors.[78] Unlike her outdoor work at Kalem, the Biograph job often kept her deskbound, "working over scenarios and interviewing people all day—and watching the hands of the clock for five." Nevertheless, the respect she received pleased her. "Every morning I went up to Mr. Marvin's office," she recalled, "to talk over plans for the next day, discuss new productions and Biograph business in general. It is still rather surprising to me how confidential these heads of both Kalem and Biograph were with a mere slip of a girl."[79] Gauntier returned to Kalem in the winter of 1908. By 1909 she was still doing her "triple job of writing scenarios, playing the leads, and helping Olcott with the directing" at Kalem's winter location in Florida. In 1910 Kalem offered Gauntier the position of director-producer of the Florida unit after Olcott's resignation, but she refused on the grounds that it was too strenuous. Instead, Gauntier spent 1911 "mastering many new secrets of the industry," such as film development, printing, tinting, retakes, titling, editing, "and other mechanical details." She never went back to the stage.[80]

THE FORMER STAGE ACTORS AND ACTRESSES who doubled in brass in the early film industry, like Gauntier, did improve dramatic quality. But the ruthless demands of the standard eight-hundred-foot to one-thousand-foot one-reel film (approximately ten to fifteen minutes) compromised story lines. If the "motion scenes" were in error, the high cost of film stock demanded that the subtitles be rewritten to fit the scenes already shot, no matter how awkward the fit.[81] Even at Kalem, a studio Gauntier considered to have "artistic leanings," director Olcott would "dance up and down shouting: 'Hurry up, folks, film's going. Grab her Jim; kiss her. Not too long, Quick!'"[82] In fact, the perception of film as a technological product (rather than a dramatic art) reached its zenith during the nickelodeon era. This view became manifest in the Motion Picture Patents Company, which attempted to monopolize the industry.

Edison created the Patents Company in 1908, after winning court cases involving several vital camera patents. To celebrate his victory, Edison created a cartel of the most important film producing companies then in existence: Biograph, Essanay, Lubin, Selig, Vitagraph, Kalem, Méliès, Pathé Frères, George Kleine (an importer), and, of course, his own company. These licensed companies paid royalties to Edison for the use of his patented designs. In return they enjoyed freedom from litigation, access to raw film stock, a dependable release schedule (through the cartel-owned distribution company), and a standardized price per foot (ten cents) for films. The Patents Company's standardized film length, per-foot pricing, and scheduled weekly output epitomized the "technological view" (as opposed to any artistic vision) of the movies.[83] This perspective worked against women's progress into positions of authority, for it drew from the masculine meanings of scientific and mechanical expertise. Yet while the Patents Company consolidated its technological approach to moviemaking, the first female film producer appeared. In October of 1910 *Moving Picture World* profiled the new Solax Company, whose "chiefest and most valuable" asset was its director-general, Madame Alice Guy Blaché.[84]

Alice Guy Blaché was neither an early entrepreneur nor an actress working her way up by doubling in brass. She was a French expatriate who brought a decade's worth of experience as a film director with her when she arrived in 1907. Although her name was not well known among Americans, Alice Guy Blaché was one of the first filmmakers in the world. It was while working as a secretary to photographer and early filmmaker Leon Gaumont that Guy first tried her hand at making movies. Like his American counterparts in the 1890s, Gaumont was more interested in camera technology than in what was actually filmed, and he shot primarily street scenes. When Guy, who dabbled in amateur theater, suggested that Gaumont film a story, he said it seemed "like a silly girlish thing to do."[85] Gaumont agreed to let Guy try her hand at filming stories as long as it did not interfere with her office duties. Sometime between 1897 and 1900 she made her first film, *La fée aux choux* (The Cabbage Fairy).[86] Decades later, Guy observed that "if the future development of motion pictures had been foreseen at this time, I should never have obtained his consent. My youth, my inexperience, my sex, all conspired against me." Even at the time, Guy was not without enemies. Once filmmaking became "interesting, doubtless lucrative," she claimed, "my directorship was bitterly disputed."[87] The Gaumont property man in particular proved irascible. Guy went to great lengths to secure finely painted sets for her feature film *La Passion*, only to find them chopped up and wrapped around the pipes in the property department to keep them from freezing. According to Guy's biographer, Alison McMahan, the man was a frustrated director who, after being allowed to direct one film, did not prove talented enough for more. It seems Guy's coworkers had little patience with this person and sided with Guy.[88]

Guy's *petites histoires*, initially made on her own time, succeeded, and she soon became Gaumont's most important director. By 1906 (the exact date is unclear) Guy's *La Passion* was one of the first multireel feature films—a two-thousand-foot passion play using nearly three hundred extras at a time when American audiences were viewing five-hundred- to eight-hundred-foot films in cramped nickelodeons. But European films were not always so highbrow; Guy learned to make movies the same way the first male filmmakers did—by shameless plagiarism. But unlike nearly all other filmmakers at the time, she did not act as her own cinematographer, lending more credence to the argument that camera operation was masculinized from the start. Guy copied Lumière films most frequently, but soon she made films her own way. McMahan argues in her groundbreaking study of Guy's extant films that she not only showed movies in which "women also worked and played in the world alongside men" but "began to articulate a female address, a layer of messages aimed at women, mostly in the form of satire on heterosexual relations." Even so, some of Guy's films were vulgar from the point of view of the American middle classes. The comedy *Madame a des envies* (Madame Has Her Cravings, 1906) depicted a visibly pregnant woman giving in to a craving for sugar, making "amusing facial contortions" as "she sucks on the phallic sugar stick in a very suggestive manner." Importantly, the focus of this film is a woman who is satisfying her own desire, not unconsciously providing visual pleasure for men in the audience as was true for so many early peep-show-style films.[89] For American reformers, however, such a film would undoubtedly underscore the moral depravity of moving pictures, and of French films in general.

Gaumont, like his counterparts in the United States, deeply concerned himself in the technical end of the business, and like several other inventors, including Edison, he developed a fairly successful talking-picture device, which he promoted from 1902 to 1913. Guy made more than one hundred sound films for Gaumont's Chronophone between 1902 and 1906, while continuing to make regular silent films. The Chronophone consisted of a separate wax recording device that played music or voice in sync with the onscreen images. As a director, Guy played the prerecorded sound while dancers danced, or actors lip-synced to their own voices. Exhibitors needed to precisely link the film and recorded sound or lose the entire effect, though when done correctly, the process worked well. Technical problems surrounding synchronization and amplification prevented it (and other inventions like it) from inaugurating the era of sound movies, but Gaumont hammered away at improvements and at potential markets. Guy even nearly secured Caruso for a *phonoscène*, but the great opera singer declined.[90]

It was the Chronophone that brought Guy together with her husband, Herbert Blaché, and then brought them to the United States. Guy met Blaché, a Gaumont employee, while shooting a *phonoscène* of a bullfight at Nimes. Guy asked Blaché to fill in for her ailing cameraman, so Blaché took his one and only turn behind the crank. He apparently overexposed the film. In any event the couple fell in love on the romantic set (the toreador dedicated his kill to Guy). After making a trip together to Germany at Gaumont's request to instruct exhibitors on the proper use of the Chronophone, Blaché imported his English father to ask Guy's family for her hand in marriage. It was 1906; Guy was thirty-three, and Blaché was twenty-four.[91]

Guy feared that Gaumont would fire her when their engagement became known.[92] Her status did change. The Blachés, with Gaumont's blessing, severed their official relationship with the French company and went to the United States in 1907 to oversee an effort by Cleveland entrepreneurs to create a Chronophone franchise. Guy-Blaché was in some ways quite traditional and seems to have temporarily given up her own career to follow her new husband. For nine months they lived off their savings, but the Cleveland Chronophone venture failed. Broke, Blaché asked his former employer for work at Gaumont's new film processing plant on Long Island. This plant would soon house Solax.[93]

The would-be Gaumont plant sparked protest from the Patents Company, which feared the plant would be used to house a new nonlicensed competitor. While most movie manufacturers welcomed the creation of the Patents Company, and with it the end of litigation over patent rights, film distributors and exhibitors took a different view.[94] The Patents Company's aim was not only to control film production but to control distribution through the General Film Company, which distributed the product of licensed companies. In 1909 the General Film Company began buying film exchanges across the nation. Within a year and a half many exchanges went out of business, and most of those remaining were owned by General Film. Understandably, distributors were angry. But so, too, were some managers of nickel theaters. Although the General Film Company offered them a regular supply of varied films at a fair price, some exhibitors resented the lack of product choice and the attempt to monopolize distribution.[95]

Disgruntled exhibitors soon had an alternative source of films. Unlicensed distributors formed their own distribution company—the Motion Picture Distributing and Sales Company (or simply Sales Company). The creation of a distribution mechanism for filmmakers outside the Patents Company led to a flood of "independent" movies made by a slew of new filmmaking concerns. By July of 1910 the Sales Company serviced about one-third of America's moving picture theaters.[96] Given the explosion of independent product, many would-be filmmakers were evidently stifled by Edison's nascent monopoly. By November the trade journal *Nickelodeon* warned that "if the multiplication of producers continues there will soon be more sellers of films than there are buyers."[97]

This is the context in which the Gaumont Long Island plant appeared in 1909. Its alleged function was to process Gaumont films for Patents Company distribution; the Gaumont company claimed that it was cheaper to do this on American soil. But the Patents Company viewed the move as "aggressive." After all, laboratories were the very foundation of a full-fledged film factory. Thus the Patents Company feared that Gaumont was secretly planning to make films in the United States for the new independent market.[98] The plant limped along for nearly a year while the lawyers for Gaumont and the Patents Company exchanged angry letters. Finally, in the summer of 1910, Herbert Blaché bought the Gaumont plant with the help of investors. *Moving Picture World*, following the dispute, asked Blaché if a new independent Gaumont-American product would soon appear. He answered coyly that "one never could tell what the future would bring forth."[99] For several months nothing happened. Then, in October of 1910, *Moving Picture World* profiled the new Solax Company. According to the trade journal Solax benefited from "Gaumont brains, Gaumont experience, Gaumont knowledge" but was not formally associated with the French company. It was run by Alice Guy Blaché, Solax's "director-general." According to Guy Blaché the Gaumont plant was "underused," and she gave into "temptation" and rented it to make a "few" films.[100] Now the mother of two-year-old Simone, Guy Blaché began making two one-reelers a week for the independent market. In 1911, while Vitagraph's business practices were deemed "entirely inadequate" by professional auditors, Alice Guy Blaché was enlarging the Solax studio at Flatbush, New York, where "the finest and most modern devices are in use" and where manager Wilbert Melville began his experiment in applying the principles of scientific management to filmmaking.[101] In the late summer of 1912 Guy Blaché's new $100,000 studio in Fort Lee, New Jersey, was "the last word in moving picture plant architecture" and efficiency. Designed to allow the product to flow through the studio in an assembly-line fashion, the new Fort Lee studio could produce twelve thousand feet of finished positive film each day.[102] Blaché supported his wife's company but remained Gaumont's U.S. representative until his contract ended in 1912.[103]

Despite *Moving Picture World*'s warm support of "Madame" in its initial announcement of the new company, Solax advertising never revealed the sex of its director-producer. Indeed, in an oblique way early Solax advertising indicated that the studio was run by a man. Solax's first advertisement introduced its mascot, "Ol Sol—the Father of Photography," a sun-faced male figure that claimed, "I've made good pictures for years."[104] Given the traditional take on the marriage, it would seem that Herbert Blaché's ego played a role, but at this point the more likely reason for Solax's apprehension was the issue of how the director's gender would be received. The Blachés and their business manager may have been unsure if Americans would buy films from a studio run by a woman. Despite her decade of filmmaking experience at Gaumont, a respected brand, few if any in the American film industry would have recognized her name. When *Motography* profiled Guy Blaché in October of 1912, it concluded its short biography by stating that the fact "that the Solax guiding star is a woman seems to be lost in the hurly-burly of business."[105]

Ironically, Guy Blaché's gender might not have mattered as much just a few years earlier, when demand for films meant distributors paid little heed to brand names. But by 1910 the supply was catching up with demand.[106] With price and length standardized and plenty of product on the market, the only factor that could induce an independent film exchange to buy one brand of film over another was the films themselves. Between 1909 and 1911, with the patents war behind them and a flattening of demand, film manufacturers necessarily shifted their focus from cameras and projectors to the quality of the films, hoping to achieve brand-name recognition and brand-name loyalty.[107] Establishing their company in the midst of this shift, the Blachés may have taken extra care to make sure that the Solax brand was received with confidence.

In February of 1911, just four months after Solax opened for business, *Moving Picture World* noted "with satisfaction the steady improvement in quality" of Solax pictures, claiming that a prominent exchange manager had declared that he was now going to give the Solax brand "preference over another make of film."[108] Although Solax had a long way to go to achieve the kind of brand-name loyalty that Biograph had already won under the direction of D. W. Griffith, Solax achieved what all film manufacturing concerns craved: brand-name recognition. But such recognition meant that Alice Guy Blaché's gender could not go unnoticed. Within a short time, recalled Guy, "I rarely passed a week without being interviewed."[109] Generally, the female director-general was treated with respect.[110] In 1912, when Solax moved into its new $100,000 state-of-the-art studio in Fort Lee, *Moving Picture World* took note: "The entire studio and factory were planned by Madame Alice Blaché, the presiding genius of the Solax Company. She is a remarkable personality, combining a true artistic temperament with executive ability and business acumen. Every detail of the making of a Solax picture comes directly under her personal supervision. She takes full responsibility for the Solax product, and, when one considers that this model factory is the result of her work during the two years existence of the Solax Company, her judgment is hardly to be questioned."[111] Guy Blaché's gender clearly did not detract from her ability to make movies. She was simply lauded as a competent filmmaker.

Despite the shift away from sheer quantity, the technical aura of the film's first decade did not completely recede. The expertise of the camera operator, the laboratory, and the quality of the film stock itself were the primary criteria used to distinguish a good film from a poor one. Thus, like its contemporaries, Solax used its early advertisements to expound on the studio's superior photography, its tinting and toning, and the longevity of its films— "Looks as New on 30th Run as the Day of Release."[112] As for genre, Solax seemed to make the typical melodramas, comedies, westerns, and military pictures.[113] The military picture was in vogue as a result of American expansionism, and Solax became the clear leader in the trend.[114] In April of 1911 Solax announced a "Big Series of Gigantic Military Productions," beginning with *Across the Mexican Line*. Interestingly, Guy promoted active female leads, and the first military picture concerned the story of brave Juanita, sent to spy on an American lieutenant. In *Across the Mexican Line* Juanita falls in love with her would-be informant, who teaches her telegraphy, a skill she will soon put to dramatic use. When a Mexican general finds them and orders the American lieutenant to reveal military secrets, Juanita "dashes away to a telegraph pole, climbs, taps the wires, connects them with her instrument and is successful in conveying the news to the American troops," thus assuring his rescue in the nick of time.[115]

McMahan, who viewed the 111 extant films of the hundreds Guy made, noted the strong tendency toward powerful women and gender transgression throughout Guy's career. Although cross-dressing was common in all early films, McMahan argues that in Guy's films it can be read as questioning the immutability of gender—if gender can be modified by a mere change of clothes, it cannot exist outside human agency. In cross-dressing films by other directors the women don men's clothes "as proof of great love," as in Griffith's 1910 Biograph film *The House with the Closed Shutters*, concerning a woman who dresses as a soldier to fight for her ne'er-do-well brother and to protect the family name. She dies in battle, and all suspect that it is the mournful sister of the war hero who lives in the house with the closed shutters, when it is really the irresponsible drunken brother. In Guy's Solax film *Cupid and the Comet* (1911), by contrast, the woman dresses as a man to pursue her own desire—her true love. In this film a father takes his daughter's clothes to keep her from eloping and sleeps with them. Unable to wake him, the daughter dresses in her father's clothes to make her escape and then locks his closet so he cannot dress to pursue her. The daughter shows up at a minister's office in male attire with her equally masculine fiancé, but the effeminate minister will not marry two men. The daughter takes off her hat to reveal her identity, but her bearded father arrives in female dress to stop the ceremony.[116] According to McMahan, at the center of Guy's films

> is the preoccupation with female agency, the connection between agency and gender construction, and the obstacles facing the development of female agency in a patriarchal society. She was quite conscious of the fact that she herself had achieved an unprecedented degree of self-realization through her career as a film producer and director; almost all of her films are addressed directly to women with the message "you can do more—here's how." The "how" usually involved creative thinking, daring action, and a sense of humor: all three qualities required by the tomboyish persona and by crossdressers.[117]

Guy addressed feminism directly in at least two films, one made while she was at Gaumont, the other for Solax. Not surprisingly, the French film took greater risks. *Les Résultats du féminisme* (The Results of Feminism, 1906) begins with a scene of "elegant" male milliners. When one walks outside with a hatbox, he is accosted by a female masher, who first stares at him, then invites him to sit, and then presses him with her advances until he cries. Another female bystander shoos the masher away, and after she comforts the assaulted man, they begin to flirt. Proper male passersby "cover their

 EXPANSION, STARDOM & UPLIFT

eyes prudishly" to avoid looking at the lovers. The scene shifts to a home in which a man wearing a housecoat sits at a sewing machine while his father irons and his mother smokes at her leisure and reads the newspaper in a chair. After the man's parents retire, the apparently kind woman from the street appears at his door and lures him away to her room, where she seduces him. A few years later the man is a harried househusband while his wife sits in a café smoking and talking with her friends. Oddly, Guy includes a scene of male nannies breastfeeding babies, truly pushing the flexibility of gender! When the househusband begs his wife to return home, she ignores him. Desperate, he throws acid in her face. The other "neglected husbands swear vengeance" and mob the cafés wherein their wives take their leisure, throwing them out. The film ends when "the triumphant men remain in the café knowing justice has been served."[118]

A tamer version of gender-role reversals appeared in Guy Blaché's *In the Year 2000* (Solax, 1912), a science fiction fantasy in which women rule and men obey. Unfortunately, the only evidence left of this film is a brief description in *Moving Picture World*, which begins by claiming that "prognostications often terrify us with visions of what will be when women shall rule the earth and the time when men shall be subordinates and adjuncts." Calling the film "serio-comic," the writer claimed that it was the "very seriousness of the purpose of the theme" that "makes the situations ludicrous" and that it was one of the most amusing films ever put out by Solax, a company known for its comedies.[119]

As INTERESTING AS THESE FILMS ARE, they were probably less shocking to audiences than one might suppose. American audiences, accustomed to cross-dressing male actors, as well as cinematic burlesque of suffragists, had seen similar films before. But McMahan's suggestion that they contained a feminist message aimed at empowering women in the audience is intriguing, especially in light of the fact that Guy seemed to go to some length to hide her gender from the trades. Like female filmmakers who followed in her wake, Guy walked a tightrope between the centuries—balancing Victorian constraints for women on the one hand with possibilities for New Women on the other. At this point in the development of the American film industry, in which films were still judged by their technical merits, Alice Guy Blaché's most important contribution was her demonstration that a woman could successfully run a studio. But her background was unique. Other women wanting their own studios would have to follow a different route, particularly as $100,000 studios became de rigueur. Fortunately, the outline of a new path to creative and financial control was becoming clear. After 1909 two unforeseen developments rumbled through the movie industry, changing every aspect of the business in ways that proved exceedingly favorable to a handful of fortunate women—the arrival of the movie star and the creation of the multireel feature film.

# CHAPTER TWO

# "To Get Some of the 'Good Gravy'"

# for Themselves

## *Stardom, Features, and the First Star-Producers*

The story of the first movie star, as told in the contemporary trade press and many subsequent histories, goes something like this: On February 19, 1910, a St. Louis newspaper reported that the popular actress known only to movie patrons as the "Biograph Girl" was dead following a tragic streetcar accident. Fans did not know the name of their favorite actress because producers withheld the names of their players to avoid paying the higher salaries film stardom would bring. But fans soon learned that the Biograph Girl was named Florence Lawrence and that she was alive and well and in the employ of a new producer because Carl Laemmle, head of the Independent Moving Picture Company (or IMP), told them so. Taking out a full-page announcement on March 12 in *Moving Picture World*, Laemmle claimed that Biograph planted the story in a fit of pique over losing its best player to the rival IMP Company. Using the actress's real name for the first time in print, Laemmle assured readers that Lawrence, "the 'IMP' girl, formerly know as the 'Biograph' girl," was "in the best of health" and that "very shortly some of the best work of her career is to be released."[1] To prove it, Laemmle sent Lawrence, her leading man, and her director to make an appearance in the flesh at two St. Louis theaters—the Gem and the Grand Opera House. The *St. Louis Times* announced the arrival time of Lawrence's train, planned a welcome party, and offered a special clip-out coupon to female fans, who would receive a "handsome photo" of Lawrence upon presenting it to the actress.[2] What allegedly happened next shocked observers. A crowd larger than the one that had greeted President

Taft a week earlier rushed at the moving picture actress. They ripped the buttons from her coat and stole her hat. According to one source, Lawrence escaped only with the help of the police.[3] Foreshadowing the now familiar hysteria of film fandom, the first American movie star was born.

Laemmle's publicity stunt was so successful it carried this creation myth to the end of the twentieth century, until the late Robert deCordova revised the history of the early star system in 1990. He discovered that the names of film actors and actresses were revealed months earlier, even Lawrence's name, and he found no evidence of the alleged original newspaper story claiming Lawrence had died. The Lawrence stunt did not invent the film star, he argued; instead, the arrival of the star system was a natural shift from the focus on the "apparatus" as the creator of film to the actors on the screen, concomitant with the rise of the narrative film.[4] However, both Laemmle's announcement in *Moving Picture World* and Lawrence's appearance in St. Louis were real. And although we cannot be sure of her reception, the reactions published by *Moving Picture World* after this publicity stunt indicate a sea change. A mere two months after Lawrence's live appearance in St. Louis, *Moving Picture World*'s "Man About Town" professed astonishment at "the interest the public has taken in the personality of many of the picture players." Letters allegedly poured into the offices of film manufacturers and exchanges, from both men and women, asking for autographed photos of their favorite leading actors. One actress claimed to have received three thousand offers of marriage just three months after the Lawrence incident.[5] By the end of 1910, *Moving Picture World*'s new "Picture Personalities" column profiled Florence Turner of Vitagraph, Mary Pickford of Biograph, and Pearl White of the Powers film manufacturing company.[6] Even if it did not invent the film star, the Lawrence incident signaled to the industry that the star had arrived.

The second major corrective from deCordova was his proposal that the film industry marshaled the emergence of the star system for its own gain, an idea that runs contrary to "[s]tandard accounts," which, he asserted, "would almost lead one to believe that the industry entered into the star system against its will."[7] Truly the film industry benefited from the box office draw of the star and the publicity of fan culture and participated in the development of both. When the stars first emerged from within the film industry's own stock companies, rather than try to stuff their employees back into the previous policy of anonymity, film manufacturers immediately set out to capitalize on their fame and to create new stars. Their ability to do so was soon enhanced by the longer feature films that showcased the stars. But there were those within the industry, particularly those with the-atrical backgrounds, who knew that although stars increased admissions, they could also demand higher wages, acquire greater control, start bidding wars, and strike out on their own as independent producers. Stars could, and did, create trouble for the established studios. And as if to validate the naysayers, the very first stars, most of whom were women, soon deserted the established studios for the freedom of the independent market. Between the emergence of the star in 1910 and the merger of Famous Players—Lasky and Paramount in 1916, a significant number of the most famous female stars entered production on their own terms, in the words of actress-producer Florence Turner, to "get some of the 'good gravy'" for themselves.[8] These were the first actress-producers of the film industry.

THE EMERGENCE OF THE STAR SYSTEM was a logical result of the changes that created the nickelodeon boom. The arrival of inexpensive permanent mov-ing picture theaters boasting frequent, sometimes daily, program changes created a demand for steady actors and a following for screen favorites. Al-though the exact numbers are often debated, contemporary observers re-ported extraordinary attendance figures. In 1909 a writer for the *Survey* estimated that 250,000 New Yorkers patronized the city's 350 moving picture theaters each day (500,000 on Sundays), and in Chicago 200,000 daily patrons attended the windy city's 345 theaters. Nationally, four mil-lion Americans were said to attend a moving picture show every day.[9] But before 1908 or so, the possibility of movie stardom seemed remote. Actors were ephemeral, hired by the day. In addition, the shooting of movies at stage distance reduced the thespian to a small and often indistinguishable figure on the screen.[10] By 1909, however, as film producers hired permanent stock companies, audiences began to recognize frequent players.[11] Early di-rectors like D. W. Griffith and Sidney Olcott worked on improving scenar-ios, editing, and cinematography, and as the camera came in closer, audiences were able to observe even subtle emotions. According to many historians these improvements created the first stars of the cinema by encouraging pa-trons to become sufficiently absorbed in the story and the personalities on the screen.[12]

In these respects film acting was becoming more like stage acting, and surely these actors of "canned drama" were cognizant of the fortunes that fame could bring. The theatrical star system originated when European stars began to tour the United States in the mid-nineteenth century. Soon after,

American stage stars appeared. For a few decades these traveling stars did not disturb the local stock companies but merely arrived for a performance or two of their own play, produced with the support of the local company, and then moved on. By the 1860s, however, stars had proliferated to the extent that "every actor of any merit, and many with nothing but sheer imprudence, strove to find a place in the firmament," while the public "fell into a state of mind in which it regarded any production as inferior that did not include a star in its cast."[13] Just as occurred later in the film industry, theater managers had to secure stars or lose their audiences, but doing so was an expensive prospect. First-class leading ladies and men earned upwards of $250 a week. By the late nineteenth century, Maude Adams, John Drew, Ethel Barrymore, and Minnie Maddern Fiske allegedly made more than $50,000 a year. Managers were forced to cut the salaries of their regular stock companies to afford the famous.[14]

As the drawing power of the star became evident in the 1860s, the unwieldy star-stock system began to be replaced by the "combination system." This system provided a clear precedent for the actress-producer of the early film industry. Under the combination system the traveling star brought not only a play to each theater but also a company of supporting actors. According to Alfred L. Bernheim, an early historian of the American theater, under the combination system the star, now a producer, "gained complete control over the production and was able to avoid all conflict of authority" by providing everything that was necessary to produce the play, down to the props and scenery. If the risks were greater for the star, so were the potential profits, as the combination companies proved extremely popular.[15]

The star system empowered women as well as men, and some female stars of the stage became managers of important theaters bearing their name. The Laura Keene Theatre presented high quality drama to New Yorkers between 1855 and 1863, when it passed into the hands of Mrs. John Wood, a comedienne famous for her male, or "breeches," roles. Louisa Lane (Mrs. John) Drew ran the famous Arch Street Theater of Philadelphia between 1861 and 1879 while raising a large family of adopted and natural children, including Georgiana Drew Barrymore, mother of Lionel, John, and Ethel Barrymore. In 1888 three female "manageresses" were said to be looking for theaters in New York, including the British actress Lily Langtry. Minnie Maddern Fiske—actress, playwright, producer, manager, director, and a fierce opponent of the Theatrical Syndicate of the 1890s—was perhaps the most renowned American actress-producer as the century came to a close.[16] Women thus had major roles in maintaining some of the largest legitimate theaters in the nation. On a lesser scale local stock companies proliferated between 1820 and 1860, and women continued to appear in management.[17]

By the early twentieth century, female managers and actress-producers could be found in all aspects of the theater. Mae Desmond and her husband began the family-oriented Mae Desmond Players in upstate New York when it became impossible for the husband and wife actors to find concurrent employment in the 1910s.[18] The Spooner Stock Company, headed by Mary Gibbs Spooner, opened in Brooklyn in 1901 to star her two daughters, Edna May and Cecil. After two years of considerable success, they moved to Manhattan, and in 1909 and 1912 the Spooner daughters each opened their own successful stock companies.[19] Some of the first film actresses, such as Cleo Madison, gained experience as independent producers in this branch of the theater, while others encountered female managers at some point in their stage careers.[20] Lotta Lawrence, mother of the original "Biograph Girl," led her own touring repertoire company until it fell on hard times in 1906, probably caused by the influx of movies into the cheap houses that once hosted the ten-twent'-thirt' shows.[21] Named after the staggered (yet inexpensive) ticket prices, the ten-twent'-thirt' shows proliferated at the turn of the century by offering the same plays as the expensive theaters, albeit a season or two later and with an emphasis on blood-and-thunder melodramas.[22]

Although less prestigious, vaudeville was even more lucrative for actresses. Just as plays seen in the more expensive theaters could be seen more cheaply at the ten-twent'-thirt' shows, scenes from popular plays could also be seen in an abbreviated form in vaudeville. A vaudeville program comprised six to eight twenty-minute variety acts, which were booked into theaters individually or as part of a traveling bill. Dramatic scenes, sometimes enacted by the original star, appeared as a single act, sandwiched between other vaudeville turns. Women performers appeared in other typical vaudeville acts also, as dancers, impersonators, acrobats, and comediennes.[23] By the turn of the century, vaudeville was the most popular commercial amusement in the nation, and women in vaudeville made tremendous salaries. Female vaudevillians received more money than male performers on every level, from novice to star. Ed Albee, who controlled the vaudeville industry with B. F. Keith, told reporters that "a minor two-girl dancing act is paid half again as much as a similar two-man act. This holds true even when the men are better."[24] Female stars received some of the highest salaries in the nation if not the world.[25] In 1906 the Herald Square Theater in New York paid

Mrs. James Brown-Potter $3,000 per week, and E. F. Proctor offered May Irwin $4,000 a week.[26] At a time when the typical female wage earner in the United States earned between $5 and $15 a week, the pay for standard acts was $200 to $500 a week.[27] Furthermore, the lower rungs offered the opportunity to produce one's own act. According to a writer for *Variety* in 1907, each vaudeville artist "must be his own manager, buy or write his own special production," and then become "his own advance and press agent; his own property man," that is, until she succeeds. Then a manager signed the artist and assumed such duties.[28] Edwin Milton Royale, a noted playwright, author, and actor, alleged in 1899 that the structure of vaudeville provided "'the open door' . . . [W]hatever or whoever can interest an audience for thirty minutes or less, and has passed quarantine, is welcome."[29] This undoubtedly hyperbolic statement points nonetheless to a space where women could achieve fame, fortune, and control.

As discussed in the previous chapter, nearly all of the first film actors and actresses came from the stage. Future film star Florence Turner did impersonations, and at age four Florence Lawrence was "Baby Flo, the Child Wonder Whistler."[30] Gene Gauntier recalled future film director Lois Weber playing in *Why Girls Leave Home; or, A Danger Signal on the Path to Folly*, future film star James Kirkwood in *The Worst Woman in London*, and Gladys Smith, before her name was changed to Mary Pickford, appearing in *The Fatal Wedding*. Gauntier herself recalled too many summers in New York, hoping for a Broadway engagement, "only to take, and gladly, another melodrama and troupe out again to the 'sticks'—the long hard road to fame."[31] Despite the anonymity of most early film actors and actresses, it is likely that given their background on the stage, they connected fame with greater creative control and greater earnings. But the kind of fame that led to stardom was elusive in the film industry before 1910. Film companies did promote their female leads, but they referred to them by their brand name—the "Vitagraph Girl," the "Biograph Girl," and the "Kalem Girl"—rather than by their given or stage name. These nicknames for specific actresses may have originated from audiences, bereft as they were of any names to place on the faces of their favorite players. But such brand-name monikers worked as advertisements, deftly forcing filmgoers, exhibitors, and exchange managers to remember the brand, not the name of the player.[32] A company's players provided a constant that could be exploited in an era when films often changed daily.[33] Film manufacturers even resurrected the lobby card, "the ancient custom of regular play houses," displaying photographs of their stock company for filmgoers to observe as they entered the theater, ensuring that patrons associated the faces of favorite actors with the appropriate brand.[34] Manufacturers strained the limits of this system by repeatedly using the real first names of their players onscreen. As the "Biograph Girl," Florence Lawrence frequently played "Flo," and Mary Pickford played "Little Mary" in many of her films. Ironically, this practice was particularly common at Biograph, one of the last studios to reveal the actual names of its players.[35]

Whether this strategy was intended to prevent full-blown stardom or was merely a stage in the process by which stardom became possible is open to debate. There does seem to be evidence that film companies desired stars at this time, as they began to import them from the stage in 1909 in an effort to upgrade the reputation of the movies. American film firms appeared to be influenced by European imports, especially from the French firm Film D'Art, which featured European stage stars in filmic versions of celebrated authors (Dumas, Zola, Balzac) as early as 1909. In the second half of that year Edison hired American stage actress Cecil Spooner to act in Mark Twain's *The Prince and the Pauper*. In September 1909 the Edison company announced with great fanfare that it had engaged French pantomimist Pilar-Morin, a respected vaudeville star who waxed forth on the connections between theatrical drama, pantomime, and the evolution of film acting.[36] A censorship crisis, which will be discussed in the next chapter, probably contributed to the emphasis on uplift, but at this point it seems clear that American film companies desired stars and thought the only place to get them was the stage.

When the popularity of Florence Lawrence, the former Biograph Girl, proved that stardom could come from the stock company, firms took action. Shortly after the St. Louis incident, Vitagraph established the first fan magazine, *Motion Picture Story Magazine*, in 1910.[37] The "Vitagraph Girl," Florence Turner, made her own live appearance just one month after Lawrence visited St. Louis. Vitagraph continued to promote Turner throughout 1910, sending her to visit various moving picture theaters with her own illustrated song:

> I'm in love with the Vitagraph Girl,
> The sweet little Vitagraph Girl,
> Each moment a picture of romance or hate,
> Her tragedy's bully, her love simply great.[38]

Although male actors became well known, the first movie stars of the highest magnitude were primarily female. Patrons favored female stars over their male counterparts on the stage and in vaudeville at the beginning of the twentieth century, but in the early movies this sexual asymmetry was pronounced.[39] It is true that Maurice Costello won an early popularity contest and had his own live appearance in 1911.[40] And Ben Turpin, John Bunny, Tom Mix, and Broncho Billy were early screen favorites. But before Charlie Chaplin and Douglas Fairbanks captured the public's imagination in the mid-1910s, it was women who achieved full-blown stardom and all the perks that went with it.[41] Why this was so is not completely clear. A critic in 1911 argued that female stars "of the Florence Lawrence caliber" seemed more numerous than male stars because they were simply better actors—it was "seldom that any male actor puts enough character in his interpretation to be remembered."[42] But it may have been her transgressiveness that made the actress memorable. Susan Glenn recently argued that the turn-of-the-century stage offered a zone in which women could develop a nascent feminism by stretching the boundaries of acceptable female behavior, such as bodily display. The moving pictures extended this zone.[43] Film theory, too, offers sophisticated analyses exploring women and spectatorship, but what is of moment here is the concrete result of female fame—the leverage to gain greater control over one's own career.[44]

FAME ALONE WAS NOT ENOUGH to allow the first female film stars to become producers. The structure of the industry needed to be configured in a way that would create that opportunity. Fortunately, after 1909, developments within the film industry did, in fact, make independent production possible, and the background of the stars encouraged them to take advantage of this moment. Assisting the would-be star-producer in the movies was the creation of relatively autonomous director units. There is some debate over the existence of a director-unit stage of production. It was one of four transitions in film production postulated by Janet Staiger: (1) the cameraman system (1896–1907); (2) the director system (1907–9); (3) the director-unit system (1909–14); and (4) the central-producer system (after 1914).[45] In contrast, Charles Musser finds a collaborative system prior to 1907, characterized by relatively horizontal relationships between a pair of filmmakers, followed by the central-producer system after 1907, characterized by larger studios with a greater division of labor and more hierarchical control. Musser argues that it was the increasing demand for narrative film production brought on by the nickelodeon boom that encouraged film companies to adopt modern management structures after 1907 in order to keep track of costs and clarify studio hierarchies. Under the central-producer system a single executive was in charge of film production, ending the fluid boundaries between crafts that characterized the early nickelodeon era. However, Musser asserts that there were variations on the theme, even well into the 1910s. Under the central-producer system a director could have "virtual autonomy" or suffer under "almost complete dominance of producer over director."[46]

From about 1909 to 1916 many production units, headed by director-producers, enjoyed a great deal of autonomy. These units were formed when film manufacturers, making more than one film at once, employed a second director and a second company, and then a third, and so forth, each turning out a one-reel film per week. By 1909 Selig had four director-producers in three locations—California, New Orleans, and Chicago—and Kalem sent a unit to Florida to work during the winter.[47]

It was the autonomy of the director units that ultimately encouraged stars to take a better offer or to plunge into the independent field. The Kalem Florida unit, for example, handled nearly every aspect of filmmaking, from creating the story to editing the film.[48] This degree of self-sufficiency and independence made it relatively easy for production units to split from the original company. Most often the seceding party did not include the entire unit but rather the creative core: the director, the star, the writer, and the cameraman, or some combination thereof. Most frequently, directors and their stars split to form a new independent company named after the star. Not only would the star's popularity provide currency in the marketplace, but male partners typically stayed in the background, probably because of the traditional belief that an actress is "more popular when she is apparently unattached to any man."[49] A star's name meant that the films would sell, which in turn attracted investors and distributors. These were the critical issues for independent companies, for the elements they left behind were financing and distribution. Fortunately, moviemaking was still a relatively low-cost proposition.[50]

Distribution, however, was often a greater hurdle. As a *Photoplay* columnist stated in 1915, anyone "with a few thousand dollars can 'make a picture,'" but placing pictures "regularly in the hands of the nation's exhibitors" was the trick.[51] It was, however, possible, thanks to the erosion of the Motion Picture Patents Company and its ultimate destruction by the courts in 1915. Although independent manufacturers like Carl Laemmle's IMP Company challenged the Trust for years, two legal developments buried the Trust in 1912: IMP's legal victory, allowing filmmakers to use cameras other than those licensed by Edison, and a government antitrust suit brought against the Patents Company that same year.[52] Independent producer-distributor alliances continually shifted between 1911 and 1915, but there was always a way to get a picture into theaters.

Given the dearth of memoirs and personal papers, just how much control these stars exercised in the creation of their independent films is nearly impossible to discover. It appears that male directors assumed production responsibilities in most companies, but it is also true, as we will see, that many of the star vehicles created for these women under their own brand names featured unusually strong heroines. In any case the increasing number of production companies named after, and created for, female stars changed the face of the production branch of the industry.

The first star to have her own company was Marion Leonard, who was thrust into the spotlight in the winter of 1909–10 when she replaced Florence Lawrence as the "Biograph Girl."[53] Leonard was a talented actress and an exceptional beauty, and she quickly joined the ranks of the first film stars. Like her peers Florence Lawrence, Florence Turner, and Mary Pickford, Marion Leonard moved from studio to studio as her fame increased and the offers improved. In 1911 she and her husband, Stanner E. V. Taylor, a director who began his career at Biograph during Griffith's reign, created the Gem Motion Picture Company with the sole aim of exploiting Marion Leonard's growing popularity.[54]

The centrality of Leonard's stardom was clear in the Gem's first advertisement to the trade, which barely mentioned the name of the company. The focal point was a portrait of Leonard, framed by a diamond ring. Leonard flirted with her admirers in the ad: "People—I am engaged!"—to the new Gem company.[55] *Moving Picture World*'s first profile of the Gem company emphasized the freedom of independent production: "Never before has Mr. Taylor had such absolute carte blanche in the development of his ideas." The journal also noted the Gem's generous, but anonymous, investors. The company boasted a brand-new studio, the latest in lighting, and a "very competent stock company" to support Leonard in her starring roles. Indeed, barely one month old, the Gem claimed it had dozens of finished negatives in its vault, suggesting a long and well-financed incubation period.[56]

Leonard played strong and independent women in her Gem releases. The first, *The Defender of the Name*, echoes D. W. Griffith's 1910 *The House with the Closed Shutters*. In this film Leonard enacted the part of a proud Confederate woman whose brother was assigned to spy on the North. The brother, observing the execution of a Union spy, turns coward and runs home, committing suicide. Leonard's character "cannot abide the shame of a coward in the family," so she completes his assignment, procuring the documents from the North under fire and returning home to place the documents on her brother's body. Thanks to her courage, the brother is believed to have died a hero. In her third release, *Under Her Wing*, Leonard played the sister of an irresponsible and chronically unemployed father. Leonard finds the "sluggard" a job in the bank where she is a trusted stenographer, only to get caught in the middle when her brother is found stealing. Leonard takes the blame rather than see her brother's family starve, but the truth is found out. In the end Leonard regains the trust of her boss, her brother leaves his family to "no regrets," and Leonard becomes the breadwinner, taking her brother's family "under her wing."[57] Unfortunately, the Gem company folded before the first film was released. According to Leonard and Taylor, the jealousy of another independent manufacturer forced them out of the Sales Company, the only major distributor of independent films in 1911. All twenty-six of the Gem negatives were sold to the Rex Motion Picture Company, also an independent manufacturer.[58] After this debacle Leonard left the screen for a year but returned when the advent of longer feature film created new opportunities for screen stars.

Florence Lawrence, the original "Biograph Girl," and her director-husband, Harry Salter, established their own company, Victor Films, in the spring of 1912. According to one source, as early as 1908 Florence Lawrence was cognizant of her growing fame and demanded a private dressing table, a weekly salary, and double the pay received by other Biograph actors. D. W. Griffith, an opponent of the theatrical star system, "could see what was happening" and "decided he must groom other young girls to insure that no one actress made him dependent upon her." Lawrence, determined to get her due, began secretly to search for a studio that would more actively advance her career. When her actions were discovered, Lawrence and her husband were fired from Biograph. Thus she was "at liberty" when Laemmle hired her for his IMP Company.[59] Salter may have been the source of the Lawrence's stage-star demands at Biograph. According to one biographer Salter's ambitions were fueled by his conclusion "that he had a little more leverage since Florence, his wife, was developing not only as the top favorite but as a very fine actress." He was certainly the cause of their departure from the IMP film company after only a year. According to personal correspondence Salter instigated the pair's abandonment of IMP for Lubin because the manager of IMP, Thomas Cochrane, got "on his nerves."[60]

Lawrence and Salter left IMP for Lubin but, after a trip to Europe, created Victor Films, with Salter as director and Lawrence as the star. Victor cleared the first hurdle for small independents when it found sufficient financing for its first film, *Not like Other Girls*, but it was stumped by distribution. Just as the film was ready for release, the Sales Company began falling apart as two of its members, Harry E. Aitken and Carl Laemmle, disputed.[61] After the Sales Company broke up, Lawrence and Salter cast their lot with Carl Laemmle's new Universal Film Manufacturing Company, where they became one of several quasi-independent companies that shared Universal distribution.[62]

At Universal each company worked as a kind of production unit. They made films independently but operated under the guidance of "a general managing producer," who was "responsible for results" and was to keep close tabs on production.[63] In return the companies of Universal, which kept their own brand names, were relieved of a great deal of the financial risk of production, and they were assured of distribution. As director, Harry Salter received $200 a week; as star, Lawrence received $500 a week, a salary comparable to that earned by stars of the legitimate stage.[64]

Although Victor Films was one of the smallest companies in the Universal family, no other Universal company boasted a star in 1912. While other Universal companies depended on genre (Bison westerns, Rex melodramas), the Victor product was promoted on the basis of Lawrence's superior acting and superior stories.[65] This combination proved successful. Starting July 12, 1912, Victor released a one-reel film for Universal every week, beginning with *In Swift Waters*, a romantic comedy that *Moving Picture World* called "an entirely perfect moving picture." Salter, as director, was praised for his "fine discretion" and "artistic sense," Lawrence for her "sweet and fresh" personality. *The Players*, their second film, also received a lengthy and positive review from *Moving Picture World*.[66] Throughout the summer of 1912 Victor continued to produce critically acclaimed light comedies and dramas.[67]

Unfortunately, once established in Victor, Salter became jealous over his wife's alleged relationship with Universal producer Pat Powers.[68] When Lawrence rebuked him, Salter fled to Europe, and Lawrence took an ill-advised year off from filmmaking at the height of her popularity.[69] While in Europe, Salter begged for forgiveness and promised that on his return Lawrence would be "Mademoiselle La Directress," but it was not to be.[70] After Salter returned, and he and Lawrence began making movies for Victor again, Salter asked Universal to equalize their paychecks so that at least on paper his wife would not be making more money than he did.[71] Although Salter and Lawrence made twenty-five two-reelers for Victor between 1913 and 1914, Lawrence never regained her former popularity.[72] In 1914 an injury on the set of *Pawns of Destiny* forced Lawrence into retirement for two years. This episode permanently severed Lawrence's connection with Victor and Universal; it severed her marriage to Salter as well. Single and forced to take in boarders, Lawrence complained bitterly to Salter that he never even tried to give back the portion of his salary that was really hers.[73] Lawrence tried several times to become an independent producer after 1916, but her continuing fame, based on nostalgia alone, could not sustain a career. Lawrence, the first American movie star, committed suicide in 1938.[74]

Florence Turner, the former "Vitagraph Girl," also became an independent producer. In the spring of 1913 she formed Turner Films, Ltd., in England with another former Vitagraph employee, director Larry Trimble. Turner cited "trade factionalism" in the United States and the open market in England as the determining factors for her move.[75] Speaking in 1919, Turner said, "It was the only thing for me to do." She wanted her own company, but she could not "fight the trust here; it was altogether too strong." Since we know others were able to make independent films, this claim seems dubious.[76] But the Patents Company's high tariffs on foreign films between 1908 and 1912 all but destroyed English producers. Thus Turner and several other American filmmakers, such as Edwin S. Porter, traveled to England after 1912 to take advantage of dark studios and unemployed technicians. So many American filmmakers went to England, in fact, that many Britons protested that Yankees were overrunning their native cinema.[77] Economic conditions determined where Turner located her company, but her motivation to become an independent producer was based on her own assessment of her power as a star. In her own estimation she was responsible for Vitagraph's superior sales, so, as she told a reporter in 1913, she was now going "to get some of the 'good gravy' for herself" while she was in her prime.[78]

Turner and Trimble established Turner Films as an independent company, but it operated under the auspices of the Hepworth Manufacturing Company.[79] Their first product, *The Rose of Surrey* (September 1913), was a two-reel melodrama featuring Turner and Jean, the former Vitagraph collie, another defecting Vitagraph star whose likeness formed the company's logo.[80] Between the summer of 1913 and the early autumn of 1914 Turner Films produced about two films a month, typically comedies, melodramas, thrillers, and scenic split-reels, which comprised some of the best films to emerge from England in the 1910s.[81] Unfortunately, Florence Turner's British company was destroyed by the first World War, and when she returned to films, she no longer enjoyed the same audience appeal.[82]

Female stars and their directors, then, were inspired to form their own companies for several reasons: they recognized their power as stars and wanted to get some of the "good gravy" for themselves, the director-unit mode of filmmaking within the studios encouraged functional independence, capital requirements were low, and independent distribution was available. Indeed, thanks to the last two factors, one did not have to be a movie star to become an independent producer. Cecil Spooner created the short-lived Blaney-Spooner Feature Film Company in 1913 with Charles E. Blaney.[83] Eleanor Gates, author and playwright, organized a feature film company in 1914 to make movies of her own books, plays, and stories.[84] The Liberty Feature Film Company, organized in the spring of 1915, was "the only producing company owned and managed entirely by women," according to *Motion Picture News*.[85] The Liberty Feature Film Company intended to make pictures that "would be desirable by the exhibitors wholly on account of their artistic merit." The company's president was Sadie Lindblom, described as the "wife of the well-known Alaskan capitalist," perhaps indicating the source of the company's capital. The company lasted for less than a year; nevertheless, it produced a respectable number of two-reelers.[86]

Politically motivated women from outside the entertainment industry also came to the studios hoping to make films to further their causes. Among the first women to do so were suffragists. Film manufacturers churned out antisuffrage comedies nearly by the dozen. Thus, the Women's Political Union, in conjunction with the Eclair Company, countered these images by producing their own brief well-received prosuffrage comedy, *Suffrage and the Man* (1912).[87] Shortly thereafter the Reliance Company released the much more important *Votes for Women*. In 1912 Reliance approached the National American Women Suffrage Association (NAWSA), proposing that they coproduce a prosuffrage film. The result was the two-reel melodrama *Votes for Women*, supervised by the NAWSA and written by three women. *Votes for Women* was truly a social problem film, contextualizing women's suffrage as a solution to a host of injustices caused by men. The story followed the political awakening of a naive rich girl, the daughter of a department store mogul and the fiancée of a powerful senator. Her cloistered world is shattered when her closest friend, a reformer and suffragist, takes her to the dilapidated tenement owned by her fiancé. There she meets two young orphaned girls and their baby sister. One labors at home in appalling conditions, sewing embroidery sold at the department store and tending to the baby, while the other works endless days as a clerk in the lingerie department. Or at least she did. Returning home in tears, she reports to her sister that she lost her job because she refused the improprieties of the floorwalker. "If women had the vote," asks the friend, "do you think that such things could go on around us unchecked? Do you think that men would be allowed to fill their pockets at the expense of the lives and toil of helpless girls and all to enrich themselves and their associates? Not for a minute!" The heroine is converted to the cause of suffrage, and after she contracts scarlet fever from her slum-manufactured trousseau, her father and fiancé convert as well. The grand finale used footage from an actual suffrage march down Fifth Avenue.[88]

*Votes for Women* received very good reviews, and for uplifters within the film industry it provided a sweet moment of victory. One of the nation's most outspoken early critics of the movies, Anna Howard Shaw, the Methodist minister who once famously referred to Chicago's nickelodeons as the "recruiting stations of vice," made her film acting debut in *Votes for Women*.[89] A second serious suffrage melodrama appeared in 1913, *What Eighty Million Women Want—?*, a four-reeler produced by the Women's Political Union and the short-lived Unique film manufacturing company. This time "the English Militant Leader" Emmeline Pankhurst and Harriot Stanton Blatch, president of the Women's Political Union, appeared together onscreen. *What Eighty Million Women Want—?* exposed the mistreatment of women under American laws and also received good reviews.[90]

Low costs and an open market encouraged men to enter independent production as well. In 1913, for example, Frederick S. Dudley and James D. Law opened the Colonial Motion Picture Corporation after securing the screen rights to several novels, and H. M. Horkheimer, a former ticket seller for a circus, created the Balboa Amusement Producing Company with capital of $7,000.[91] In 1914 Ivan Abramson, manager and producer of Yiddish theater, created Ivan Film Productions, and L. Frank Baum, author of *The Wizard of Oz*, created the Oz Film Manufacturing Company to make movies based on his books.[92] Between 1912 and 1915 actors Hobart Bosworth, Mack Sennett, James J. Hackett, King Vidor, and Carlyle Blackwell all formed companies to make films starring themselves.[93] It is impossible to arrive at an accurate count of these companies, since most of them came and went within a matter of months, but at least 105 companies were begun by men between 1910 and 1915, whereas fewer than 10 were begun by women.[94] But numbers hardly tell the whole story, for most of the women who went into independent production were the biggest stars in the industry. Unlike the legions of companies begun by men, typically bereft of a proven star, the independent companies created by top female stars constituted a threat to the established studios. And the studios were angered by the defection of their stars. As Johnson Briscoe of *Photoplay* asked, speaking from the point of view of a studio head, were not the studios "the very ones responsible in the first place for the actor's popularity?"[95]

❧

THE STUDIOS WERE IN A QUANDARY. Audiences clearly preferred films with stars in them, but how could manufacturers hold on to their stars when the independent market beckoned? This situation worsened for manufacturers after the introduction of the multireel film, soon referred to as a "feature" film to distinguish it from one- and two-reelers. A multireel film, which could last for an hour or more, allowed stars to appear in entire plays rather than abbreviated drama. Importantly, feature films introduced the kind of star vehicles offered by the stage.

Established studios experienced difficulty adjusting to the feature film. Although some European manufacturers made multireel films before the nickelodeon era, the American film industry resisted them. Thanks partly to the practices of Edison's Patents Company, producers, distributors, and exhibitors were locked into a rigid system based on the one-reeler. Film producers made a specified number of reels per week, which were sold to distributors as a package at a constant price per foot. The single reel, or at most, a two-reel film, constituted a single unit. Any film longer than two reels threw off the entire system of distribution, as well as customary exhibition practices. Exhibitors used one-reelers to create mixed programs—a comedy, a scenic, a drama, and so forth. A multireel film would take this power away and perhaps drive away patrons who preferred variety. In addition, features could not pay for themselves under the rigid price-per-foot method. Features cost more to produce per foot than shorter films because they tended to require better actors, directors, costumes, and sets.[96] And since American exhibitors simply split features, showing them one reel at a time, distributors did not see why they should pay more per foot for a feature than for any other kind of film.[97]

The challenge of features became unavoidable in 1911 and 1912, when a few showmen, such as Adolph Zukor, bought the rights to European-made features and exhibited them in their entirety in rented legitimate theaters, where patrons willingly paid as much as $2 to see them.[98] Using the road-show method, distributors barnstormed from legitimate theater to legitimate theater with a feature film instead of a play, sending an advance man to promote the film, hiring a special projectionist, and providing music and narration. The road-show method was costly, but because patrons were charged admission comparable to a stage play, potential profits were enormous. Huge profits could also be made through the states' rights method of distributing features, which sold exclusive rights to distribute features within a specific territory. The states' rights method allowed the distributor, once he or she had bought the rights, to exhibit the film any way they wished, either through special exhibition akin to the road show or more simply in nickelodeons or tent shows, where usually more than a nickel was charged. Almost three hundred features were exhibited using these methods between 1912 and 1914.[99] Film stars, no doubt recognizing the potential of the feature film as a star vehicle, were given yet another reason to desert the established studios for independent production. Granted, feature film production was more expensive, but features could still be made for a few thousand dollars, and they were extremely lucrative.

Helen Gardner was the first film star to enter independent production with the specific intent of making features. It was not that her original studio, Vitagraph, refused to make multireel films. On the contrary, Vitagraph became the first U.S. company to do so in 1909, and Gardner's own claim to fame was her performance as Becky Sharp in Vitagraph's three-reel version of *Vanity Fair* (1911).[100] Rather, Gardner's motives for entering independent production were simple: equal parts profit, fame, and control. In June of 1912 a notice in *Moving Picture World* announced that Helen Gardner, "the beautiful and talented leading woman who appeared as Becky Sharp," was organizing a company to release five-reel "feature plays" on the independent market. Charles L. Gaskill acted as director-producer-scenarist of the "Helen Gardner Picture Players," and Charles L. Fuller promoted the New York company much like a theatrical advance man.[101] Although Gardner, like Marion Leonard and Florence Lawrence, did not officially direct or produce, her stardom provided the raison d'être for the Helen Gardner Picture Players. Each of the half dozen or so features was written specifically for Gardner, and she probably received a percentage of the profits, something actresses working for established film manufacturing companies did not receive. Her first release was a six-reel version of *Cleopatra*, copyrighted as *Helen Gardner in Cleopatra*. Critic Louis Reeves Harrison claimed that her performance was "one of the greatest ever shown on the screen."[102] Importantly, *Cleopatra* proved to be more than just a critical success: it made money at the box office, too.[103]

Another early star-producer encouraged by the possibilities of the feature film was Marion Leonard, returning from a year's absence from the screen after her Gem company folded in late 1911. She and Stanner E. V. Taylor, her director and husband, joined the Monopol Film Company to much fanfare in December of 1912. That month the entire unit, consisting of some twenty supporting actors and actresses, as well as cameramen and laboratory workers, boarded a train for the West decorated with huge streamers "bearing the name of Marion Leonard and the Monopol Film Company." At Monopol Leonard earned an unheard-of salary of $1,000 a week.[104] But Taylor and Leonard left the company in the late summer of 1913 to create the Mar-Leon Corporation. Their first film, *In the Watches of the Night*, resembled the first Gem releases in that Leonard played a woman who "saved the good name of her husband and the father of her child."

Indeed, it seemed that the more independence enjoyed by Taylor and Leonard, the stronger the female characters.[105] Of the five features made by Mar-Leon between November of 1913 and June of 1914, at least three concerned strong and complex female characters that triumphed over vicious men. The heroines of Mar-Leon productions are interesting and sophisticated women; although the means to their triumphant ends are not particularly moral, they are rewarded for cleverness and for being in control of their destiny. In *A Leaf in the Storm* Leonard played a woman driven by widowhood to consider becoming a taxi dancer and perhaps a prostitute. Changing her mind once she is inside a dance hall, she is prevented from leaving by a "cadet," but through a series of fortuitous events the cadet is kicked out, and Leonard meets a wealthy woman slumming in the dance hall. They become friends, and Leonard accompanies her on a trip abroad. When their ship sinks, both women find themselves on an island with a previously shipwrecked sailor. The wealthy companion proves to be an alcoholic, carouses with the sailor, and has passed out when a rescue boat finally arrives. Leonard sees her chance and assumes the wealthy woman's identity, spending months fooling the woman's relatives in Europe and living the high life. When the real woman appears, Leonard feels ashamed and leaves a note of confession, but an infatuated and wealthy nephew follows her to America, and they marry.[106] In *Mother Love* (June 1914) Leonard played a musician whose jealous husband steals her violin and torments her so terribly she leaves home, only to become a musical success. Eventually she reveals her husband's cruelty, shocking him into an early death, after which she blithely resumes her musical career in the city.[107] After the summer of 1914 Mar-Leon disappeared for reasons unknown, and Leonard faded into obscurity.

In December of 1912, at the same time that Leonard was reviving her career at Monopol, the "Gene Gauntier Feature Players" appeared, named after the popular film actress. The inspiration for this company may have come from its director, Sidney Olcott, after Kalem refused to renew his contract at the same salary level. Ironically, Kalem claimed that it could not offer Olcott a raise because of financial constraints caused by the new independent competitors.[108] Information in the trade magazines, however, indicated that a new independent film distributor, Warners' Features, persuaded the director-star partnership of Olcott and Gauntier to form their own company. Warners' Features was a system of states' rights exchanges begun by two of the not-as-yet famous Warner brothers. It took its profits as a percentage of rentals rather than charging manufacturers a lump sum up front. Warners' Features offered feature filmmakers states' rights distribution "practically free."[109]

Gauntier and Olcott were ready for independent feature production. They were already making features for Kalem, including the well-known (and controversial) five-reel life of Christ, *From the Manger to the Cross* (1912). Accustomed to working independently, the Olcott-Gauntier unit made movies in Florida, Ireland, and Egypt, far from the studio's home base in New York City.[110] Thus the Gene Gauntier Feature Players were the autonomous Olcott-Gauntier Kalem unit gone independent. The personnel was nearly identical, including Jack Clark, Gauntier's husband, who played leading male roles.[111] The new company even settled on the St. John's River in Jacksonville, Florida, precisely where the unit spent many winters while working for Kalem.[112] And as at Kalem, most of their scenarios were penned by Gauntier.[113]

The Gene Gauntier Feature Film Company thus began with more human capital than most independents. Calling the company "big news," writer George Blaisdell of *Moving Picture World* reminded his readers that Gauntier "had much to do with the making of pictures other than the work before the camera," noting that it was her idea to make *From the Manger to the Cross* and praising it as the most ambitious film ever attempted.[114] The company lived up to this promise when its first three-reel feature for Warners' distribution, *A Daughter of the Confederacy*, made it to the screen. The story revolved around "Nan, the Girl Spy," and was based on a character Gauntier had created in 1909. The story proved so popular it resulted in a two-year series. According to H. C. Judson, a *Moving Picture World* critic, the film "gives us the old taste exactly, but is bigger in every way." In this story "Nan" captures a Union spy and brings him into headquarters, amidst "a great deal of suspense that keeps the spectator on knife's edge." After many adventures Nan watches as the two armies clash, soon throwing herself into the battle in a borrowed uniform, rallying the Confederate troops "to final victory."[115] After twenty weeks of making films in Jacksonville, including *The Mystery of Pine Tree Camp*, *In the Power of the Ku Klux Klan*, and *The Little*

*Rebel*, the Gauntier Feature Players headed for New York to look for a studio in the city that would be of "sufficient capacity for the staging of the class of pictures they desire to make."[116] But the company took off for Ireland in August 1913 to make films on location, which it had also done with great success at Kalem.[117]

Although the Gauntier Feature Players created popular films and plenty of good publicity, production trailed off in the winter of 1913–14.[118] In July 1914 a writer for *Moving Picture World* indicated that the company was moving into a new studio in a converted church at 515 West 54th Street in New York. But that same month Gauntier left for Europe, allegedly "on doctor's orders," telling *Moving Picture World* that she hoped to regain the "strength she lost in the previous winter."[119] The outbreak of war drove Gauntier back to the United States in August, but she apparently continued her vacation on the West Coast until February of 1915.[120] In the spring of 1915 Gauntier rented her studio to another company while she, Clark, and several members of their troupe went to California to make three-reelers as a unit for Universal.[121] Although Universal boasted more women in direction and production than any other studio, Gauntier returned to New York by the summer of 1915, apparently after a disagreement with Sidney Olcott. In an interview in her apartment, located only a block away from her studio on 54th Street, Gauntier told Mabel Condon, a *Photoplay* reporter, that housekeeping was "per-r-r-fectly delightful"; in other words, she was in no rush to return to the screen. Although not making films at the moment, Gauntier claimed to still have her own company and that a separate company for her husband, Jack Clark, was in the works.[122] *Motion Picture Magazine* indicated Gauntier was returning to the screen in 1917, but by 1918 *Photoplay* described her as having been "retired for several years."[123] Her husband still played leading roles, and Sidney Olcott continued to direct.[124]

As is becoming clear, one of the constants of the female star-producer was her partnership with a man. While it seemed relatively easy for a star to parlay her power into independent production, most female star-producers did not attempt to do so without a male partner. Many were already partnered with a male director because of the dynamics of the collaborative system of production. Others turned to financiers and husbands. When these partnerships failed, it was the female partner who disappeared. The disintegration of the Gene Gauntier Feature Players, for example, occurred at precisely the same moment as the disintegration of the personal relationship between Gene Gauntier and Sidney Olcott. Whether this breakup was the cause or the result of the failure of the company is unknown, but what happened to the partners afterward was typical. Sidney Olcott went on to enjoy a long career as a director.[125] But Gene Gauntier's career went into decline, despite the fact that she was hired by Universal, the most inviting studio for a woman filmmaker in the mid-1910s.[126] Even while partnerships were intact, the very reasons why female actor-producers needed male partners made them vulnerable to manipulation. Although it was true that most of these partnerships were grounded in creative collaboration, such as that between actress and director, it was also true that a male partner was needed because men could more easily gain entry to the business end of making and selling movies. Again, the stage offered a precedent. According to an actress-manager in 1888:

> [B]ehind every successful lady manager stands a man, to whom she owes her financial prosperity, at least. The theatre is divided into two principal departments. The stage is devoted to art; the front of the house to business. On one side of the curtain there is a studio; on the other a bank. The ladies may manage one: but the men are best fitted for the other. Laura Keene would not have made money without Mr. Luetz; Mrs. Wood without Mr. Duff, Marie Wilton without Mr. Bancroft . . . and so on . . . [T]he positions of management and responsibility claimed for women shall not be the sole management and sole responsibility. In theatricals, as in society, the partnership of men and women is best. Ladies should be supreme upon the stage and men in the box-office.[127]

Even after the turn of the century traditional gender boundaries made it difficult for women to develop the required relationships with businessmen and investors. The spaces in which business relationships were nurtured, a men's club or a meeting, were still highly masculinized. Even in the heyday of Solax, Herbert Blaché told Alice Guy not to attend meetings with her distributor because she would embarrass the men who "wanted to smoke their cigars and to spit at their ease while discussing business."[128] This attitude limited the control female partners could exercise over the proceedings of their company.

One extreme example is Grandin Films. Although the details are sketchy, this company was created not so much to further the career of Ethel Grandin as to satisfy the financial greed of two scheming brothers. Raymond C. Smallwood created Grandin Films, Inc., in 1914 to exploit the nominal stardom of his wife, Ethel Grandin. His partner was his brother, Arthur N. Smallwood.[129] The undercapitalized company went into receivership by June of 1915, but film importer and financier George Kleine decided to enter into an agreement with Grandin and her husband, Raymond, who were to make two-reelers at their own expense for distribution by the General Film Company. The deal soured when Kleine discovered that Raymond Smallwood lied about the company's outside sources of capital. This outside capital proved to be "no more than a hope," and Kleine, having advertised the series of films, was forced to advance money to the pair to protect his own reputation. Worse, Kleine discovered through "gossip" that the two ready-for-release films that the Grandins brought to the original agreement actually belonged to Raymond's brother Arthur, who threatened to use an injunction to keep Kleine from releasing them. Kleine was in an "equivocal position" and claimed damages in court. The Grandins actually produced several films under these circumstances, but by November of 1915 Kleine refused to advance them any more money, and the Grandin company ground to a halt. Grandin, of course, may have been a willing accomplice in the messy financial dealings of her husband and brother-in-law, or she may have willingly looked away. Kleine, however, saw her as a victim. In his deposition he concluded that Raymond Smallwood was "a trickster within his limited mental capacity," but "the woman is a hardworking actress" who "probably trails along without being individually vicious."[130]

It is certainly no surprise that some of the new movie stars, perhaps especially minor stars like Grandin, would be vulnerable to manipulation. But a pattern of manipulation and subsequent failure emerged even among the most famous female film producers. There is much evidence to indict Harry Salter, Florence Lawrence's husband and partner, of deliberately manipulating his wife's career to suit his own purposes. Herbert Blaché also appears under a cloud of suspicion. Shortly after his Gaumont contract expired, Blaché took control of Solax, dissolving his wife's company in the fall of 1913 and founding Blaché Features. Although Alice Guy Blaché directed half the films for this company, she lost her studio and her autonomy.[131] Blaché added insult to injury by running off to Hollywood with one of his actresses in 1918. The machinations of her husband seemed clear, but biographer Alison McMahan rehabilitates Blaché's reputation by claiming that the demise of Solax came not at the hand of Herbert Blaché but rather as a result of the vicissitudes of independent distribution. The breakup of the Motion Picture Distributing and Sales Company in 1912 (into the Sales Company, the Mutual Film Corporation, and Universal) diminished the number of theaters acquiring Solax films and diminished Solax income. When Blaché took over as president of Solax in 1913 following the end of his Gaumont contract, it was not necessarily to take power away from his wife. As president, Blaché tended to financial matters, leaving Guy Blaché to do what she liked best—direct and produce. Guy Blaché's staff voted to take a 50 percent pay cut to help Solax survive, but it was no use. Solax's acting company disbursed in 1913. After creating his own film distribution company (Film Supply Company of America), which McMahan claims he did to save Solax, Herbert Blaché created Blaché American Features, naming Alice Guy Blaché as vice president. For a year, from September 1913 to August 1914, the Blachés alternated directing duties, and some features were released under the Solax brand name. In 1914 the Sales Company offered to buy Solax for $200,000 in stock and put the Blachés on a joint salary of $600,000 a year. Believing that their Solax plant was worth more than stock, they refused, only to find that the war in Europe created a coal shortage, which cost them both heat and light, and that in any case filmmakers were deserting the East Coast (where their studio was located) for Hollywood.[132]

Whether or not Blaché was the cause, Guy Blaché's fortunes continued to decline. In 1914 she became a "director for hire," making a white slavery film, *The Lure*, for the Schubert Film Manufacturing Company. She enjoyed no control over the producer, cast, or script and made little money because of a poorly negotiated deal. In September 1914 Herbert Blaché created U.S. Amusements Corporation, naming Alice Guy Blaché as a director along with five men, most of whom the couple had worked with before. The company was to use the Solax plant and was to control its own processing and distribution, but this venture failed. The Blachés directed and produced films for a company called Popular Plays and Players between 1915 and 1917, each making six features. Once again, Guy Blaché was just "a director for hire." It was for this company that she directed the famed actress and occasional screenwriter Olga Petrova in a series of films: *The Tigress* (1914), *The Heart of a Painted Woman* (1915), *The Vampire* (1915), *My Madonna* (1915), and *What Will People Say?* (1916). Petrova starred in other films for Popular Plays and Players, some coproduced by Guy Blaché and Blaché. Petrova believed that she could do better and founded her own company, Petrova Pictures, in 1917.[133]

Alice Guy Blaché's fame and fortune had begun to slide. "In short," she recalled, "everything went from bad to worse."[134] Although she directed seven features in 1916, by the fall of 1917 Guy Blaché left New Jersey for

North Carolina after her children fell ill. Herbert Blaché, on his own, took up with an actress and in 1918 left for Hollywood with her. Guy Blaché returned to the Northeast and lived in an apartment for a year in New York, where she made her last film, *Tarnished Reputations, or a Soul Adrift* (1920), for Léon Perret, an ex-Gaumont actor now with his own production company. Blaché, passing through New York when Guy Blaché was ill, took pity on his wife and invited her to live in California, although not with him. A distraught Guy Blaché sued for divorce but won only a small alimony. Meanwhile, she presided over the auctioning of the Solax plant, which fell into bankruptcy. In 1922, with Solax in bankruptcy and her divorce final, Guy Blaché moved back to France with her children. She claimed sexism made it impossible for her to pursue directing in France. Returning to the United States in 1927, Guy Blaché could find no copies of her films to help her find employment, even at the Library of Congress. Leon Gaumont acknowledged Guy Blaché in a 1954 speech he delivered in Paris, entitled "Madame Alice Guy Blaché, the First Woman Filmmaker," and a group of filmmakers including Georges Sadoul and Charles Ford began including her in the canon, but when she died in New Jersey in 1968 at the age of ninety-five, there were no obituaries calling attention to the passing of the first female filmmaker in the world. She was all but forgotten.[135]

Although McMahan's argument somewhat rehabilitates Herbert Blaché, it does not change the pattern. Despite the openness of the industry, it was always easier for individual men to find work as directors and producers than it was for individual women. Once male-female partnerships failed, women's careers typically failed as well. Herbert Blaché, Sidney Olcott, and Larry Trimble went on to enjoy long careers, whereas their respective female partners, Alice Guy Blaché, Gene Gauntier, and Florence Turner, disappeared. But some female filmmakers soon found that they enjoyed a distinct advantage over their male counterparts. Although millions of Americans visited nickelodeons every week, not all citizens were pleased at the arrival of the movie theaters. When Progressive reformers, many of them clubwomen, questioned the impact of the movies on American society, the presence of women behind the camera suddenly took on new meaning.

Nickelodeon front lined with baby carriages, indicating the popularity of films among women (undated). *Courtesy of the Academy of Motion Picture Arts and Sciences.*

Undated portrait of filmmaker and star Gene Gauntier.
*Courtesy of the British Film Institute.*

Undated portrait of filmmaker Alice Guy Blaché.
*Courtesy of the British Film Institute.*

Florence Turner in a still from an unidentified film made by Turner Films, Ltd.
*Courtesy of the British Film Institute.*

Celebrating fraternity in an illustration from the *Screen Club Souvenir Programme,*
1st Annual Ball (April 19, 1913). Comedian John Bunny at center.
*Courtesy of the New York Public Library, Performing Arts Division.*

Frightened rural wife (Lois Weber) calls her husband at work as an intruder breaks in—a triple-screen effect from *Suspense* (1913), demonstrating Weber's filmmaking talents. *Photo courtesy of the British Film Institute.*

An intruder peers through a keyhole at his potential victim (Lois Weber) in Weber's *Suspense* (1913). *Courtesy of the British Film Institute.*

Portrait of Lois Weber (c. 1912).
*Courtesy of the British Film Institute.*

In this still from Lois Weber's *Hypocrites* (1915) the figure of "The Naked Truth" holds
up her mirror to expose the secret illicit behavior of the suitor while Gabriel looks on.
*Frame enlargement courtesy of the Library of Congress.*

Still from *Ruth Roland, Kalem Girl* (1912), demonstrating the future serial heroine's boxing skills. *Photo courtesy of the British Film Institute.*

Telegrapher Helen (Helen Holmes) sits at her post in this still from *The Hazards of Helen* (c. 1915). *Photo courtesy of the British Film Institute.*

Mabel Normand and Charlie Chaplin in *Mabel's Busy Day* (1914).
*Photo courtesy of the British Film Institute.*

Mabel Normand in a publicity still for *Mickey* (1918).
*Photo courtesy of the Academy of Motion Picture Arts and Sciences.*

Mr. and Mrs. Sidney Drew, producers of genteel comedies (c. 1917).
*Photo courtesy of the British Film Institute.*

# CHAPTER THREE

# "So Much More Natural to a Woman"

## Gender, Uplift, and the Woman Filmmaker

In the fall of 1907 a grammar school principal advised the ladies of the Woodlawn Women's Club of Chicago that "no properly conducted home" would let its children attend the city's nickelodeons. Those "devil's apothecaries," the educator claimed, must be "starve[d] out of existence."[1] The agitation over moving picture theaters continued the following summer. In Grand Rapids, Michigan, Mrs. S. Gauw gave a speech entitled "The Evils of the Moving Picture Theaters."[2] In Kansas City the Franklin Institute, "looking to better the morals of the children and youths," began an investigation of the movies shown "by the cheap amusement companies throughout the city."[3] In Salt Lake City school principal Evelyn Reilly "took a stand in favor of censorship."[4] Communities across the country questioned the nickelodeons that dotted the landscape and drew children as effectively as the Pied Piper.

From a middle-class point of view, particularly among conservative Americans, the problems of the nickelodeon were many. Popping up in the same general context as a commercial amusement already under close surveillance—the penny arcade—urban nickelodeons were perceived by some observers as a menace. According to Gene Gauntier a New York nickelodeon, circa 1906, was

a dingy, odorous little hold, about the size of the average shop on Main Street. Outside it blazed with electric lights of low candlepower, which made up in quantity what they lacked in quality. The entire front was usually plastered with glaring sheets of pictorial paper and strips of canvas proclaiming the current atrocity of crime and adventure. Inside were benches, kitchen chairs, or, in the more luxurious, hard wooden opera seats. Admission was a nickel. The pictures were jumpy and dingy, running five to eight hundred feet, and thrown upon an oblong of more or less white canvas, generally ornamented with sagging folds and an occasional rent.[5]

Penned in 1928, this backward glance served to measure how far the industry had come and was thus at least somewhat exaggerated, yet it reflected the assumptions made by conservatives. Nickelodeons, claimed some, were dirty little firetraps, strewn with peanut shells, rocked by noisy crowds, and steamy with the effluvium of the urban masses. Within them, "impressionable" audiences of children and immigrants were exposed to deleterious scenes of violence and crime, young men and women exchanged improprieties, and neighborhood pockets of working-class solidarity flourished.[6]

Since nickelodeons were generally profitable, the decentralized film industry did not, at first, form a unified response to its critics. This changed after the mayor of New York City shut down the city's 550 nickelodeons on Christmas Eve in 1908. A variety of political reasons inspired the mayor, but the specific reasons were immaterial.[7] A police censorship board already existed in Chicago. Which city might be next? As W. Stephen Bush of *Moving Picture World* concluded in the immediate aftermath, the question was not whether there would be censorship but "whether to have censorship done by the police or by the industry itself."[8] The very real possibility that both films and theaters would be controlled by the state inspired producers and exhibitors to unite in 1909 under the rubric of *uplift*. Writers for the new trade journals such as *Moving Picture World*, *Motion Picture News*, and *Nickelodeon* led the effort to make moving pictures acceptable to the American middle classes. From 1909 to the mid-1910s, the various factions of the American film industry were united in their effort to make moving pictures respectable, an effort that was by its very nature gendered. Women in the audience, women as exhibitors, and women as filmmakers became politicized as the film industry mobilized to improve its reputation. Already present, women in the American film industry now became assets to be exploited and models for men to follow. At this moment women became the most celebrated exhibitors and filmmakers in the American cinema.

When the mayor of New York City temporarily revoked all the nickelodeon licenses in the city, closing them down during one of their busiest seasons, it came as a shock. Acting quickly, the Moving Picture Exhibitors Association of New York initiated the creation of a voluntary national censorship board. It made sense to locate such a board in New York because most film manufacturers were located there and imports came through its harbor.[9] Just three months after the closure of New York's nickelodeons, a new National Board of Censorship began evaluating films. Within a few years the board reviewed approximately 85 percent of all films shown in the United States.[10] If a film passed, it received from the board a seal of approval that appeared prominently in its opening frames. From the manufacturer's point of view, submitting films to the board was good business: the board's seal of approval might entice better-heeled audiences to attend moving pictures, and the film industry as a whole might stave off state censorship.

The National Board of Censorship was a quintessential Progressive accomplishment: it was a regulatory agency aimed not at punishing businessmen but at directing their energies toward the public good. To organize and run the board, the New York exhibitors turned to a prestigious and liberal Progressive ally, the People's Institute, which successfully censored and produced live entertainments. The People's Institute, in turn, was introduced to the movie question by the Women's Municipal League, which asked for its help in conducting a citywide survey of working-class amusements in 1907.[11] Their joint report, published in early 1908, asserted that the physical conditions of most of Manhattan's nickelodeons were dangerous, but the problems were "subject to immediate remedy." More important, there "was a seed of genuine drama and living human interest, which had been planted in the hungry soil of human need and would grow."[12] Created under the auspices of the liberal People's Institute, the National Board of Censorship began reviewing films in March 1909. In keeping with the ideology of female moral superiority, the majority of the volunteer censors were middle-class women. Lee Grieveson found that by 1912, 57 of 75 censors were women, and by 1915 the figure had risen to 100 out of 115.[13]

The belief in the moral superiority of women can be traced back to the Jacksonian period, when both industrialization and the extension of the franchise created an ideology that men were suited to capitalism and politics, a rough-and-tumble public sphere ruled by self-interest, whereas women, believed to be naturally pious and pure, maintained the morality of the nation through the spiritual management of the private sphere of home and family.[14] It was primarily the belief in spiritual superiority that legitimated women's actions outside the private sphere before the Civil War, notably in campaigns against alcohol, slavery, and prostitution. But after the Civil War, when a new generation of women used the essentialist notion of female moral superiority to publicly combat poverty, male drunkenness, child labor, and the contamination of food and drink, religious motivations began to be replaced by a focus on maternalism. All women were alleged to enjoy special insight as mothers or potential mothers, and since this insight was denied to the men who made public policy, male politicians were forced to give the female voice a place of privilege regarding issues concerning family life, which could and did extend to working conditions, education, poverty, and public health. Indeed, the fact that women could not vote placed them above politics, further enhancing their clout during an era of acknowledged political corruption. This was what made the antisuffrage argument persuasive to some women—if women gained the vote, they would lose the power they enjoyed as members of a separate sphere characterized by selfless morality.[15]

At the start of the twentieth century many women observed that the social problems under their purview were too large to be addressed through local efforts, and they began to look to government to adequately address their concerns. Maternalist female reformers spearheaded legislation addressing tenement conditions, sweatshops, clean milk and water, sanitary conditions at schools, and playground construction even prior to gaining the vote. In doing so, as Paula Baker asserts, women "domesticated" politics by bringing their traditional concerns to the attention of legislators, concerns that reached the national stage during the Progressive Era. Baker argues it was women's lobbying, in part, that created the new definition of political liberalism as a philosophy that held the government responsible for addressing the ills of industrialization, immigration, and urban expansion.[16] Thus a gendered political liberalism informed the Progressive Era, and a feminized approach was therefore deemed necessary to uplift the cinema. Male censors did exist, but they were mocked as being effeminate. In 1911 *Moving Picture World* described male censorship advocates in Baltimore as "childless but careful bachelors."[17] The industry could not, and did not, question the need to reform or the presence of female censors.

Under the strategy of uplift the film industry began to manipulate gender to meet its own ends. The first place a specifically gendered policy of uplift appeared was in the trade journals' pleas to "play to the ladies." Attracting women to the audience for the purpose of morally cleansing a commercial amusement was not new; it had been used with success in theater and vaudeville in the nineteenth century. Impresarios in these venues created ladies' matinees and gave away hams and silk dresses to entice women to attend their entertainments.[18] Vaudeville in particular needed to overcome the masculine associations of the concert saloon. By making women feel comfortable and welcomed, vaudeville became the most popular family amusement by the 1890s. As late as 1906 *Variety* asserted that "a healthy condition is best evidenced by several theaters on the circuits having their matinee audiences almost entirely made up of women."[19] When *Nickelodeon* urged exhibitors to "play to the ladies" in 1909, it was a coded dictum that was understood to mean uplift, as well as audience expansion.[20]

The women who attended the nickelodeon, however, did not necessarily fit the profile of the female patron idealized by uplifters: a respectable middle-class, middle-aged matron. Female moviegoers were often drawn from working-class and immigrant communities, and, just as important, they were often young. As Shelley Stamp found, the "ladies" who were meant to distance the cinema from tawdry cheap amusements did not behave in the subdued manner of the middle class. They chatted with their friends, smooched with their sweethearts, cheered transgressive heroines, applauded cinematic suffragists, and soaked up salacious white slavery films that allegedly warned them that going to the movies was dangerous in the first place.[21]

Exhibitors found it necessary to instruct women in respectable behavior. Slides asked ladies to remove their hats and reminded them to tell the management if they were "bothered" by male patrons. Advertisements that showed women in refined dress attending the movies suggested that respectable dress and manners were expected. Mrs. McGabb, a recurring "boisterous, talkative woman" caricatured in *Moving Picture World*, offered "a gentle corrective to the practice of talking through pictures."[22]

It is doubtful that the contradictions of the imperative to "play to the ladies" surprised industry insiders. The assumptions on which the "play to the ladies" dictum rested were removed in time and space from the "moviestruck" girl. The belief that the presence of women was morally cleansing rested on the understanding that women existed in the private sphere, above the corrupting influences of the public sphere. This earlier ideology still had cultural currency, and was still employed by female moral reformers and one strain of suffragists. By the end of the nineteenth century, however, the foundations of this ideology were countered on a daily basis by the presence of a younger generation of women in the public sphere of higher education, work, politics, and commercial amusements. The typical female nickelodeon patron was behaving as a New Woman, with far greater access to, and comfort in, the public and heterosocial spheres of work and pleasure than previous generations had experienced. New Women challenged the physical and philosophical boundaries of maternalist reform by achieving a new modernized independence that weakened the separate-spheres ideology on which female moral superiority rested.[23] Furthermore, most urban nickelodeon fans were from the "lower orders." They were workers and immigrants, and the daughters of workers and immigrants, removed by age, class, and often ethnicity from the middle classes the industry sought to attract.[24]

The film industry handled these contradictions by instructing the women who already attended the cinema in the codes of middle-class respectability and by attempting to lure even more women to the movies. Like the stage impresarios before them, nickelodeon managers began using baby shows, giveaways, and special ladies' matinees to encourage female patronage.[25] As exhibitors began to move into larger and more expensive theaters, female patronage continued to be cultivated.[26] In the fall of 1910 the new $60,000 Princess Theater of Milwaukee offered luxuries designed to appeal specifically to women. "Girl" ushers showed patrons to their seats, and "retiring rooms" for women, staffed by attendants, ensured cleanliness and comfort.[27] As the trade journals looked for best practices among exhibitors, female theater managers were particularly highlighted. As we have seen, female exhibitors existed prior to the rise of the nickelodeon, and given the acceptance of women in small business, there were probably hundreds of female exhibitors. But it was only in the context of the uplift movement that they gained the attention of the industry. The first was Jane Addams of Hull House fame, who opened a nickelodeon in 1907 to demonstrate what a morally healthy nickelodeon might look like. Addams's theater showed the sorts of films praised by liberal reformers as demonstrating the educational potential of the movies—travelogues, actualities, and narrative films with strong morals. Within three months Addams's theater closed because she could not find enough suitable films nor draw a large enough audience.[28]

Addams was an anomaly. When the uplift movement began in earnest in 1909, the industry's trade journals looked for regular exhibitors who managed to present clean shows while keeping their audience. Few exhibitors were interviewed more often than Mrs. Edward H. Clement of Boston, who was profiled by *Nickelodeon*, *Moving Picture World*, and *Motion Picture News* between 1910 and 1913. In 1910, according to *Nickelodeon*, Mrs. Clement's Boston Bijou Theater boasted a "wholesome condition" and "spick and span" attendants and had successfully raised the price of admission to twenty cents.[29] *Moving Picture World* attributed her success to her gender: "As we have said before, and as we say again this week elsewhere, the influence of good women in the moving picture field is of incalculable advantage and value." Mrs. Clement was contrasted to the "ignorant, obstinate commercialism of short-sighted, money-grabbing" exhibitors, who "unfortunately rul[ed] the roost."[30] In 1913 Clement's refined management style was once again applauded by *Motion Picture News*:

> The ticket-taker is a quiet refined girl—not the giggling, gum chewing kind . . . [T]he man at the ticket receiving box is most gentlemanly; the woman in charge of the coat room is a lady who might well grace a drawing-room; the ushers, all women, are courteous and thoughtful, and the show itself is beyond possible criticism. There is no comedian, with his disgusting and stale bowery jokes; the illustrated song is not twanged out to a rag-time tune . . . [A]s to the pictures themselves, not one is presented that has not some strong lesson to teach.[31]

A year earlier, Mrs. Clement confirmed the industry's efforts to reform (and her leadership) when she read a paper entitled "Standardizing the Moving Picture Theater" at a meeting of the Massachusetts State Conference of Charities.[32] Other laudable female exhibitors included Mrs. A. C. M. Sturgis. Under the title "How a Woman Plans to Run a Theatre," *Motion Picture News* noted that Mrs. Sturgis's theater management included accomplished pianists, a clean house, careful selection of films, and giant exhaust fans into which perfumed deodorizers were to be sprayed on the hour. "It is attention of this nature that is a strong point in a woman's management of a picture house," she claimed. "I mean to do all I can to interest the women in the Lafayette, and I am inviting suggestions from my patrons as to what they want."[33]

As noted in the coverage of Mrs. Clement's Boston theater, the presence of female employees inside the theater might also create the desired atmosphere. Not all did so—the "giggling, gum chewing" girl in the box office was already a stereotype—but refined women employed as lecturers during the live portion of the show could set a high tone. In the estimation of *Nickelodeon* many live acts were "vulgar" and "distasteful to the majority of cultured people."[34] In their effort to educate exhibitors regarding appropriate choices, the trade journals sometimes suggested highbrow female acts, such as Miss Russell, a lecturer, and the "celebrated dramatic reader" Lora Bona, whose repertoire included *A Midsummer Night's Dream* and *An Evening with Tennyson.*[35] The editor of *Moving Picture World* added, "We want to see women of her refined intellectual type acting in the uplift of the moving picture, because we think such personalities as hers would be instrumental in attracting the better classes of the community to the picture house."[36]

WHAT OF THE FILMS THEMSELVES? It was *The Unwritten Law: A Thrilling Drama Based on the Thaw-White Case* (Lubin 1907) that inspired a crusade against the movies by the *Chicago Tribune* and ultimately the creation of a police censorship board in that city. It concerned the sensational 1906 murder of famed architect Stanford White by Harry K. Thaw. Five years earlier White had "pursued" Evelyn Nesbit, who was then a sixteen-year-old chorus girl, and allegedly drugged and raped her in his decadent Madison Square Garden apartment. Thaw, married to Nesbit for three years, cried, "He ruined my wife!" when he fatally shot White. Newspapers emphasized the sordid details regarding White's previous affairs and the sexualized décor of his apartment, which included a red velvet swing and a bedroom lined with one hundred mirrors. When the film was released, it was already deeply intertextual in a way that would encourage critics to characterize the cinema as a "school for crime."[37]

Although films depicting crime and violence worried reformers most, critics were also dismayed by the abundance of slapstick and broad humor during the nickelodeon era. By the end of the century, refined humor marked middle-class self-control, while the unpredictable and often pointless humor of slapstick, common in vaudeville and in short filmed comedies, raised the specter of anarchy in an era of rapid immigration and urbanization.[38] Imports were often criticized. Although self-consciously high-class French imports were inspirational to the uplift movement, French comedies made light of marital infidelities, child endangerment, and gustatory experiences (such as eating rats and cats) that middle-class Americans found repulsive. After the closing of New York's theaters, the trade journals began to list subjects to be avoided: sex, crime, violence, and any scenes that degraded religion.[39] And rehabilitating the movies meant more than relief from censorship headaches. According to banker A. H. Giannini the film industry was unlikely to draw investors unless it improved its product: "the poor stories, the inexperienced director, the caliber of the cast, the incompetent title-writer—all these factors were not calculated to awaken an intense interest in the public. The banker, of course, was not attracted to this business."[40] Cultural illegitimacy was costly.

The hiring of stage personnel, a shift away from competition based on technology to competition based on dramatic quality, and the emergence of the uplift movement combined to encourage producers to seek cultural legitimacy by looking to the stage, to "art," for inspiration. Critics like Louis Reeves Harrison of *Moving Picture World* argued hopefully that the "motion pictures are almost insensibly operating as a training school for all matters pertaining to the stage drama save what is spoken."[41] Between 1907 and 1910 the Vitagraph Company produced enough "quality" films based on classic sources (Dante, Shakespeare, the lives of Napoleon, Washington, and Moses) to establish a reputation as a highbrow producer.[42] Actors directed by D. W. Griffith at Biograph (1909–13) exhibited the restrained "verisimilar" or naturalistic style of acting that was standard on the legitimate stage rather than pantomime, which was still prevalent in lowbrow melodramas. Griffith understood that as the camera moved closer to the players, large and unnatural gestures were no longer necessary. A simple look could convey meaning.[43] As Roberta Pearson notes, this new acting style played directly into the hands of uplifters, as the restrained style was associated with the " 'better' classes."[44] Film acting as a whole improved its reputation, as evidenced from a change in language; the verb *acting* supplanted *posing* after 1907.[45] Now that films could convey subtle psychological states, they could sensitively handle controversial topics; thus film content shifted as well. The first film viewed by the volunteers of the National Board of Censorship was Biograph's *A Drunkard's Reformation* (1909), one of several temperance dramas made during 1908 and 1909.[46]

Social problem films were nothing new in the 1910s. Brief films depicting labor strife, alcoholism, judicial corruption, and the like were common even prior to 1907.[47] But the 1908–9 temperance cycle validated the quintessential maternalist moral reform of the late nineteenth century, the core issue of the Woman's Christian Temperance Union (WCTU). Although the WCTU inspired women to go beyond temperance through its "Do-Everything" policy, cinematic attention to this issue underscored what Lee Grieveson calls "a gendering of the moral address of cinema." By 1909, he argues, the actions of the film industry were gendered in three ways: (1) by the influence of procensorship female reformers, (2) the fact that women volunteers did most of the actual censoring for the voluntary National Board of Censorship, and (3) by the industry's filmic response of male "conversion narratives" (such as temperance films and "save the family" dramas). In fact, Grieveson argues that the focus on achieving middle-class respectability through the gendered strategy of reformist discourse shaped not only film content but also film form. It was the censorship crisis that encouraged the industry to move from a "cinema of attractions" model to a narrator system to better tell stories that can demonstrate that the cinema can be an educational and even moral influence on its audiences.[48]

The most controversial social problem genre of the decade was the white slavery film. A panic over "white slavery," which referred to the drugging and kidnapping of young white women in order to force them into prostitution, began in 1909, following an article published in *McClure's Magazine* that claimed that immigrants in New York City ran the largest slave trafficking ring in the world with the help of Tammany Hall and corrupt police officers. (Anna Howard Shaw's 1910 description of nickelodeons as the "recruiting stations of vice" echoed the sentiment of many: the dark, little theaters made work easy for such predators.)[49] In 1913 white slavery reached Broadway in plays such as *The Fight* and *The Lure* but not without controversy. Reformers debated the appropriateness of the subject for a play, and police arrested actress Cecil Spooner between performances of *The House of Bondage*, but as would be true for white slavery films, police activity encouraged even larger crowds to see for themselves. Feminist reformers, such as suffragist Carrie Chapman Catt, supported the white slavery plays, as they would later support the films.[50]

The first white slavery film was *Traffic in Souls* (Universal, 1913), allegedly inspired by Mrs. S. M. Haggen, president of the Immigrant Girls' Home in New York City, who asked Universal writer Walter MacNamara to produce a film to educate immigrants about the potential dangers of white slavers. Universal contextualized the film as a reform document by claiming it was based on the "Rockefeller study" of the white slavery problem, released in the summer of 1913, and described the film as a "Truthful picture-sermon." The potential that this film might act as a warning to young girls, especially immigrants, encouraged the liberal National Board of Censorship to consider passing the film, but unsure of what to do, the board invited social reformers to submit their judgments. The film was ultimately passed (with minor editing), but the board wondered if the fictional story, geared to sell, was truly educational.[51]

No doubt benefiting from Broadway publicity, *Traffic in Souls* was a sensation. It was soon followed by *The Inside of the White Slavery Traffic*, a more graphic film that depicted the inside of a brothel but more authentic as well, connecting prostitution to low wages. Samuel H. London, who wrote the film, had worked within Rockefeller's Bureau of Social Hygiene and for the Justice Department, a fact stressed in publicity surrounding the film to underscore the authority of his voice and to stress the realism of the film. Endorsements from feminists Carrie Chapman Catt and Charlotte Perkins Gilman were further evidence that London sought educational goals, as was the name of his production firm, the Moral Feature Film Company.[52] After the board reviewed the film for the first time, it delayed making a decision until its next general meeting. Meanwhile, the film opened, and the New York City police raided the Park Theater, confiscating the film. The board finally rejected the film, but it was too little too late for more conservative reformers, and in any case the voluntary board did not have the power to prevent exhibition. The theater, granted an injunction, enjoyed huge crowds thanks to the headlines created by the police raid. As the film moved to other theaters, police raids followed. So, too, did temporary injunctions allowing the shows to continue. The injunctions soon expired, however, and the police drove the film out of New York City.[53] But there were more white slavery films to come.

The white slavery cycle would appear to represent everything the film industry was trying to prevent. Shelley Stamp argues, however, that the white slavery cycle actually enhanced the status of the moving pictures. First, the "white slave films announced cinema's 'arrival' as a major entertainment form in New York" by encouraging upwards of fifteen legitimate theaters to convert to moving pictures in order to show the blockbuster films. In addition, these theaters advertised their showings in the Sunday papers in the section set aside for stage productions, a first for the film industry. Second, Stamp argues that the film industry's vocal condemnation of white slavery films allowed it to demonstrate it was even more stringent than the waffling National Board of Censorship.[54] In fact, writers for the trade journals wrote in hindsight, after the police activity, that the films were working against the general goal of cultural legitimacy. This inconsistency underscores the fact that cultural legitimacy was never an end in itself but a means to smooth the way toward growth and profit. With profit the primary motive, the film industry pursued often contradictory strategies to achieve its goal, even within the framework of the uplift movement. White slavery films pushed the film industry dangerously close to realizing long-term damage for short-term profit, but the cycle burned itself out in early 1914.

THE WHITE SLAVERY FILMS may not have convinced all observers that they were serious social problem films, but the fact that they were features, rather than one- or two-reelers, bequeathed them a measure of respectability. The first longer feature films that began to appear in 1910 drew from literary classics, as did the Vitagraph films, but now their full-stage-play length allowed for the further development and validation of the cinema as a theatrical art. Inspired by the 1911 commercial success of three imported Italian cinematic "spectacles"—*Dante's Inferno*, *The Crusaders; or, Jerusalem Delivered*, and *Odyssey*—W. Stephen Bush of *Moving Picture World* urged American producers to copy their formula.[55] As mentioned in the previous chapter, the French firm Film D'Art produced a series of features starring celebrated European actors and actresses, culminating in the appearance of theatrical luminary Sarah Bernhardt reprising her role as *Camille* for the screen in 1912.[56] These better-quality films proved an appropriate accompaniment to the increasingly ornate theaters.[57] It was at this time, too, that noted writers began penning scenarios and noted actors from the stage began making their appearance onscreen, each one lauded by uplifters as evidence that the movies were approaching the legitimacy of the stage.[58] These developments culminated in the creation of Famous Players Film Company in 1912, created by Adolph Zukor to produce films in which American stage stars played highbrow roles on the screen. Although most directors were male, the "artistic" feature film highlighted the allegedly natural abilities of female directors. As the uplift movement continued, essentialist claims for feminine artistic sensibilities meshed with assumptions of female morality to create a powerful context for women filmmakers. Women who were alert and sympathetic to the uplift movement could now situate themselves as the very ideal of what a filmmaker should be.

In 1911 a writer for *Nickelodeon* lamented the difficulties of making films based on literary classics. They were "the hardest kind to present," he asserted. "They demand an expensive outlay of costumes and scenic effects, deep and careful research into the manners and customs of the era depicted." Furthermore, they demanded "faultless photography, and above all a Producer . . . who shall . . . possess the eye of an artist and the mind of a poet."[59] Three years later, Alice Guy Blaché described herself, and all women, as naturally suited to the demands of the artistic feature film. Women "can make such splendid use of talents so much more natural to a woman," said Guy Blaché in 1914, "and so necessary to its perfection": "In the arts of acting, music, painting, and literature, woman has long held her place among the most successful workers, and when it is considered how vitally all of these arts enter into the production of motion pictures one wonders why the names of scores of women are not found among the successful creators of photodrama offerings."[60]

Trade discourse increasingly referred to Guy Blaché as the ideal film-maker. In 1912 Louis Reeves Harrison described her as a woman of "enlightenment and superior breeding" when paying an extended visit to her Solax studio on behalf of *Moving Picture World*. Harrison observed Blaché's direction of the comic opera *Fra Diavolo*, a "massive feature" of precisely the type assumed to draw well-heeled customers.[61] The fact that she was a woman—and a European woman at that (since the first major features were imports)—played a large role in her being singled out. Indeed, the change since 1910, when Solax hid her sex, was remarkable. In the context of the uplift movement, being a woman was now an asset. Interestingly, no one, and certainly not Guy Blaché, mentioned at this point that she had been responsible for at least a few of the objectionable early French imports while she was still working for Gaumont.[62] Rough edges remained, but now they were excused on account of her foreign birth. "She is French," reported an industry journalist in 1912, and "her foreign training and the knowledge of what would 'go' abroad has been the innocent cause of an occasional shock when she tried to bring home a strong moral lesson."[63] A 1915 review of *The Heart of a Painted Woman*, written by uplifter W. Stephen Bush, noted Guy Blaché's "strange tendency to ultra-sensationalism of the morbid order," such as the use of a hypodermic needle as well as gunplay and an execution scene.[64] Such scenes might not have been tolerated had she been a male filmmaker. In 1914, when Guy faced a censorship inquiry over *The Lure*, a film version of the white slavery stage play, a female censor stood up and said, "I think that only a woman could treat it with this delicacy. I think that Madame Blaché has succeeded very well." The film passed unanimously.[65]

No female filmmaker, however, satisfied the industry's ideal more than the talented Lois Weber. The story of her life, as repeated in trade magazines, emphasized her middle-class credentials and the religious devotion that traditionally empowered female reformers. Weber was born in 1879 into a religious, middle-class Pittsburgh home, where her parents encouraged her to explore the arts, and at age sixteen she became a concert pianist, though a bout of stage fright ended her career in less than a year. Back in Pittsburgh Weber sang and played the organ for the Church Army Workers, a Salvation Army–type missionary society aimed at rescuing prostitutes. After her father died, Weber's uncle encouraged her to try a performing career again, this time on the musical stage. According to Weber she "was convinced that the theatrical profession needed a missionary," and her uncle "suggested that the best way to reach them was to become one of them" (an alleged motivation that relieved her of the taint of immorality that still attended professional stage actresses). In 1904 Weber earned promising reviews for her performance in a touring company of *Why Girls Leave Home*. In 1905, at age twenty-six, she married Phillips Smalley, the stage manager of the company, and gave up her career to travel with her husband's touring company. After two years of waiting in hotel rooms and writing freelance moving picture scenarios, Weber had had enough. In 1908 she offered her services to the American branch of the Gaumont company, the same company that brought the Blachés to America. After the 1908 theatrical season Smalley joined his wife at Gaumont.[66]

Between 1908 and 1912 the filmmaking team of Weber and Smalley moved from studio to studio and worked for a time with Edwin S. Porter at his Rex company.[67] In 1912, when Rex became one of several production units releasing under the Universal brand, Weber and Smalley took over, writing, directing, and playing the leads. The couple, now in Los Angeles, made one two-reeler a week for Rex/Universal.[68] During this period Universal's production units enjoyed nearly complete creative control, and it was at Rex/Universal that Weber achieved her reputation as a serious social uplifter and as the leading partner in the Weber-Smalley unit.[69] The key to Weber's success was the fact that she did not take a simplistic approach. Her films did not talk down to the working classes that frequented the movies but typically questioned middle-class viewers about their own pretensions. And although "true" sentiment in Weber's films was found among families who lived modestly, her films included Parisian fashions, popular dances, sumptuous surroundings, and immoral behavior.[70] An early example of Weber's ability to advocate modesty and maternalism while presenting alluring images of sexuality and abundance was *The Spider and Her Web* (1914), "a virile story of ultra-modern highlife, showing aspects of love, passion, vanity and fear as they are under the thin stratum of lackadaisical sentiment and false tradition."[71] The film ostensibly asked, "Does the end ever justify the means?" but contextualized the question through the story of an alluring vamp. The "Spider," played by Weber, attracts intellectual men to her salon and then leads each to ruin. A scientist eventually administers a potion to the "Spider," which harmlessly imitates grave illness, thus frightening her into mending her ways and adopting a baby. When she discovers the trick, the "Spider" tries to return to her old habits but cannot. "The orphaned baby has

EXPANSION, STARDOM & UPLIFT

worked her salvation," asserted the *Universal Weekly*. Thus the vamp learns the value of "real" sentiment through motherhood. Among Weber's other notable early films were *The Jew's Christmas* (1913), depicting the evils of prejudice, and an adaptation of *The Merchant of Venice* (1914), inherently uplifting because it was Shakespeare.[72] Weber even managed to overcome the developing social barrier between the emergent film colony and Los Angeles society by addressing the Woman's City Club with a lecture titled "The Making of Picture Plays That Will Have an Influence for Good on the Public Mind."[73] By 1914 a writer from *Photoplay* claimed that what Weber and Smalley "have done for the uplift of the artistic end of moving pictures is too well known to bear repeating."[74]

Weber's films found their mark because they reflected the generational conflict of the era. As Nancy Cott has argued, in the 1910s an older generation raised on the ideology of woman's sphere collided with the younger generation, which tended to embrace the freedoms of the New Woman and the emergent consumer culture.[75] Weber's life was an expression of this generational divide: she was a stage performer and a Church Army Worker, a filmmaker and a middle-class matron, a childless advocate for birth control who "radiates domesticity."[76] Her own status was that of a New Woman by virtue of her career, but it clashed with her equally public status as Mrs. Philip Smalley, wife and collaborator. Weber shared credit with the Oxford-educated Smalley, the grandson of abolitionist Wendell Phillips.[77] Some observed an equal partnership between the two; others did not. As early as 1913 a writer directly traced the "source" of the duo's films to "the fertile brain of Lois Weber."[78] Historian Anthony Slide, who interviewed eyewitnesses, found contradictory statements. A prop boy said, "Phillips Smalley did nothing—he just sat around the set." A child actor claimed that Smalley "chased every woman on the lot including my mother," adding that Weber and Smalley had public "arguments and shouting matches." But Mary McLaren, one of Weber's favorite actresses (who worked on the same set as the child actor), claimed that the filmmakers were respectful to one another and that "it was a pretty 50–50 proposition."[79] Shelley Stamp's recent and careful analysis of the publicity attending the Weber-Smalley partnership reveals a more consistent image. According to Stamp, industry writers focused on Weber's marriage to Smalley to confirm her status as a solidly bourgeois married woman, precisely the desired audience demographic. But at the same time that publicity emphasized her modest, middle-class home and her role as Mrs. Philips Smalley, it also highlighted the outlines of a modern companionate marriage, as she and her husband were photographed time and time again working together, or at least next to each other, in domesticated settings. She argues that this discourse did not "domesticate" Weber but rather "blurred the boundaries between work and leisure" and presented a model of a "nonhierarchical marriage" based on mutual interest and support. The end result helped to uplift the industry while "creating not only a legitimate place for women in the industry, but a privileged one."[80]

Armed with the imprimatur of the film industry, Weber began making a series of highly controversial and highly successful social problem films. With her own credentials beyond reproach, Weber not only criticized the very class of patrons the movie industry desired, but she also antagonized censorship advocates at the very height of a censorship flurry in 1914 and 1915. Weber left Universal in 1914, probably because the studio balked at feature-length productions, and made her first major feature film, *Hypocrites*, for Bosworth, Inc.[81] *Hypocrites*, made in 1914, starred an allegorical figure entitled "The Naked Truth," who repeatedly appears holding up her mirror of truth to the hypocrisy of modern life.[82] There is nothing unusual about such heavy-handed moralizing in the silent era. What was atypical was the fact that the figure of the Naked Truth was played by, in the words of one contemporary reviewer, "a naked girl, about 18 years of age."[83] In fact, she was played by Margaret Edwards, winner of a gold medal as the "most per-fectly formed girl in the world."[84] Although shot as a double exposure and thus transparent, when Edwards faced the camera, spectators were clearly confronted with what we now refer to as full frontal nudity.

Onscreen nudity was rare in the silent era, but it was not unheard of. What requires explanation is that this particular film, with its highly un-usual full-frontal shots, was not only tolerated in most cities but was cel-ebrated as an artistic, cultural, and moral landmark for the film industry. Elites such as William Randolph Hearst and the heads of Progressive orga-nizations such as the Mother's Club, the Drawing Room Club, and the Flor-ence Crittendon Home attended the film during its Broadway run and thus observed the figure reveal more than just middle-class hypocrisy, yet they did not publicly flinch.[85] It is true that Ohio banned the film, and Boston audiences could only view it after clothes were painted on the Naked Truth frame-by-frame, but elsewhere it drew critical acclaim in its original form. *Moving Picture World* gushed over the "dignified attention and warm support of the press," claiming that the film "created more comment and aroused more enthusiastic press notices than almost any other motion picture yet shown in New York." The *New York Times* called it "artistic," and the *New York*

*Evening Journal* said it was "the most startlingly satisfying and vividly wonderful creation of the screen age."[86] Released prior to Griffith's *Birth of a Nation*—often regarded as the film that finally legitimated the movies for American middle classes—*Hypocrites* attracted the well-to-do by the thousands. One writer noted "that the highest price seats are sold out every performance and the low price seats have to be forced."[87]

Barely a decade past the nickelodeon boom, Weber's allegorical nude challenged the moving picture's new middle-class audiences and elite censors to prove their own cultural status by accepting it as art rather than smut. The most popular piece of American art before the Civil War was Hiram Powers's 1844 sculpture of a female nude, *The Greek Slave*. Thanks to the artist's barnstorming promotion, *The Greek Slave* defied lewd interpretation to the point where scores of well-to-do Americans bought small-scale reproductions for their parlors. "By the end of the century," argues Joy Kasson, "*The Greek Slave* had become synonymous with respectable, even staid, taste."[88] After the Civil War, as lowbrow and highbrow culture separated, the acceptance of the nude became more firmly class-based. Elite Americans who feared that they would be mocked by Europeans for prudish attitudes began cultivating a taste for the European salon nude, buying both original paintings and photographic reproductions. At the same time, censors, particularly the New York Society for the Suppression of Vice, under the leadership of Anthony Comstock, worked to rid American society of anything that would encourage licentiousness. The class nature of the nude-as-art emerged in high relief following Comstock's 1887 arrest of an elite Fifth Avenue art dealer, Roland Knoedler, for selling photographic reproductions of French salon nudes. Although the arrest four years earlier of a store clerk selling precisely the same photographs caused no stir, Knoedler's arrest inspired an outcry. The *New York Times* described Knoedler's gallery as "a respectable house that has furnished respectable citizens with good pictures for more than a generation" and that "the best proof" of the gallery's innocence was that its pictures were "not sought by persons of vicious life or inclinations." As sociologist Nicola Beisel observes, "a picture's obscenity depended upon the social class of those who sold it or viewed it." It was one thing to censor an unknown store clerk but quite another to censor a tastemaker to the denizens of Fifth Avenue. In the court of public opinion the interpretation of the salon nude as sophisticated, rather than morally degenerate, won—at least for consumption among the upper classes.[89] Weber's figure of "The Naked Truth," in fact, bears an extreme likeness to French academic painter Jules Joseph Lefebvre's allegorical painting *Truth*

(1870), which depicted a young, voluptuous full-frontal nude holding aloft a small mirror, a conventional symbol of truth. This painting was his great initial success, after which he was made an officer in the Legion of Honor and became a member of the Academie des Beaux Arts, frequently exhibiting his works in the Paris Salon.

This lavish, four-reel feature began with a still portrait of the director (notably without her husband), signed, "Sincerely, Lois Weber." Thus, audiences were told that they were about to be treated to a personal message from the writer-director-producer of the film. This kind of introduction—the still shot in contemporary clothes—was frequently used to introduce the stars but rarely the director. Weber then introduced the cast, each appearing in the same thronelike chair. Then viewers see the white Gates of Truth open and the figure of "Naked Truth" walk through, exposed over the original shot and therefore transparent, yet fully discernable. To introduce the plot and to remind the audience, who have just seen a completely naked woman, that this is a moral film, a typed page appears that reads, in part, "The text of my sermon this morning is taken from the 23rd chapter of Matthew, 28th verse: 'Even so, ye outwardly appear righteous unto men, but within ye are full of hypocrisy and iniquity.'" The congregation fidgets as the minister speaks; chatting and yawning, a choirboy reads a newspaper, but a few pay close attention. At the end of the service, top-hatted capitalists, who just congratulated the minister on a great sermon, gather outside the church and agree to ask for his resignation, though none want their name known. Back in the church the crestfallen minister finds the newspaper in the choir area and slumps into the thronelike chair seen in the opening introduction. He glances at the paper's front page. Under a large reproduction of a painting depicting the naked female figure of Truth, the title reads "Why the Truth Has Startled Wicked Paris," a reference to "the Most Talked-About Picture in the New Paris Salon." The minister looks to the heavens with concern and despair and apparently sleeps. In a double exposure, what looks like his spirit clothed in monk's robes arises over his unconscious body. The monk is now outside, pointing the way up a steep path, the "narrow way" to truth. The congregants from church stroll down the wide, easy lane, only a few trying to climb the narrow path. A wealthy capitalist refuses to leave his heavy gold behind. The women give up after soiling the hems of their dresses. Only a repentant praying woman and a choir girl with a crush on the minister make it most of the way. The Monk, now at the top of the hill, tells the cavorting figure of Truth, "Since my people will not come to you, come to my people."

The Monk, now called Gabriel (that is, the messenger of God), is then seen in a medieval monastery, where he fasts while others feast. In solitude he works on his statue of Truth. When he is finally ready to unveil his masterpiece, the abbot arranges a fete for the occasion. Royalty, artisans, and peasants assemble, along with monks and nuns. An apprehensive Gabriel watches from the woods as the abbot unveils the statue. The naked statue of Truth shocks the multitude, and the mob attacks Gabriel, spearing him through the heart. Two women (those who tried to climb the narrow path), now representing Magdalene and the Virgin Mary, mourn the fallen monk. In many ways this time-traveling interlude is awkward, but it is a brilliant anticensorship message that places the potential censors of the film in the position of the barbarous and simpleminded medieval mob that killed the Christlike Gabriel.

The film now shifts to the present day, where the Naked Truth, accompanied by Gabriel, holds up her mirror to politics, society, love, modesty, and the home. In "Politics" three top-hatted capitalists sit behind a politician giving a heated campaign speech before a sign that says "My Platform Is Honesty." When Truth holds up her small mirror, the picture dissolves and the audience sees the candidate taking bribes from his supporters. In "Society" Gabriel and the Truth enter the luxurious living room of a society couple giving a party. The hostess tells them that "Truth is welcome if clothed in our ideas" and gives Gabriel a sheer scarf, which he rejects in disgust. In "Love" an earnest suitor gives flowers to his sweetheart, but in Truth's mirror the audience sees him sharing a jolly drink and a kiss with a woman of questionable morals and gambling and fighting with men. In "Modesty" a group of chilly young people on a beach wrapped in cloaks seem innocent, until the Truth shows how on a warm day they expose their arms and legs in suggestive swimwear (one doubts that the irony of a nude pointing this out was lost on contemporary audiences). Finally, in the "Home" we see a young family gathered around their daughter's sickbed as a doctor shakes his head. As Truth holds up her mirror, the audience sees an open box of candy labeled INDULGENCE and an open book entitled SEX. The young mother gives the little girl piece after piece of candy while the father, engrossed in his own reading, fails to notice his young son looking at SEX until it's too late.

As the film ends, Gabriel and the Naked Truth go back through the white gates, apparently into heaven. Back on Earth the congregants find their minister slumped in his chair. The next day's headline reads "PROMINENT MINISTER EXPIRES IN CHURCH—After preaching a sermon on Hypocrisy, it was unfortunate that he should be found with a newspaper in his hand. The congregation was much shocked."[90]

*Hypocrites* was scheduled for release on October 19, 1914, but was held back as the National Board of Censorship debated Weber's use of nudity. It is true that the National Board of Censorship, created by the film industry and run by liberals, was typically tolerant, but only six months earlier protest from more conservative reformers inspired the House of Representatives to open hearings regarding the possibility of federal censorship.[91] This was precisely what the film industry wanted to avoid. By approving *Hypocrites* at this moment, the National Board of Censorship, and the film industry it supported, took an enormous leap of faith. None of the advance advertising mentioned the use of a nude figure, which impressed both those inside the film industry and its potential critics.[92] As one reviewer noted—Ohio and Boston aside—"even the most fastidious can find nothing offensive to carp at, it has been so well handled."[93] (It caused another critic to state, perhaps tongue in cheek, that after seeing the film, "you can't forget the name of Lois Weber.")[94]

*Hypocrites* was an instant success when it opened on January 20, 1915, at the Longacre Theater in New York.[95] When it was announced that the film had to be moved to a different theater to make way for a new show, patrons paying fifty cents to one dollar a seat filled the Longacre to capacity, earning the theater $5,000 in a single week.[96] By the end of its run, *Hypocrites* earned $119,000 in domestic sales, or six and one-half times its negative cost of $18,000.[97] Was Weber conscious that patrons might be flocking to theaters to see a naked woman rather than to hear her sermon? She answered that in 1917 when she said, "Let those who set themselves up as idealists chatter as much as they please about their art, the commercial side cannot be neglected. We're all in business to make money."[98] But despite Weber's remark, she was sincere about proving that moving pictures were an art form. In an article entitled "The Greatest Woman Director," an interviewer claimed that by conversing with her, "you become imbued with her enthusiasm, you believe in the high mission of the screen, as she does; you look at the whole 'industry' from a new point of view—that of an art—and you feel inspired with some reflection of her soaring ideals and standards."[99]

*Hypocrites* made Lois Weber a household name. The typically budget-conscious Carl Laemmle welcomed back Weber and Smalley in the spring of 1915, allowing them not only to make features but giving them nearly complete creative freedom and distributing their films under Universal's most prestigious brands.[100] Most of the films Weber made for Universal were also social problem films. *Scandal* (1915), the sequel to *Hypocrites*, showed how gossip can lead to destruction.[101] *Hop, the Devil's Brew* (1916), was an antiopium tract made in collaboration with the customs bureau.[102] In *Shoes* (1916), a film based on a Jane Addams story, a struggling dime-store clerk with a profligate father is forced to "sell out" her virtue to secure a necessity.[103] *Saving the Family Name* (1916) exposed the hypocritical prejudice of the well-to-do against actors and actresses.[104] *Idle Wives* (1916) offered a movie-within-a-movie: "a community of dissatisfied husbands, wives, sweethearts, and parents" saw themselves—and their possible futures—played out on the screen, causing all to take stock of their lives.[105]

By 1916 Weber was "the greatest woman director," according to *Moving Picture Stories*, and one of the top six directors in the entire industry (as "Lois Weber and Phillips Smalley"), according to the *New York Dramatic Mirror*.[106] Weber earned this acclaim as a result of her previous work in conjunction with the release of *The Dumb Girl of Portici*, the cinematic version of the nineteenth-century opera, starring Universal's biggest coup, the famed Russian ballerina Anna Pavlova. Pavlova became interested in filmmaking after a visit to Universal City, and as she was considering the purchase of the Boston Opera Company, she was interested in the money as well (her filmmaking fee amounted to nearly $500 an hour).[107] According to one writer, Laemmle assigned Weber to handle Pavlova's film because "only a woman would understand a woman."[108] Filmed at great expense and starring an internationally renowned prima ballerina, *The Dumb Girl of Portici* brought Universal, and the film industry, a great deal of prestige, despite the fact that the disjointed and nearly danceless film pleased neither critics nor fans.[109]

Weber's greatest fame came with more daring films. In 1916 she made *The People vs. John Doe*, a film against capital punishment based on an actual case in which the wrong man was blamed. The *New York Times* called it "the most effective propaganda in film form ever seen here," and *Variety* was equally ebullient in its praise.[110] Even more controversial were a pair of films Weber made in 1916 and 1917, which indelibly linked her name with one of the most controversial subjects of a highly politicized decade—birth control.

In 1916 the birth-control movement in America peaked, thanks to Margaret Sanger's impending trial for obscenity for sending contraceptive information through the U.S. mail. The primary theme of the American birth-control movement in the 1910s was the denunciation of class privilege: contraceptive information was a readily discussed "secret" among wealthy women and their doctors but was denied to the poor women who needed it most.[111] According to Shelley Stamp, in depicting a wealthy socialite who ruins her chances for motherhood by undergoing too many illicit abortions, *Where Are My Children?* played into elite fears of race suicide as opposed to the working-class sensibility of Margaret Sanger's crusade.[112] In any case *Where Are My Children?* supported the dissemination of literature about birth control, at the same time upholding the idealized mother and denouncing abortion.

The hero of the film (Tyrone Power) was a child-loving but childless husband, a district attorney who unsuccessfully defends a doctor whose only crime was publishing a book on contraception. While the lawyer pines for a child, his wife is seen smoking, drinking, and arranging a party. When she finds herself pregnant, a friend tells her that if she is "determined to evade motherhood," she knows a doctor who can help. They visit the doctor, and a small angel is seen going into heaven. "One of the 'unwanted' ones returns," reads the title, "and a social butterfly is again ready for house parties." The husband eventually discovers what has been happening. The wife repents, but she has made too many visits to the abortionist to "wear the diadem of motherhood." The film ends with her bitterly imagining the family that might have been.[113]

Although the traditional family is upheld, and the abortion itself implied only in masked, allegorical terms, *Where Are My Children?* was extremely controversial. A committee of eight from the National Board of Review (formerly the National Board of Censorship) received the film in March of 1916 but were unable to reach a clear decision. Convinced of the film's merits, but not ready to give the film its standard seal of approval, the board gave four special showings before prominent New Yorkers, including representatives from the Merchant's Association, the Bar Association, the American Social Hygiene Association, and the Neighborhood Worker's Association. Representatives from Boston, Philadelphia, and other cities were also present. The results of polls taken at these showings encouraged the board to pass the film as long as it was shown only to adults.[114] Interestingly, *Moving Picture World* critic Lynde Denig noted that the film did not reflect the class realities stressed by Margaret Sanger. It was working-class women, bereft of knowledge regarding birth control, who most likely ended up at an illicit abortionist's, not the society woman. Other observers made similar remarks, claiming that the film confused abortion and birth control, whereas the "voluntary motherhood" campaign under Sanger wished to limit abortions by disseminating birth control information.[115] Weber's handling of the issues is somewhat confusing. It is true that the society wife takes the blame, and it would be true that under the eugenics ideology of the era this would be the family that should reproduce. But her brother, who impregnates the daughter of the housekeeper, who subsequently dies by the hand of the abortionist, is also condemned. Birth control information could have been accessed by the elite wife, but she takes what is portrayed as the easier way (which also elicited complaints), and it is the working-class woman who ultimately loses her life.

Not surprisingly, *Where Are My Children?* raised the ire of some local censorship boards. Pennsylvania censor Ellis P. Oberholtzer called it "unspeakably vile" and banned it from the state altogether.[116] Nevertheless, the film proved popular, no doubt thanks to the controversy it engendered. According to Universal, opening night filmgoers blocked Broadway traffic, and *Variety* reported a similar scene at Boston's Majestic Theater, where two thousand people had to be turned away.[117]

Lois Weber embodied the ideal director at the height of the uplift movement, between 1909 and 1916. Her films did not simply provide safe fodder for mixed audiences but rather challenged the film industry and censors to redefine moving pictures as a medium for mature audiences, as well as immigrants, workers, and children. Her middle-class, religious background, her apparently sturdy bourgeois marriage, and her embrace of maternalist reform allowed Weber to make films that perhaps no male filmmaker dared. More than any other filmmaker, Weber both fulfilled and expanded the definition of what women might do for the industry, given their allegedly innate morality and insight into social causes.

IN SPITE OF THE ATTENTION PAID TO WOMEN in the industry (who were always a distinct minority), or perhaps because of it, the men of the industry began gathering in all-male clubs and trade associations. As we have seen with the creation of the Static Club, this was the era of professionalization, which often began with the establishment of a fraternal order. The first such organization was the all-male Screen Club, formed in New York in 1912 for producers, directors, actors, authors, cameramen, and "photoplay newspaper men." Its purpose was to "raise the industry to the highest status of respectability and dignity." Modeled on the all-male Friars Club, it was also a social organization, and to this end it opened a clubhouse on West 45th Street with a library, a smoking room, a dining room, and a café. Soon there were similar clubs in Chicago and Los Angeles, the latter leasing a three-story building that "will have practically every convenience common to men's clubs."[118] The similarity between the Screen Club and a men's club did not end at its physical amenities, as the club's anthem attests:

> Members of the Screen Club we
> Linked in true fraternity
> Brothers marching hand in hand
> A loyal and faithful band.[119]

Camaraderie and fraternity were as important to the Screen Club as was the "advancement and preservation of the motion picture art."[120] In 1914 photos of the new clubhouse exposed spaces clearly encoded as masculine: a darkly paneled grill, billiard room, officers' room, and secretary's quarters with a highly visible spittoon. And if there was any doubt, a caricature included in a pamphlet commemorating the club's second annual ball depicted "Noon Hour at the Screen Club": a dozen men enjoying cigars and drinking from glasses conspicuously labeled "milk."[121] Hardly the model of uplift, and antithetical to the legacy of female reform, the Screen Club was off-limits to women; so, too, were any benefits to membership. In 1913 *Moving Picture World* reported a "movement among the women engaged in the producing end of the industry to form an organization of their own on similar lines," but it never coalesced.[122] As similar trade associations formed, only the Photoplay Author's League, established in 1914, welcomed both men and women.[123]

But the uplift movement did not encompass the entire industry. The only thread tying all members of the film industry together was the bottom line. And not all women chose to present themselves within the framework of domesticity. Although the feature film dealt with serious issues and was geared to attract a better-heeled crowd, there was still an audience of immigrants, workers, and children to please. While D. W. Griffith and Lois Weber filled Broadway theaters, a group of female filmmakers working in the trenches of "short" films drew public attention by defying traditional notions of womanhood. As Weber carefully groomed her womanly image, other female filmmakers contradicted middle-class notions of femininity in slapstick comedies and gender-defying serials. Clearly modern, could these "New Woman" filmmakers of the 1910s withstand the industry's changing political winds?

INTERLUDE

# WOMEN IN SERIALS &
# SHORT COMEDIES
# 1912–1922

As longer feature films came to define Hollywood's premiere product, women filmmakers achieved some of their greatest successes in the short-film format. Two-reelers were typically exhibited, often along with live acts, before the feature film to create an evening's entertainment. Two genres dominated the short-film scene: hair-raising serial thrillers, which starred the "serial queens" of the 1910s, and slapstick comedies. Both forms promoted an image of women who were, much like young women in the audience, comfortable in the heterosocial spheres of work and commercial amusements. Women onscreen, however, transgressed normal boundaries through heroic and daring action or anarchistic comedy that turned gender roles upside down. These rich New Woman fantasies were often created by the women who starred in them, offering women in the audience yet another layer of identification. But it was not only women who enjoyed these genres; they were among the most popular films of the mid-1910s.

As a result of its unique format and consistent portrayal of the New Woman onscreen, the short film followed a somewhat different trajectory from the feature film's. The entry of women into serials and comedies will reflect the trends discussed in Part I, and the exit of these women from filmmaking will foreshadow Part II, particularly with regard to the increasing control of the central producer and the financial hazards of the independent field. But the demise of the woman filmmaker in these genres was not due to the encroaching studio system. Serials and short comedies did not follow the feature into the most luxurious first-run theaters at the end of World War I but were relegated to subsequent-run theaters, where they became the staple of children's matinees and the bane of a new wave of censorship advocates.

CHAPTER FOUR

# The "Girls Who Play"

## The Short Film and the New Woman

The New Woman appeared on the screen in the 1910s primarily in two forms: as a serial queen—one of the "girls who play with death"—and as an irreverent comic spoofing the conventions upheld by the guardians of public morality.[1] Short one- and two-reelers in the form of suspenseful cliffhangers and brief comedies not only survived as added attractions in the age of the feature film, but at mid-decade they proved at least as popular as the features. Largely freed from expectations of uplift attending feature films, short films could explore the sensuous, rebellious, and athletic traits associated with the generation of young women sitting in the movie theaters.[2] In a new genre, the serial thriller, strong, self-reliant heroines mastered pistols and Packards, wrestled criminals and wild beasts, and saved their fathers and boyfriends from certain death. To a degree not seen in longer films, women initiated the fun in short comedies by eluding chaperones, visiting cabarets, flirting with strangers, and making fools out of bosses, boyfriends, and husbands. Quick-witted and independent, New Woman heroines and comediennes ran away from home, donned scandalous clothing styles, and defied authority. Significantly, many of these New Woman–style heroines were the product of female producers. More than any other genres of the silent era, the serial and short comedy of the 1910s suggest that women filmmakers, as a group, contributed to an alternative vision of gender onscreen. But by 1922, when censors rejected the New Woman–style serial heroine and the slapstick comedienne, female directors and producers of short films disappeared as well.[3]

THE TERM *New Woman* first applied to women in the middle classes who ventured into the public sphere, such as female reformers, women in business, college students, and athletes.[4] But as we have seen, by the beginning of the twentieth century many young working-class women had also ventured into the public sphere, both to work and to enjoy the new mixed-sex venues of amusement parks, dance halls, and nickelodeons, where they flaunted new styles and new behaviors.[5] As Ben Singer convincingly argues, they probably also read the work of yellow journalists who regularly wrote of courageous women, such as housewives who fought off intruders, and "plucky girl reporters" who scaled bridges, drove speeding cars through city streets, and dressed as men to sample urban places deemed off-limits to respectable women.[6]

This daring New Woman also appeared in stage melodramas, setting a direct precedent for the daring onscreen heroine. Always on the lookout for action-filled sequences, filmmakers regularly plundered the blood-and-thunder melodramas of the nineteenth century for their action-filled climaxes of crimes, murders, and rescues. One of the most popular of these plays was *Under the Gaslight*, which premiered in 1867 and may have been the origin of the venerable last-second rescue of a victim bound to the railroad tracks. In *Under the Gaslight* the victim is male, the rescuer female, the latter allowed to breach nineteenth-century gender constrictions by the legitimate need to save her family.[7] As stage melodramas grew more and more sensational, requiring real horses, trains, water, explosions, and burning buildings, the athleticism demanded of women expanded as well. *Edna, the Pretty Typewriter* (1907) required the heroine to "jump from the roof of a building to the top of a moving elevated train" and then leap from one speeding car into another.[8] Since the first professional film actors and directors hailed from the stage, it is no accident that they brought the tricks of sensational melodrama with them.

Films mimicking the female heroics of the stage melodrama appeared in a few pre-1908 films such as *The Girl from Montana* (Selig, 1907), in which the pistol-packing star saves her boyfriend and captures a band of horse thieves.[9] By 1911 a subgenre soon to be exploited countless times by serial producers began with Griffith's *The Lonedale Operator* (Biograph, 1911). *The Lonedale Operator* resembled Porter's *The Great Train Robbery* (1903), except the telegraph operator at the Lonedale railroad station was a woman. Notably, she was every bit as brave in the face of armed robbers as Porter's male hero. A year later Biograph's *A Girl and Her Trust* again concerned a brave female railroad telegrapher attacked by robbers. In this case the heroine threw herself onto a boxcar to catch the thieves, leading to a suspenseful chase.[10] But the moving picture that most clearly presaged the serial film was Kalem's *The Girl Spy* (1909). The story concerned the true exploits of Belle Boyd, a Confederate girl who disguised herself as a soldier to spy on the North. Written as a single film by Gene Gauntier, who also starred in the leading role, *The Girl Spy* became a two-year series as a result of popular demand. Complaining that she was "tired of sprains and bruises and with brains sucked dry of any more adventures for the intrepid young woman," Gauntier finally "married her off and ended the war." That should have been the end of the girl spy, but "Not so!" According to Gauntier, popular demand forced the addition of *A Hitherto Unrelated Incident of the Girl Spy* (Kalem, 1911).[11]

Another example of the fascination with female athleticism that attended the New Woman and her translation to the screen occurred in 1907. That year Boston police arrested Australian swimming champion Annette Kellerman for sporting a form-fitting bathing suit on a local beach. Both athletic and feminine, risqué and wholesome, the tremendous appeal of the Kellerman-style suit began a cycle of diving girl films, which exploited the suit's obvious attractions while emphasizing physical fitness as an ideal for women. Mabel Normand donned a Kellerman suit in Biograph's *How She Triumphed: An Argument in Favor of Physical Culture* (1911) and *The Diving Girl* (1911), before becoming the premier New Woman–style comedienne, and future serial queen Ruth Roland displayed her diving ability while wearing a Kellerman suit in *Ruth Roland, Kalem Girl* (Kalem, 1912).[12] *Ruth Roland, Kalem Girl* offered a patchwork of vignettes to illustrate the athleticism of this future serial star. The film begins with Roland in a fashionable silk dress, driving a convertible, and entering her dressing room at Kalem, where she smells a vase of fresh flowers and is attended by a maid. After highlighting her fashionable femininity, the bulk of the film advertises Roland's tomboyish athleticism through a series of shots: Roland pummeling a punching bag in a men's gymnasium, wielding a rifle outdoors and holding up evidence of her hunting skills (ducks), catching a huge fish on a lake, riding a galloping horse on a beach, canoeing in a sailor suit, diving three times off a tall pier in her Kellerman suit, and taking off in what looks like an open airplane. The film ends with Roland once again in feminine dress, rejecting the offers of three wealthy suitors.[13]

ALTHOUGH THE SERIAL QUEEN APPEALED TO MALE VIEWERS, she was created with a female audience in mind.[14] Until 1912 it appeared that only the exhibition branch of the industry specifically catered to women. Even *The Girl Spy*, for all its feminist overtones, was simply one of several local stories Gauntier "borrowed" while her Kalem unit worked on location in the Deep South.[15] The first filmmaker who publicly created a film especially for women was, somewhat ironically, Thomas Edison. When McClure's, publisher of *Ladies' World* magazine, approached Edison about "a plan whereby the moving picture could be worked in conjunction with the [serialized] story," Edison agreed to take part.[16] Although technically a series, Edison's *What Happened to Mary* (1912) is considered by most scholars to be the first movie serial. The hallmarks of an early serial were the deliberate continuation of characters and collaboration between filmmaker and publisher, in which episodes appeared simultaneously in print and on the screen. Edison-McClure's *What Happened to Mary* pioneered both of these motifs, which clearly played to the female audience through its association with *Ladies' World* and its character of "Mary," played by Mary Fuller. "Mary" was an ordinary girl, a nineteen-year-old in "an old print gown" who was not so very different from the average young woman in the theater, that is, of course, until she discovered the truth about her past: as a baby, she was left on a doorstep with a note promising $1,000 to her guardian if she was married off successfully. Realizing that her stepfather planned a distasteful marriage for her, Mary decided to take fate into her own hands and boarded a train out of town, intending to gain her rightful inheritance for herself. With this simple act "Mary" began the trend of proactive serial heroines that would distinguish the genre for almost a decade.[17]

When the loosely connected stories forming *What Happened to Mary* proved to be an enormous hit, imitators immediately sprang into action, copying both the continued-story format and the independent heroine. A few months after *Mary* debuted, Selig and the *Chicago Tribune* collaborated on a chapter play entitled *The Adventures of Kathlyn* (1913). The first true cliffhanger serial, *The Adventures of Kathlyn* tested the credulity of spectators with the story of a young American woman who inherited a throne in India. Like Mary, Kathlyn was a "girl without fear."[18] Confronting lions, tigers, and sexually aggressive natives, Kathlyn exhibited, according to the intertitles, "not a sign of that natural hysteria of women."[19] Even more popular than *What Happened to Mary*, *The Adventures of Kathlyn* inspired a "Kathlyn" waltz, cocktail, hairstyle, and hat.[20] Over the next two years serials flooded movie screens across America. The fantastic popularity of Pathé's *Perils of Pauline* (1914) anticipated the enormous profits of *The Million Dollar Mystery* (1914), which earned well over that amount at the box office. By early 1915 managers of the largest theaters in Elizabeth, New Jersey, put serials on the regular weekly program, and Indianapolis reported that "serial pictures seem to be the big drawing cards in all the houses."[21] In 1916 Pearl White, the serial star who began her career in *The Perils of Pauline*, handily beat out Mary Pickford and Lillian Gish in a fan magazine contest to determine the most popular moving picture star.[22]

At the center of serials of the 1910s was a woman of action, a woman who "saved the hero's life nearly as many times as he rescued her."[23] *The Perils of Pauline* began when Pauline, the adopted daughter of a rich industrialist, spurns the proposal of his biological son, Harry. She wants to become a great writer before settling down. To help her get the experience she needs, the father promises Pauline a trip around the world. Suddenly, however, he dies. He has willed Pauline half of his fortune but only if she lives long enough to wed Harry. Otherwise the fortune will go to the father's villainous secretary, Owen. Owen spends the next nineteen chapters trying to do away with Pauline before she weds Harry. A trusting Pauline assists Owen's many attempts to trap her, not the least by continuing to put off Harry while she engages in adventures to enrich her writing career, from ballooning, to movie acting, to racing thoroughbreds and automobiles. Harry does indeed come to the rescue time after time, but Pauline does quite well rescuing herself on occasion.[24] Universal's answer to Pauline, the similarly intrepid *Lucille Love, Girl of Mystery* (1914), proved her mettle while being "wrecked in an airplane, shanghaied aboard a ship, shipwrecked in the South Seas," and "washed up on an island peopled with beast men." Lucille usually rescued herself. "Never before in anybody's picture," declared Universal historian I. G. Edmonds, "has a hero—defining the word as the one who gets the girl —had so little to do."[25]

The serial queen of the 1910s was clearly not the virginal heroine of a D. W. Griffith film. She was a New Woman fantasy—athletic, courageous, intelligent, and popular. But the film industry handled this new character with apprehension. This attitude stemmed partly from the experimental nature of "playing to the ladies." Faced with an unprecedented number of female patrons, and surprised at the sudden popularity of the serial queen, the industry rather clumsily attempted to refine the serial heroine to appeal to more genteel, "feminine" tastes. Informed by trade journals that women were interested in clean movies, free from violence, sex, and vulgarity, some filmmakers logically concluded that a gentrified serial heroine would be even more popular than the stunt-driven, wild-animal wrestling Kathlyn and Pauline.[26] To this end Reliance publicized *Our Mutual Girl* (1913) as "strictly a women's series." A Pygmalion tale of a rural girl launched into New York society, *Our Mutual Girl* featured high society, prominent suffragists, a female senator, and even an appearance by Otto Kahn, of "the famous banking house of Kuhn, Loeb & Co." Through fifty-two episodes *Our Mutual Girl* learned how to dress in the latest fashions and mimic the "little mannerisms and personalities" necessary to gain acceptance into the "smart set."[27]

Despite these attractions, *Our Mutual Girl* did not cause much of a stir; it certainly did not turn Norma Phillips, its star, into a household name. Nevertheless, two years later Reliance paid respected authors George Randolph Chester and Lillian Chester $25,000 to write *Runaway June* (1915), a serial that deliberately avoided "the chase element and blood and thunder." Reliance launched an enormous publicity effort, including advertising campaigns in the *Ladies' Home Journal* and *Saturday Evening Post* and a national beauty contest giving winners from each state a free trip to California aboard a "woman's special train." But *Runaway June* was not a runaway success.[28] Danger, action, and suspense provided the nuts and bolts of a successful serial, precisely the "chase element and blood and thunder" missing from *Runaway June*. The few highbrow serials that did succeed at the box office, like *The Ventures of Marguerite* (Kalem, 1915) and *Who Pays?* (Balboa, 1915), did not eliminate villains and gunplay.[29]

It was clear that the rough-and-tumble serial heroine made for box-office success, but this presented a dilemma for the film industry; the combination of strong, assertive women and violence was as dangerous as it was appealing. Consider the provocative plight of Kathlyn Williams in *The Adventures of Kathlyn* (Selig, 1913):

> You will see her bound by fanatical natives on top of a great funeral pyre with the flames creeping ever nearer her helpless form.

> You will see her tied with thongs in a tiger trap as human bait for the bloodthirsty beasts of the jungle.

> You will see her swim for life to escape a maddened water buffalo in the black waters of the Bengal River.

In fact, time after time, in scene after scene, Kathlyn takes her life in her hands and walks grimly up to the very jaws of death in order to portray with life-like realism the scenes necessary to make *The Adventures of Kathlyn*.[30]

Or consider episode 13 from a later Pearl White serial, *The Lightning Raider* (Pathé, 1918). When Pearl sees her enemy place a vial of germs in white roses ready for delivery, she determines to become a "rescuing angel" and secures a list of intended recipients. She finds that the first bouquet adorns the table at "The Yearly Banquet for the Society for Anthropological Research." Six or seven male academics with pince-nez and beards anticipate Professor Absolom's paper on "The Inferiority of the Female Brain Cavity." Just as the professor begins, "From the natural timidity of the female, I deduce—," White bursts into the room, pistol in hand. As the men look on in fear, she grabs the flowers, rummages through them searching for the vial, and keeps the gun pointed at the academics. She does not find the vial but laughs at the frightened men as she runs outside.[31]

Torn between exploiting the lurid (and lucrative) aspects of the serial heroine and offending both potential middle-class audiences and cultural authorities, the film industry attempted to contain the sexual and feminist qualities of the serial heroine. Industry insiders accomplished this containment in two ways: through contrived motivations onscreen and through ameliorating publicity for the star offscreen. Onscreen, an absurdly large number of serial heroines were orphaned heiresses whose guardians were inept or corrupt, thus granting them the means and the moral imperative to begin their adventures. Freed from parents and guardians through no ambition of their own, the serial heroines had no choice but to bravely forge onward. In addition serials sometimes portrayed the perils of the serial heroines as just deserts for their excessive independence. In episode 1 of *The Ventures of Marguerite* (Kalem, 1915), for example, the heroine's curiosity entices her into the villains' trap: "We were familiar with your adventurous disposition, my dear, and our little affair at the restaurant was simply a subterfuge to get you here!" Feminine assertiveness did not always go unpunished.[32]

Offscreen, publicists carefully balanced the independence of the serial star with womanly interests. As presaged in *Ruth Roland, Kalem Girl*, studio publicists wanted audiences to believe that these athletic New Women really performed their daring onscreen stunts (and many did), but they also wanted to reassure the public that the serial heroines happily observed the bounds of feminine propriety.[33] When a *Photoplay* reporter first met Kathlyn

Williams in 1914, he expected to "find a dashing, mannish woman arrayed in more or less masculine attire" but instead saw before him "a decidedly womanly lady, quietly but tastefully dressed and one whose charm is increased by a beautifully modulated voice."[34] The attempt to describe serial heroines as feminine without diminishing their aura of athletic prowess and bravery approached the schizophrenic in the 1910s. A *Photoplay* writer reported in 1915 that Helen Holmes of *The Hazards of Helen* (Kalem, 1914–17) liked "pretty gowns" but added that she could "burst the sleeves of any of them by doubling up her biceps."[35] A writer for *Motion Picture News* claimed that Ruth Roland—star of the serial *Who Pays?*—rides "a horse as if she were born in a saddle. She runs an automobile. She boxes, fences, plays tennis, football, and baseball. She swims, swings Indian clubs and bowls. She is a crack shot with rifle or pistol, and has won a number of prizes at archery." But he also described her as "adorable," "dainty," "sweetly vivacious," and "exquisitely feminine." In fact, she was "all girl."[36]

THE NEW WOMAN COMEDIENNE received the same treatment. "She rides like a Centaur, swims like a fish," and has "muscles as strong and springy as cold-rolled steel," wrote one critic of Mabel Normand, but that same critic also described her as a "compelling" beauty.[37] In the work of Mabel Normand, the wholesome athleticism of the serial heroine and her tendency to find trouble blended with the flirtatious rebellion of the working girl. While no other comediennes came close to matching Normand's popularity, other studios tried to copy "Keystone Mabel," creating a multiplication of New Woman–style comedies starring sassy working girls, disobedient daughters, and flirtatious wives.

Like the serial heroine, the New Woman slapstick comedienne challenged traditional gender roles. Slapstick purposefully violated the bounds of refined middle-class behavior, finding humor in the daily chaos of an urbanizing, heterosocial society, where both men and women inhabited the public realm of work and leisure, where strangers could and did wink at each other in the park, and where the forced intimacy of city life often meant watching men and women flirt, argue, and even resort to blows. More pointedly, it allowed female characters to reduce male authority figures—bosses, boyfriends, husbands, and policemen—to an equal or lower level.[38] Office workers across the United States, for example, no doubt enjoyed Fay Tincher as "Ethel" the stenographer, who brought pandemonium to the office with her garishly striped dress, spit curl, and chewing gum in a series of films produced for Mutual. City dwellers and rural townsfolk alike enjoyed Louise Fazenda's Universal comedies, which usually found her playing a rebellious farm girl trying to elope against the wishes of her father.[39]

For critics the violation of conventional feminine behavior was sometimes too much. "Deliver us, O Lord, from the woman who attempts comedy," invoked one motion picture critic in 1914. Notably, however, this critic made an exception for Mabel Normand.[40] Before Normand, female comediennes (versus those who played the "straight" role) were typically unattractive and abrasive.[41] Henry Jenkins dubbed the anarchistic style of physical humor "New Humor," and it was particularly controversial when women performed it. Laughing out loud was a sign of a lack of self-control among the nineteenth-century bourgeoisie. By the beginning of the twentieth century a "New Humor" emerged out of the rich context of urban, interethnic life and was developed on the vaudeville stage. Even newspapers began to carry comics and humor columns for the first time, multiplying readership sometimes by half. Women, once again lumped under the rubric of the middle-class matron, were assumed, since they were morally superior to men, to be without a sense of humor. Some astute observers noted that for a dependent married woman, laughter could be construed as criticism of male superiority and was best kept confined. Thus the humorous woman was the antithesis of the refined beauty, which was assumed to be the goal of all women. When humorous women emerged on the vaudeville stage, they were seen as dangerous "wild women," threatening the sanctity of the middle-class home.[42]

The beautiful Mabel Normand embraced the New Humor. According to Mack Sennett, the head of Keystone Studios, Normand essentially played herself. When asked what made Normand funny, filmmaker Hal Roach replied that "you knew that if a guy kicked her, she'd kick him back."[43] In most of her films Normand appeared as a working-class girl in a tattered dress who loved to play practical jokes and have a good time.[44] In comedy after comedy, from *Mabel's Dramatic Career* (1913), *Mabel at the Wheel* (1914), *Caught in a Cabaret* (1914), *Mabel Lost and Won* (1915), and *Mabel, Fatty, and the Law* (1915), Normand deftly defied authority, met the physical challenges and humiliations of slapstick, and still remained an attractive, respectable, American girl. A former model for Charles Dana Gibson, creator of the idealized woman of the turn of the century, Normand was literally a Gibson Girl thumbing her nose at middle-class conventions.[45] In *Mabel's Married Life* (1914), codirected by Mabel Normand and Charlie Chaplin, Normand and Chaplin begin the film as a calm, mannerly couple, relaxing in a public park; but after Normand suspects her husband of flirting with another woman, she becomes incensed, and Chaplin has to pull her from the other woman's throat. Miffed, Chaplin storms off to a local bar and returns home drunk to find what he thinks is another man in his apartment. Chaplin takes a swing at the suspect, and the "man" swings back, but the stumbling Chaplin soon realizes that his nemesis is a punching dummy and that he's been had by his wife. Normand, hearing the blows, comes running out of the bedroom in pajamas to laugh at her foolish husband.[46]

Normand's freedom to create her onscreen persona emerged from the same collaborative filmmaking that created the star-producer of the longer feature film.[47] At Keystone Studios, all employees within a production unit—actors, director, and crew—hammered out ideas for scenarios, which were turned into scripts by the writing staff and submitted to Mack Sennett for approval. Even after approval each unit was free to improvise within certain limits during the shooting. As we have seen, this kind of collaboration made everyone in the unit a kind of generalist and a potential filmmaker, and it was not unheard of for writers or actors to find themselves directing or supervising a unit.[48] Yet such fluidity might also set up a conflict of authority within the production unit, and as woman director, Normand was particularly vulnerable.

Mabel Normand's nemesis was the inexperienced Charlie Chaplin. With only a few months of directing experience under her own belt, Normand supervised the twenty-four-year-old Chaplin in a film starring both of them: *Mabel at the Wheel* (1914). Chaplin wrote in his autobiography that he "doubted her competence as director" because she was "only twenty, pretty and charming, everybody's favorite." In one scene, where Normand hosed down a street to cause the villain's car to skid, Chaplin wanted to insert his own bit of business: to accidentally stand on the hose, look down the nozzle in a puzzled manner when the water stopped, unconsciously step off the hose, and get squirted in the face. "But she shut me up quickly," he recalled. " 'We have no time! We have no time! Do what you're told.' That was enough, I could not take it—and from such a pretty girl." Chaplin replied, " 'I'm sorry, Miss Normand, I will not do what I'm told. I don't think you are competent to tell me what to do.' "[49]

Chaplin blamed his difficulties on Normand's youth, gender, and even her attractiveness, but he experienced the same problem with Henry Lehrman, his first director at Keystone. Chaplin didn't understand that his vaudevillian gags took too long for the fast-paced world of filmmaking. Sennett stood behind Normand during this incident, intending to fire the

"Girls Who Play"

temperamental Chaplin by the end of the week, but Chaplin's first films were popular. Chaplin stayed at Keystone long enough to begin directing his own films, which proved humbling.[50]

Since Normand was engaged to Sennett during her Keystone career, one might cast a jaundiced eye at her activities behind the camera, but both contemporary critics and historians laud Normand's competence as a filmmaker. When Sennett left Biograph to form Keystone in 1912, he took Normand not only to be Keystone's leading lady but to help run the studio. According to *Photoplay* writer James R. Quirk, Sennett was "the first to proclaim her capabilities as a creator of situations and the important part she has played in the tremendous task of organizing the forces of the company."[51] As soon as demand necessitated a second production company, Sennett entrusted it to Normand's supervision, and by December of 1913 Normand began directing her own films.[52] In 1916, even after Chaplin and Roscoe "Fatty" Arbuckle arrived at Keystone, critic Julian Johnson gave Normand the credit for "bulwarking of all the Keystone comedy with her own slender shoulders," claiming that "Normand knows more about screen comedy, and has made better screen comedy, than any woman actively photographed." Years later, film historian Kalton Lahue argued that the films Normand directed alone "were far superior in construction to those in which [Sennett] appeared or directed."[53]

No OTHER WOMAN IN COMEDY ACHIEVED Normand's popularity or her power behind the screen, but it was not unusual for the serial heroine of the 1910s to both write and coproduce her own starring vehicles. One of the most ubiquitous serial queens was Helen Holmes, daughter of a Chicago railroad official and friend of Normand's from their prefilm modeling days. Normand secured Holmes her first acting job at Keystone, but it was as the star of Kalem's *The Hazards of Helen*, a long-running serial about a railroad telegrapher, that Holmes gained fame.[54] The seemingly endless 119-chapter serial began with a scenario penned by Holmes herself, entitled "The Girl at the Switch."[55] During a 1916 interview Holmes told a journalist that "if a photoplay actress wants to achieve real thrills, she must write them into the scenario herself," because male writers will not have a woman perform stunts "they wouldn't do themselves." Holmes continued to write many of the scenarios in which "Helen," the brave railroad telegraph operator in the far West, proved herself able to run a train as well as any man. Indeed, "Helen" even overcame sexual discrimination on the job in episode 13, in which Helen, "the night [telegraph] operator at Ferndale," bravely fought thieves only to find herself fired because she was a woman. She is informed via telegram that "Effective to-day: Male operators only will be assigned to Ferndale Station on account of recent robbery." At that moment Helen spots the crooks jumping a freight train. Determined to catch them, she climbs a bridge and drops down onto the top of the moving train, engaging one of the crooks in hand-to-hand combat. As Helen grapples with the crook, he falls into a river, but Helen pursues him and successfully wrestles him to shore. The episode ends when Helen triumphantly reclaims the stolen cash from the crook's pockets.[56]

At the height of the *Hazards* success in 1915, Helen Holmes and her director-husband, J. P. McGowan, left Kalem and the *Hazards*. They set up an independent film production company and enjoyed complete creative control under their own "Signal" brand.[57] At Signal, Holmes and McGowan made a string of extremely successful railroad serials beginning with *The Girl and the Game* (1915). In this serial the brave "Helen" saves her boyfriend and father from a train wreck, saves her boyfriend from a burning locomotive, saves the railroad from financial ruin, recovers the payroll from thieves, saves her boyfriend and a male friend from another train wreck, rescues a male character from a lynching, captures ore thieves, saves two men from a mine cave-in, recovers more stolen money, and uncouples a freight train to prevent a "terrible wreck."[58]

While Helen Holmes starred in railroad thrillers, Grace Cunard created mystery serials for Universal. A former Biograph actress under D. W. Griffith, Cunard wanted creative control from the beginning of her film career. After teaming up with Francis Ford (the elder brother of the soon-to-be-famous director of westerns John Ford), she joined Universal.[59] At Universal Cunard and Ford enjoyed a production unit of their own and a great deal of latitude. Cunard wrote all the scenarios and shared directorial duties with Ford in a collaborative partnership typical of early filmmakers.[60] After Cunard and Ford had made several successful one- and two-reelers, Universal chose their unit to develop the studio's answer to Pathé's instantly successful *Perils of Pauline*. Cunard and Ford revamped a two-reel western entitled *Lucille Love, Girl of Mystery*, and turned it into a fifteen-chapter installment play. Episodes were churned out mere days before release, but in spite of the inconsistencies of this hastily planned serial, *Lucille Love*, now a spy story, was an enormous hit.[61]

The success of *Lucille Love* led to a serial-making career for Cunard and Ford that came to full flower in *The Broken Coin* (1915), one of the most popular serials of the decade. Cunard, who thought that "ingenue parts" were "insipid," wrote the part of newspaper journalist Kitty Gray for herself. The adventures begin when Gray gains possession of a half-coin that, when matched with its partner, would reveal the location of a fortune. Universal studio head Carl Laemmle good-naturedly played her editor, and he unleashed a mountain of publicity for *The Broken Coin*, including a twenty-two-page press book, "teaser" ads in newspapers, souvenir buttons, mirrors, postcards, and, of course, broken coins.[62] Wearing several hats—leading actress, writer, codirector, coproducer, and film editor—Cunard now earned a four-figure salary every week, based on her $450 a week actor's pay, plus a 25¢ bonus for each foot of finished film, and 10 percent of the net profits for writing and codirecting.[63]

By 1916 Cunard and Ford were Universal's biggest box-office draw. During 1916 and 1917 they embarked on *The Purple Mask*, one of the most implicitly feminist serials of the decade.[64] *The Purple Mask* was the alias of debutante Patsy Montez, who traded in her party dress for a leotard, tights, shorts, mask, and cape to lead the Apaches, a band of male crooks, through the mansions and gutters of Paris. *The Purple Mask* stole from the rich and gave to the poor, especially wronged women. In one episode the "Mask" steals an expensive necklace from a bride and sells it back to the groom, giving the money to the mother of his illegitimate child. In another episode Patsy and her gang recover a fortune gathered through a "nefarious practice"; they then use the money to "establish a home for unfortunate girls."[65]

BETWEEN 1915 AND 1916 women filmmakers made some of the most popular images on the screen: Normand at Keystone, Cunard at Universal, and Holmes at her own Signal company. All were one- to two-reelers, and unfortunately, the changing status of the short film and the changing mode of production made it much harder for these women to continue working within the larger studios after 1916. Given their heightened status, first-run exhibitors, who managed increasingly ornate theaters, became cautious about what kinds of films they exhibited.[66] Slapsticks and serials came under special scrutiny.[67] In an article giving advice to freelance writers, Captain Leslie T. Peacocke informed his readers that the problem with the movies was that they were controlled by uneducated "financial potentates" with little taste, who entered the film business "when any sort of production was avidly seized by a public greedy to be amused." The public was "more discerning now," and the "old time thrill—the falling over cliffs, the automobile accidents; the fighting in barrooms and over stairs . . . have lost their power to thrill."[68] The serial and the slapstick comedy, drawing cards at even the best houses a year or so earlier, were relegated to the second-, third-, and fourth-run markets. If they appeared in major theaters, it was only at the children's matinee.[69]

Serial makers responded immediately to these changes in exhibition by turning once again to gentrification in the hopes of gaining a foothold in the first-run theater's regular program. An "evening's entertainment," after all, included not only the feature but also an array of short films. Surely a highbrow serial could take its place among the added attractions at even the best theaters.[70] George Kleine was one of the first to pursue this strategy with *Gloria's Romance* (1916). Kleine hired Broadway star Billie Burke to star as "Gloria," along with the same production team responsible for the wildly successful *Million Dollar Mystery*. With a budget six times larger than that of *Million Dollar Mystery*, *Gloria's Romance* seemed a sure bet, and some first-run theaters paid $1,000 a week just for the privilege of renting it. But despite high production values, good venues, and a Broadway star, *Gloria's Romance* failed at the box office.[71] A few months later Louis B. Mayer tried to launch a highbrow serial, *The Great Secret*, with financial backing from Boston investors. He convinced two major stars, Francis X. Bushman and Beverly Bayne, to play the leads. But outside of Boston *The Great Secret*, which premiered in January 1917, was not a great success.[72] By 1919 the term *serial* was so firmly associated with the cheaper theaters that J. P. McGowan, now at Universal, suggested that his latest serial be called a "thirty-six-reel feature."[73]

While the highbrow serial appeared to be a contradiction in terms, it was virtually guaranteed that patrons of a first-run theater would see a short comedy. Unlike serials, short comedies successfully split into highbrow and lowbrow in the mid-1910s—highbrow representing genteel comedies, which were exhibited in first-run theaters, and lowbrow designating slapstick comedies, which were, like the serial, relegated to the cheaper theaters.[74] The most famous proponents of highbrow comedy were Lucille McVey Drew and Sidney Drew, who began producing the "Mr. and Mrs. Drew" comedies immediately after their marriage in 1914.[75] Working out of their own unit at Vitagraph, the Drews specialized in what came to be known as the "polite" or "refined" domestic comedy, or in current terminology, the situation comedy.[76] The Drews managed to be funny without throwing pies, taunting the police, or getting into fistfights, and they were "always well dressed." Unlike the Keystoners, the Drews found humor in the small misunderstandings afflicting the financially comfortable: the wife with

"Foxtrotitis," the "man whose pride of ancestry makes him an insufferable bore," the woman with a closet full of clothes but *Nothing to Wear*.[77]

Fox-trotting aside, Mrs. Drew was not a New Woman. Indeed, the gentle Drew films never really challenged traditional institutions of any sort. This was apparently fine with Lucille McVey Drew, for it was she who produced and directed the Mr. and Mrs. Drew comedies; Sidney Drew allegedly preferred drinking Manhattans. This unequal division of labor was a little strange, not because of Mrs. Drew's gender but because of her relative inexperience. Her husband belonged to the Drew-Barrymore theatrical clan, and when he left the stage for the screen in 1911, he brought with him a wealth of dramatic experience as an actor and producer, quickly becoming a director at Vitagraph. Why he gave up these duties is a mystery. Perhaps his theatrical experience working with his famous mother, the actress-manager Louisa Lane Drew, instilled in him a respect for women's abilities. Being significantly older than Mrs. Drew, he may have been taking it easy. Or, perhaps, alcoholism impaired his ability to work.[78]

Nevertheless, Sidney Drew articulated the agenda of the Mr. and Mrs. Drew comedies: to provide sophisticated comedy for the literate middle classes. Drew assailed the majority of filmmakers, claiming they regarded "motion picture patrons as fools," possessing "the mental equipment of a child seven or eight."[79] One of Sidney Drew's last films as a solo director before his marriage was the extraordinary *A Florida Enchantment* (Vitagraph, 1914), in which women happily become sexually aggressive men by swallowing a magic seed, immediately leering at other women, groping at them, and even kissing other women passionately on the lips.[80] But as a member of "Mr. and Mrs. Drew," Sidney Drew would never make such a controversial film. Through their refined domestic comedies the Drews became the epitome of highbrow comedy-makers, providing the ideal product for first-run theaters. In 1918 Metro, their distributor, advertised that just by showing the Drew comedies an exhibitor could improve his theater's reputation.[81] Even the notoriously strict head of the Pennsylvania Board of Censors, who found all other comedies reprehensible, gave the Drews his hearty approval.[82]

The segregation between highbrow and lowbrow comedy exerted a conservative influence on American comedy production. In the light of new exhibition practices comedians refined their antics for a better-heeled audience. Harold Lloyd shed his broadly drawn "Lonesome Luke" character to become "a middle-class, white-collar worker whose only comic prop was a pair of glasses." He also moved from shorts into feature films. In 1920 Fatty Arbuckle left his own Comique company to make features for Paramount.

Hal Roach even toned down the slapstick in comedy shorts starring Will Rogers, Laurel and Hardy, and Our Gang to make them palatable for first-run audiences.[83] What was also being removed was the subversiveness endemic to slapstick, which allowed comediennes like Mabel Normand to assault the gender status quo. The gentrification of highbrow comedies and the relegation of serials and slapstick to the cheaper theaters was a blow to the New Woman onscreen, yet it was not fatal, and the short film did not need the first-run theater to be lucrative. Films featuring the New Woman–style heroines and comediennes now being made for the subsequent-run theaters and matinees continued to make money—a lot of money. They made enough money, and cost enough money, for studio heads to begin taking them quite seriously.

In 1916 *Photoplay* estimated that whereas the average feature cost between $5,000 and $25,000 to produce, the cost of making a serial ran between $100,000 and $500,000. Although the serial figure may have been closer to between $45,000 and $80,000, the price was still quite steep. The reason for these higher costs was obvious. A typical feature was only five reels long; a serial could be fifteen to thirty-six reels long. The era of scrambling to finish serial installments days before release was over; producers, anxious about their investment, required that the whole plot be worked out from start to finish before the cameras began rolling. Given the greater demands of holding a thirteen-chapter serial together, only the most reliable and talented writers, directors, and stars would do, and salaries rose accordingly. Even after the negative was "in the can," serials posed far greater risks than feature films because they cost so much more. "An enormous amount of money is risked," noted one contemporary, "and there is no chance to even up if it fizzles."[84] Little wonder, then, that Pathé allegedly spent $500,000 on newspaper and billboard advertising to support the six serials it released in 1916.[85]

The reappraisal of the serial's worth changed the mode of production. Central producers absorbed the responsibilities previously allowed to director-producers; it was the central producer who assigned scripts to directors, chose the cast, and kept tabs on costs. New specialists on the studio lot, such as continuity writers, editors, location scouts, and wardrobe personnel, further eroded creative control from the former director-producer.[86]

Short-film producer-directors, such as Helen Holmes and J. P. McGowan, Grace Cunard and Francis Ford, Mr. and Mrs. Sidney Drew, and to a certain extent Mabel Normand, responded to the efficiency experts and the concomitant loss of creative control by leaving the established studios for independent production. Observing changes in exhibition as well as production, many of them also attempted to make the move from the short-film world of physical comedies and stunt-driven serials into more genteel feature films. But unless allied to a strong financing and distributing company, these independents were generally financial disasters.

The first salvo launched against the new mode of production came from Grace Cunard and Frances Ford. Cunard and Ford left Universal in 1916 after "a fatal clash with that new habitue of the studios—scientific management." Universal's biggest stars at the time, they were enticed to return to the studio after a compromise, inspiring *Photoplay* to proclaim, "Score one for art over efficiency." However, on their very next project, *The Purple Mask* (1916–17), Ford and Cunard were put under the supervision of Carl Laemmle's nephews, the Stern brothers, who "constantly interfered with production." After completing *The Purple Mask* the Cunard-Ford filmmaking partnership ended. Cunard continued working for Universal, appearing in *Elmo the Mighty* (1919), a serial starring Elmo Lincoln, but she never directed or produced again. By contrast Francis Ford went on to enjoy a long filmmaking career, working for a short time as a director at Universal and then leaving for independent production in 1918, thus replicating the pattern in which male partners enjoyed long careers after their female partners "retired."[87]

Although Cunard never became an independent, Helen Holmes and Helen Gibson—two of the top serial queens of the 1910s—launched their own companies between 1915 and 1920. Helen Holmes may have chafed at bureaucracy as much as Grace Cunard. Even after creating the long-running *Hazards of Helen*, Kalem required Holmes to send all scenarios to Kalem's East Coast office for approval before shooting. In 1915, while still churning out *Hazards*, Samuel S. Hutchinson, president of the American Film Company, a partner in the Mutual distributing organization and investor in small independents, approached Holmes and McGowan, offering them financing, distribution, and creative control. It was this offer that allowed them to create their own company. They accepted and made several successful serials (as well as features, which were less successful) under their own Signal brand between 1915 and 1917.[88]

When Signal closed at the end of 1917, despite its financial success with *The Girl and the Game* and its other serials, *Photoplay* announced that "Helen Holmes is tired of dodging locomotives and wants to go in for more thoughtful stuff." Yet the likely reason for the breakup of Signal was marital discord. By the early fall of 1918 Holmes and McGowan filed for divorce.[89] By the next summer J. P. McGowan was working for Universal as a director, assigned to upcoming serial star Eddie Polo, and Helen Holmes was on the East Coast, where independent producer S. L. Krellberg created the S. L. K. Serial Corporation for her.[90] S. L. K. made highbrow serials, providing "mystery and romance" rather than railroad stunts. In *The Fatal Fortune* (1919), S. L. K.'s first release, Holmes played a newspaper reporter "drawn from one adventure into another" among opulent settings from New York mansions. But despite its allegedly "hair-raising situations" and luxurious settings, the typecast Helen Holmes did not do well in the role of a "girl reporter."[91]

In 1920 Holmes decided to take matters into her own hands, forming the Helen Holmes Production Corporation and signing a financing and distribution deal with Warner Brothers. Holmes entered the agreement with the understanding that she would be making serials, but the cash-strapped Warners convinced her to star in a feature, *The Danger Trail* (1920), to raise enough revenue to finance her serial productions (the reverse of an earlier trend, whereby the profits of cheaply made serials were used to finance features). When Holmes finally got started on her first serial, *The Tiger Band*, Warners asked her to advance $5,000 to smooth over production difficulties, and they dragged their heels in repaying her. Holmes had to take Warner Brothers to court to get the rest of her money. The Warner brothers asserted that Holmes caused her own problems by absenting herself from the set and refusing to ride a horse or get her gown wet, although they did admit her work stoppage was related to the $5,000 owed. Because her distribution agreement was tied up in litigation, *The Tiger Band* was released under states' rights distribution, with little advertising or promotion. It did not do well. After the Warner Brothers fiasco Holmes gave up filmmaking but continued to act, mostly in character roles, where, in the words of Kalton C. Lahue, she "did little more than smile at the camera and execute an occasional stunt."[92]

"Helen" Rose Gibson, who replaced Helen Holmes in Kalem's *Hazards of Helen* in October of 1915, followed a career path that paralleled Holmes's—including a foray into independent production and financial disaster. Taking the stage name "Helen" to preserve the "reality" and continuity of the series, Gibson continued to succeed in railroad dramas after *Hazards of Helen* but decided in 1920 to star in features produced by her own company, Helen Gibson Productions, to be released through Associated Photoplays Corporation. Gibson began production on the provocatively titled *No Man's Woman*. When the film was nearly finished, she signed a ninety-day note for $1,500 to complete the intertitles. When delivery of *No Man's Woman* from

Glendale to New York was delayed, Gibson lost everything, including her company. She never produced again.[93]

Although not quite as lucrative as the serial, the comedy short was also subject to the new efficiency experts. In 1916 Vitagraph fired most of its stock company and began the "jobbing" system, paying employees by production rather than yearly, and it introduced written contracts to protect itself from the power of the stars. It was at this moment that Mr. and Mrs. Sidney Drew fled to Metro.[94] At Metro the Drews continued to produce short domestic comedies for $90,000 a year, becoming independent producers the next year for their own V. B. K. Corporation, distributed by Paramount.[95] As independents, the Drews slowed production to one or two comedies a month, until Sidney Drew's sudden death in the spring of 1919.[96] According to one writer the "unprecedented demand for clean, wholesome comedies" led Lucille McVey Drew to continue making films.[97] Less than a year after Sidney Drew's death, the Pathé exchange sold what was to be a series of six to eight two-reel comedies made by McVey Drew based on the "After Thirty" stories penned by Julian Street, a popular short story writer for wide circulation magazines like *Saturday Evening Post* and *Colliers*. Despite working out of her own New York studio, and despite the apparent popularity of her first efforts, it is unclear if McVey Drew made more than the first two films, *The Charming Mrs. Chase* and *The Stimulating Mrs. Barton*.[98] McVey Drew then began writing, directing, and producing a series of short domestic comedies for Vitagraph. Most promising was her direction of the Vitagraph feature film *Cousin Kate* (1921), starring one of Vitagraph's biggest stars, Alice Joyce. *Moving Picture World*, in its column on promotion and exploitation, told exhibitors that "Mrs Drew's clever work is as valuable to you in reputation as Miss Joyce's well-founded popularity as a star." Lucille McVey Drew's filmmaking career, however, was also curtailed by early death. After an extended illness, she died in 1925 at the age of 35.[99] The Drews' brand of comedy and filmmaking continued briefly in the careers of Mrs. and Mrs. Carter De Haven, who filled the empty slot in the Paramount program after the untimely departure of the Drews. After making ten "newlywed" comedies for Paramount, the De Havens left to become independents, releasing comedies through First National. Although little is known of the participation of Mrs. De Haven behind the camera, it seems significant that the couple's First National studio was in Carter De Haven's name only. They made one feature film for First National distribution, *Twin Beds* (1920), a relatively risqué comedy concerning a drunken neighbor who mistakenly crawls into the wrong bed, but the Carter De Haven studio disappeared within a year. Within two years the De Havens were making short comedies for the Film Booking Offices of America, and they disappeared from filmmaking altogether by the mid-1920s.[100]

Even in lowbrow comedy actresses left the safety of the studios for independent production. Flora Finch, the skinny, nagging onscreen wife of John Bunny, went independent without much success in 1917, as did Alice Howell, Keystone's unkempt "scrub lady," who did little better.[101] The most famous slapstick comedienne to go independent was Mabel Normand herself. In all likelihood changes in production methods also influenced Normand's decision to leave Keystone. Normand enjoyed a great deal of creative control at Keystone between 1912 and 1915, but when Keystone merged with the Triangle Film Corporation in 1915, the autonomy enjoyed by the Keystone production units ended. Continuity scripts became mandatory, and the legendary improvisational style of Keystone production withered. Sennett himself, who had been a loose central producer of sorts, now became an administrator. Roscoe "Fatty" Arbuckle, searching for more money and more artistic control, left Keystone the same year as Normand.[102]

The reason Normand gave the press for her departure was her desire to be taken more seriously as an actress. "I wanted better pictures," she recalled. "I was getting tired of grinding out short comedies to bolster up programs in which other stars in other companies, as well as our own, were featured in pretentious films and were paid far more than I was." After her Keystone contract expired, she created the Mabel Normand Feature Film Company in 1916 with Sennett's fellow Triangle partner, Thomas Ince.[103] In an interview many years later, Normand recalled only Sennett's role in the making of *Mickey*, the first and only feature made by the Mabel Normand Feature Film Company. In her words Sennett "let" her do *Mickey*, "let" her have her own studio, and "let" her pick out her own director and cast. The truth behind the creation of the Mabel Normand Feature Film Company, however, was somewhat different. In 1915, after Normand caught Sennett in a compromising position with one of her closest friends, she disappeared from the screen. Although publicists told the press that Normand suffered from a mysterious ailment, at least one historian ascribes her disappearance to Sennett, who allegedly beat her in a drunken rage after she confronted him about the affair.[104] After her "recovery" she left for Keystone's Fort Lee, New Jersey, studio, thousands of miles away from Sennett. When her Keystone contract expired in May of 1916, she went back to Hollywood, this time with a new Triangle contract under Thomas Ince, "which recognizes her desire to play light dramatic roles."[105]

The quasi-independent Mabel Normand Feature Company, financially supported and guaranteed distribution by the prestigious Triangle organization, seemed secure, but *Mickey*, her first production, was cursed from the start. After Normand fired two directors, Sennett himself arrived to coproduce the film. Remarkably, their working relationship remained intact, but to ensure her independence, Sennett built "The Mabel Normand Studio" several miles from Keystone.[106] By the time that *Mickey* wrapped, it was seven reels long and cost between $125,000 (Sennett's estimate) and $300,000 (*Photoplay*'s estimate) to produce.[107] Even after the film was completed, its release was delayed when the director "kidnapped" the last two reels for nonpayment. Caught in the chaos of Triangle's own collapse, *Mickey* sold for the paltry sum of $175,000. The Mabel Normand Feature Film Company dissolved, and her studio went to Sennett. When finally released in August of 1918, *Mickey* was road-showed by exhibitors in first-run theaters with its own original score. It proved a monstrous hit, eventually grossing $18 million.[108] Looking back nine years later, a writer for *Moving Picture World* asserted that no other comedy had yet "made the same tremendous popular hit." "*Mickey* became an epidemic," he recalled. *Mickey* "hats, dresses, clothes, and pretty nearly everything else" filled thirty-seven storefront windows in one town alone.[109]

After *Mickey*, Normand never again attempted independent production. In 1917 she joined Samuel Goldwyn, where she became his most lucrative star in a long string of feature comedies and where she specialized in tormenting Abraham Lehr, the self-described "head of production and a sort of efficiency expert." According to Lehr, Normand had "an amazing indifference to hours and routines and costs and system." She would go out on the town nearly every night, arriving late for work the next morning, sauntering tardily across the set singing "some taunting song." Even when she was on the set, Normand would hold up production by playing the piano and singing French songs with her director, Victor Schertzinger. When Lehr tried to change her director, she "threw a fit." After a falling out with Goldwyn in 1921, Normand went back to work for Sennett, who was releasing comedy features through Paramount.[110]

WHILE THE NEW WOMAN–STYLE COMEDIENNE faded from the short-film genre, because of the departure of Mabel Normand and the gentrification of the short comedy, the demand for the strong, self-reliant New Woman serial heroine received an unexpected boost in 1916 through the popular "war preparedness" serials such as *The Secret of the Submarine* (Mutual, 1916), *Liberty, a Daughter of the U.S.A.* (Universal, 1916), and *Pearl of the Army* (Pathé, 1916).[111] The popularity of the self-reliant heroine increased even further when the United States finally entered the war. According to film historian Kalton C. Lahue working-class women and children on the home front appreciated the serials, which offered resourcefulness onscreen as a "sheer escape."[112] By 1919, at war's end, trade journals trumpeted the exuberant claims made by Pathé and Universal—the two major firms that relied most heavily on serials—that the installment play would soon be wholeheartedly accepted by even the most opulent first-run theaters. Serials were far better in terms of production values and plot than was true before the war, and they even scored a few prestigious first-run venues during the war. But despite the high hopes among serial producers that at last they were on the verge of confirmed first-run acceptance, the war years proved to be an exception. By 1920 the serial was firmly entrenched in the subsequent-run theater and the children's matinee.[113] Universal and Pathé cornered this still-lucrative postwar serial market by sticking to the old blood-and-thunder format that appealed to the cheaper theaters. Postwar advertisements for Pathé serial star Ruth Roland claimed that she was the "dynamic favorite of the masses." Universal's Carl Laemmle banked on small-town and subsequent-run theaters, offering them a complete balanced program of serials, comedies, and B-grade features.[114]

Despite the changing audience, the New Woman serial heroine still sold movies, especially Ruth Roland for Pathé and Marie Walcamp for Universal. Marie Walcamp, who began her career in 1916 when she was described as "a girl who grins at being abducted from a forty-mile-an-hour passenger train through the window and onto a galloping horse's back," sustained Grace Cunard's New Woman style after the war. Walcamp's greatest postwar success was *The Red Glove* (1919), directed by Helen Holmes's former husband and serial-making partner, J. P. McGowan. Universal supported *The Red Glove* with an enormous publicity campaign, which included the claim that Marie Walcamp chose the story herself.[115] Far more popular, however, was Ruth Roland, who began her career in early New Woman–style series before joining Balboa in 1914, where she starred in the unusually successful highbrow serial *Who Pays?* After *Who Pays?* Roland starred in *The Red Circle* (Balboa, 1915), in which she played a society girl cursed by a birthmark that caused her to lead a secret life of crime. Now a bona fide serial queen, Roland hit her stride during the war years in *Hands Up* (Pathé, 1918), in which she portrayed a journalist nearly sacrificed by a tribe of native Southern Californians.[116]

After *Hands Up* Roland was the biggest box-office draw in the world of serials. At this opportune moment she allegedly took the reins in *The Adventures of Ruth* (1919), a serial made by Ruth Roland Serials, Inc., and distributed by Pathé. Pathé had successfully handled her product since 1915 and knew that Roland was a box-office guarantee. In a full-page advertisement in *Wid's Film Yearbook* for 1919–20, Roland stated, "I wrote the story—and personally supervised the taking of every scene. *The Adventures of Ruth* is my own—in pep, in class, in action—in romance! I *know* you will like it."[117] *The Adventures of Ruth* featured Roland as an heiress whose father was killed by the mysterious "Terrible Thirteen," and Ruth must solve the murder. Roland did indeed conceive the story, but Pathé's Gilson Willets wrote the scenario, and George Marshall directed. *The Adventures of Ruth* was followed by *Ruth of the Rockies* (1920) and *The Avenging Arrow* (1921), all made by Ruth Roland Serials, Inc., and released by Pathé.[118]

THE NEW WOMAN—STYLE HEROINE appeared to be doing well at matinees and in second-run theaters after the war. In fact, the New Woman—style heroine began to appear in that most masculine of genres—the western. Fay Tincher of *Ethel*, the gum-chewing-stenographer series, was successfully reinvented as a cowgirl in 1919 by comedy maker Al Christie. As "wild and woolly" lariat-looping "Rowdy Ann," "fresh from the open range," Tincher appeared in a series of western comedies such as *Dangerous Nan McGrew* and *Go West, Young Woman*. In these films she donned a ten-gallon hat, furry chaps, and a six-shooter. Also arriving in 1919 was newcomer Texas Guinan, who first appeared in two five-reel western thrillers (originally intended to be shorts) made by the independent Frohman Amusement Corporation.[119] Guinan, who was nearly thirty and an actress of the old school of pantomime, seemed an odd choice for starring roles, but Frohman publicized her as a woman who "can handle a gun, roll a cigarette, and boss a mob of cowboys all at one fell swoop." She thus fit the ideal of the western woman, "no less brave or daring than a man and at the same time portraying her womanly side."[120] In *The Gun Woman* (1918) Guinan played a dance-hall proprietor cheated out of her money by a con man. When the sheriff tells her that the con man has been terrorizing a nearby town, she confronts the villain, pumps a bullet into him, and concludes the film by rejecting the sheriff's proposal. Billed as "the female Bill Hart" (referring to silent star William S. Hart) in *The She Wolf* (1919), Guinan shot two men to save a young woman from a false marriage.[121]

After making dozens of two-reel westerns, Guinan created Texas Guinan Productions in 1921. She planned to produce a series of two-reelers, as well as a few features.[122] Like many actress-producers Guinan probably exercised choice over story and director—her choice for the latter being none other than Francis Ford, Grace Cunard's former serial partner.[123] But despite the announcement that her first two-reelers were receiving wide distribution, Texas Guinan Productions did not go far past its first releases, *The Two-Gun Woman* (1921) and *I Am the Woman* (1921). One historian attributed this failure to a lack of capital, alleging that Guinan and her partners incorrectly assumed that she would win a $200,000 lawsuit against a previous producer, Bulls-Eye. After closing her company in 1922, Guinan left for New York, where she enjoyed greater success as a nightclub entrepreneur, gaining fame for her greeting, "Hello Suckers," a reference to the high prices she charged for watered-down bootleg liquor. In 1928 Guinan returned to the screen to star in *Queen of the Night Clubs*, but she generally left Hollywood behind.[124]

Difficulties in financing and distribution explain why, after World War I, New Woman–style independent filmmakers like Helen Holmes, Helen Gibson, Ruth Roland, and Texas Guinan went out of business, but these difficulties do not explain why the New Woman–style comedienne and serial heroine soon disappeared from the screen altogether. Although comedies were becoming more sophisticated, slapsticks were still being made, and even as the audience changed, the New Woman–style serial heroine still pleased audiences in films made by Universal and Pathé.[125] Clearly there was an additional factor, outside of the audience and the studios, that led to the sudden disappearance of the New Woman from comedies and serials. Ironically, this factor turned out to be the General Federation of Women's Clubs.

Two apparently related developments stunned the movie business between 1919 and 1922: a sudden outcry against the movies by censorship advocates and a frightening decline in attendance. While a sharp postwar recession slowed box-office receipts, a frightened industry linked poor attendance with renewed criticism of the movies. In September of 1921 *Moving Picture World* surmised, as did many other observers, that the public was simply "shunning both themes and personalities it doesn't like."[126] Because of their peculiar history in terms of content and audience, the serial and the slapstick comedy received the full wrath of reformers.

What sparked a renewed censorship campaign were newspaper stories after the war that blamed a sharp increase in juvenile crime on the serial thriller—the staple of Saturday matinees.[127] Although some judges realized young offenders found it convenient to blame the movies, the frequent claim by juveniles that they learned how to commit robberies and even cause train wrecks from seeing these crimes enacted on the screen made for frightening headlines.[128] Within a matter of months a new censorship movement gained momentum, and the focus of criticism spread. Reformers turned their attention to all audiences exposed to crime films—not only children but the adult audiences who patronized the second-, third-, and fourth-run theaters.[129] In the atmosphere of the postwar Red Scare the combination of such movies and the masses sent a danger signal, and reformers wanted "to do away with the possibility of display of low films in poorer sections."[130] According to Ellis Paxton Oberholtzer of the strident Pennsylvania Board of Censors, "The crime serial is, perhaps, the most astounding development in the history of motion pictures." Serials were "meant for the most ignorant classes of the population with the grossest tastes," and they flourished in "mill villages and in the thickly settled tenement house and low foreign-speaking neighborhoods in the big cities."[131] Female reformers even began blaming their working-class servants for the exposure of middle-class children to the serial. "How many mothers who send their children with their nurses for an afternoon of sunshine in the park know they are sitting in the darkened movie theater taking in a daring thriller?" asked Mrs. J. A. Storck of New Orleans, adding that "after all, it is to nursemaids that the thrilling serial really makes its strongest appeal."[132]

For the New Woman serial heroine the postwar censorship movement proved particularly troublesome. Men in crime serials were dangerous enough; serial heroines, women who fought criminals and even became criminals themselves, were beyond the pale. To Oberholtzer the serial heroine was unnatural. "If I should travel the country over," he proclaimed, "I should not know where to find women who conceal revolvers in their blouses." To make matters worse, the trials and tribulations of the young, attractive serial queen at the hands of the villain were innately sexual. Although serials were mostly "shooting, knifing, binding and gagging, drowning, wrecking, and fighting," Oberholtzer maintained, "always obtruding and outstanding is the idea of sex."[133] As cities across America banned the serial in 1921, the treatment of the serial heroine, particularly her being bound and tortured by the male villain, became a central offense. Some censorship boards "refused to permit even the laying of a hand upon a woman character."[134]

As the censors winnowed out the dangers of the serial genre, they soon extended their net to lowbrow comedy, where once again they found the incendiary combination of sex and crime in a genre associated with children and the working classes. Indeed, slapstick was just as much to blame as serials for the alleged rise in juvenile crime, for "the policemen and every other officer of the law has been so much caricatured that by this time they must be beyond the bounds of young people's respect." Like the serial the slapstick comedy portrayed women in a dangerous manner: either scandalous in costume (the Sennett "Bathing Beauties") or vulgar in behavior. According to Oberholtzer slapstick comedies were full of young women who "will put aside any delicacy of feeling which may have descended to them from their grand-dames, and will hazard their lives and limbs as well as their reputations to outdo the men in gross performances for creating laughter."[135]

Although clubwomen were active in censorship campaigns before World War I, they assumed the dominant role in the postwar drive for movie censorship. While a few male reformers were quite vocal, clubwomen were the most important cultural authority challenging the movies after the war.[136] Morris Ernst and Pare Lorentz, who published their study of film censorship in 1930, derided the clubwomen as "ladies at play" but acknowledged that "the movie producer is taking no chances. He doesn't want to test their power. While he was doing it he would lose too much money."[137] Thanks to the lobbying efforts of clubwomen and female reformers, censorship bills aimed at eliminating crime films were introduced into thirty-two state legislatures between 1919 and 1922, and in January of 1920 three bills were introduced to Congress that proposed to bar films "purporting to show or to simulate the acts and conduct of ex-convicts, desperadoes, bandits, train robbers, bank robbers, or outlaws."[138]

Faced with dozens of state censorship bills and a newly enfranchised majority, a truly frightened Hollywood wondered if women held the future of the industry in their hands. In November of 1920 *Moving Picture World* published a warning that highlighted the fact that women could now vote and that they and their children "comprise the major portion of our audiences. *They* have made motion pictures popular—not the men." The article advised honesty. "The moment you hear censorship is coming up in your state, *get in touch with the women, the women's clubs and organizations.* Tell them your story frankly; *tell them the truth,*" which was that "'outlaw' pictures" still existed but were only a fraction of the total output.[139]

By the time this article appeared, the industry had already taken action. In June of 1920 the National Association of the Motion Picture Industry (NAMPI), a trade association that claimed to control 90 percent of the movie business, asserted that it would support the General Federation of Women's Clubs and its efforts to protect children, and the association kept its word. In March of 1921 the NAMPI adopted a set of resolutions to show that the industry could and would cleanse itself without the help of state or federal censorship. Producers were expected to refrain from exaggerated "sex appeal," white slavery, the underworld, drunkenness, narcotics, ridiculing authority, gratuitous violence, scenes that "are vulgar and portray improper gestures, posturing, and attitudes," as well as salacious subtitles. Offenders were to be expelled from the organization.[140] To underscore the industry's seriousness, William A. Brady, president of the NAMPI, began a trip through the nation's heartland in the fall of 1921 to convince women's groups that Hollywood was doing a good job and to gain their cooperation in fighting state censorship—a tactic that was soon to be copied by Will Hays.[141] For New York it was too little too late. In 1921 New York became one of the few states to create a censorship board. Notably, clubwomen demanded that a woman be named to the board, and they got their wish.[142]

Faced with reform from the inside and the continuing threat of state censorship, filmmakers were left with little choice but to comply. For the makers of slapstick comedies and serials, complying with the NAMPI resolutions necessitated drastic changes. In September of 1921 Hallroom Boys Photoplays, Inc., introduced a "censor-proof bathing girl comedy." It announced that the new Hallroom Boys comedies series was "absolutely clean in every respect—sacrificing laughs if need be, if the scenes showed any tendency to become vulgar."[143] This announcement illustrated just how difficult it was to make a "clean" slapstick comedy. The portrayal of women in slapstick was now a minefield, and it was simply safer for men to play the broad physical comedy routines while women stayed in the background. The slapstick comedy, which once turned gender roles upside-down, became the domain of male clowns—Buster Keaton, Harold Lloyd, Laurel and Hardy, and Larry Semon; women played the straight roles or provided ornamentation. The women who starred in the New Woman comedies of the 1910s either disappeared or found other types of roles. Mabel Normand played in toned-down feature comedies for Goldwyn and Sennett. Louise Fazenda, the slapstick farm girl with the "obnoxiously homespun curl plastered in the middle of her forehead," was fortunate enough to win good roles in feature dramas throughout the 1920s. Fay Tincher, the closest female star to Normand in both beauty and ability, became famous between 1923 and 1928 for her straight portrayal of the wife in the "Andy Gump" series (based on the newspaper comic strip) but failed to find work after the series ended.[144]

Serials were equally difficult to make in the new style. A 1922 screenwriter's guide noted that the new censorship policies made it nearly impossible to create a serial around a female heroine, "and the irony of it is that the feminine star is always the most popular."[145] By the time that the NAMPI resolutions were released, Universal and Pathé had already retooled the serial format.[146] After the war both companies experimented with the idea of a male serial hero.[147] In 1919 Universal released *Elmo the Mighty*, starring Elmo Lincoln; *The Cyclone Smith Stories*, starring Eddie Polo; and *The Midnight Man*, with James J. Corbett.[148] Inspired by the jaunty Douglas Fairbanks, Pathé released *Bound and Gagged* (1920), a serial produced and directed by former Pearl White director George B. Seitz, who also starred in the serial as "a new type of serial hero—a young American, typically energetic and intrepid, whose utterly care-free attitude of mind and ability of body brought him face to face with situations he had no part in bringing about" but which, of course, he met with "thrilling heroism."[149]

Like the slapstick producers, Pathé and Universal turned to the male lead after 1921 as a safer alternative to the serial heroine, whose "perils" were too easily construed as sexual. Two weeks before the NAMPI formally announced its resolutions, Pathé trotted out *The Adventures of Bill and Bob*, gentle nature comedies starring eleven-year-old Boy Scouts from Glendale who embarked on fishing and hunting expeditions. Uplifter Epes Sargent of *Moving Picture World* greeted this news happily, announcing that "Pathé has found something new in serials . . . There is no panting heroine hanging by her neck—or her heels—from one week to the next."[150] Six months later Universal announced its own "departure" in serial production, beginning with the "censor proof" *Winners of the West*, an educational thriller based on the life of John C. Fremont. "We are making *Winners of the West*," announced Fred J. McConnell, head of the new serial exploitation department at Universal, "so that every censor and every parent will have to say: 'I see no harm in showing this to children.'" Universal continued the masculine, "natural" theme with *The Adventures of Tarzan* (1921), which perhaps pleased New York censor Mrs. Eli T. Hosmer, who stated that she liked "thrills of an outdoor type because they are more educational."[151]

These changes in serial policy did not immediately do away with the portrayal of the strong female serial heroine, for Ruth Roland continued to be Pathé's leading attraction in serials made by her own company, although they were now characterized by what were described as "clean-clean-clean" outdoor themes. *The Avenging Arrow* (1921) was advertised as a "true-blue story of adventure and romance, without a shadow of the underworld in it."[152] But while Ruth Roland, a proven star and a relatively genteel heroine, enjoyed box-office success, her career wound down.[153] When Roland closed her own company in 1922 to make serials supervised by Hal Roach, it may well have been because of Pathé's move away from serial heroines.[154] After the spring of 1921 neither Pathé nor Universal, the most important producers and distributors of serials, promoted a New Woman–style heroine. Instead, stuntman Eddie Polo, strongman Elmo Lincoln, "Cinema Colossus" Joe Bonomo, and the "censor-proof" Charles Hutchinson were promoted to stardom. The serial, which presented the most feminist characters of the 1910s, lost its complex plots and its New Woman heroines and turned instead to safe but cartoonlike male heroes such as "Tarzan" and Houdini.[155]

The NAMPI resolutions and the continuing threat of censorship in the state legislatures spelled doom for the New Woman onscreen, but in a case of spectacularly poor timing, her fate was sealed when Hollywood experienced a series of scandals that implicated, among others, comedienne Mabel Normand and serial star Juanita Hansen. The scandals began when slapstick comedian Roscoe "Fatty" Arbuckle was arrested for the rape and murder of actress Virginia Rappe at a wild party. After his first trial ended in a hung jury, women's groups began demanding the removal of Arbuckle films from the theaters. A year later Arbuckle was acquitted, but the Arbuckle scandal proved what many had suspected for years—Hollywood was innately immoral. Meanwhile, in February of 1922, a month into Arbuckle's second trial, director William Desmond Taylor was murdered in his apartment, and Mabel Normand was implicated.[156] Normand was exonerated, but her reputation suffered irreparable damage at the hands of the sensationalistic press, which speculated at the time that Normand—whose penchant for high living and tasteless gags was covered thoroughly in the papers—was a drug addict (which was apparently true). A year later police arrested Normand's chauffeur for murdering a man with Normand's gun, and a year after that Normand was named in a divorce suit. Moralists demanded that Normand's films be banned. This was effectively the end of Normand's career; Normand's health declined rapidly, until finally tuberculosis and pneumonia took her life in 1930.[157] She was thirty-eight years old.

Just as the scandals broke in the winter of 1921–22, movie attendance fell precipitously, and Hollywood naturally believed that a disgusted public was engaging in an impromptu boycott. The film industry responded to the scandals of 1920 to 1922 by creating a new trade organization, the Motion Picture Producers and Distributors of America (MPPDA). In 1922 the MPPDA appointed former postmaster general Will Hays, an old-style Protestant from Indiana, to convince America and the world that Hollywood could reform itself. Although like its predecessor, the National Board of Censorship, the "Hays Office" ultimately proved too lax for some, at first a frightened Hollywood scrambled to prove that its potential for good outweighed its temptation toward evil.[158] Unfortunately, the serial heroine and New Woman comedienne were in the crosshairs when the industry purged itself of controversial material. Anonymous mailings to the Hays Office, like the photo of Ruth Roland under which an angry writer scribbled "a teacher in the school of crime," underscored their indictment.[159]

IRONICALLY, THE CLUBWOMEN who freed the impressionable working-class and juvenile audiences from the New Woman may have wondered what they had wrought. With the departure of the New Woman from the screen in comedies and serials there was little to counter the kinds of women portrayed in films pioneered by Cecil B. DeMille, purveyor of what Pennsylvania censor Oberholtzer called the "sex rot."[160] In the eyes of clubwomen the heroines of these "sex plays" may well have been far worse than the New Woman serial heroine, whose virginity, at least, was never in question. But there was little they could do about it. The dominant voice in film regulation reverted to the industry after the establishment of the Hays Office. The near-paranoia that killed the serial heroine and comedienne disappeared, but there was no one to resurrect her. The men who created her had gone on to other kinds of films, and the women who created her disappeared altogether.[161]

# PART TWO

# "A Business Pure & Simple"

## The End of Uplift and
## the Masculinization of Hollywood,
## 1916–1928

In the mid-1910s the industry shifted away from the goal of cultural legitimacy and the uplift strategies designed to secure it and moved toward a model that prized business legitimacy. This shift ultimately marginalized the woman filmmaker.

Thanks to feature films and more comfortable theaters, by mid-decade moviegoing was becoming a middle-class as well as a working-class pastime. But the mature themes that attended some feature films, particularly those concerning controversial topics such as white slavery, led to a second censorship crisis in 1915. Within a single year Congress debated a bill calling for federal censorship, and the Supreme Court decided that the moving picture industry was "a business pure and simple," meaning that films were not entitled to protection under the First Amendment. The reform films that had been promoted during the uplift movement as evidence of shared middle-class values were now shunned by an industry wary of state censorship and increasingly concerned with the export market. Lois Weber's serious message films, once the ideal vehicle to legitimize the screen for the middle class, lost favor. Instead, feature films offering entertainment and fantasy, epitomized by director Cecil B. DeMille, became the new Hollywood ideal as the industry searched for legitimacy on Wall Street and in markets across the nation and the world.

Changes in filmmaking practices after 1916 also marginalized the woman filmmaker. The flexible work culture drawn from the theater, which allowed women to gain production experience, ended, and a business model ensued. Doubling in brass ceased as spiking film costs and scale of production led to the tightening of the central-producer system and an increasingly rigid division of labor. Masculine trade associations multiplied and expanded, creating geographic boundaries that eliminated businesswomen from networks of fellow producers, directors, and potential investors. When Wall Street investors began to fund the vertical integration of the studios that became known as the "majors," these large producer-distributor-exhibitor networks squeezed smaller, independent producers off the screen. Many of the male filmmakers in the independent sector found work within the new studio system, but the masculinized business culture of the majors meant that even experienced female directors and producers found themselves standing outside the studio gates. Before the studio system could emerge, however, there was one last extended moment when a significant number of women entered independent production.

# "The Real Punches"

## Lois Weber, Cecil B. DeMille, and the End of the Uplift Movement

When Cecil B. DeMille came to Hollywood in 1913 to become director-general of the Jesse L. Lasky Feature Play Company, he wore a mantle of cultural legitimacy similar to Lois Weber's. *DeMille* was a well-known Broadway name, as Cecil B.'s father, Henry C. DeMille, a former minister, wrote the self-consciously respectable society plays that made theatergoing a regular activity among the urban middle class and upper middle class in the late nineteenth century. The young DeMille's brother William continued in their father's footsteps and was by 1913 also a well-known playwright. Cecil B. DeMille represented precisely the sort of theatrical pedigree the film industry desired at that time. After his father died when Cecil was an adolescent, his mother, Beatrice, parlayed her late husband's plays into a highly successful career as a theatrical agent, and Cecil B. gained the rights to film plays by his father's former collaborator, the esteemed Broadway producer David Belasco. Belasco himself would have fit the uplift movement in the American cinema, as he was known as the "apostle of art" and wore a clerical collar on a regular basis. Cecil B. DeMille absorbed Belasco's sense of mission, stating, for example, that "to be afraid to develop a message in a story is to miss a great possibility."[1] In this he sounded very much like Weber. Weber's career, however, was about to go into decline, whereas DeMille's was just beginning.

The change that ultimately made Weber and an earlier definition of uplift passé began, ironically, when the film industry faced another censorship crisis. In 1915 a bill to create a federal censorship bureau began working its way through Congress, and the Supreme Court struck down the claim that moving pictures were protected under the First Amendment. Rather than bolstering the project to make movies serve the desires of middle-class reformers, these developments destroyed the uplift movement as it was known between 1909 and 1916. In an era before film ratings systematically denied entry to younger audiences, the agenda of cinematic reformers, which included juvenile delinquency, birth control, white slavery, and a host of other controversial issues, was deemed too dangerous for mixed audiences, and the idea that films were protected speech was no longer a plausible line of defense. Although the threat of federal censorship was averted for the time being, leadership within the industry began to turn away from social problem pictures and to advocate, instead, simple entertainment as the best use for the Hollywood movie. But even as the social problem genre faded, gendered questions remained. Did successful film directors need "manly" authority, or did directing require a "woman's touch"? Did the movies require the masculine "punch" of a man, or did women's "heart interest" pack the theaters?[2]

THE UPLIFT MOVEMENT REACHED A CRESCENDO in 1914–15. By 1914 trade magazines already lauded Lois Weber as an uplifter, and in 1915 she released *Hypocrites.* As we have seen, Weber, like many other social problem filmmakers, legitimized her controversial films by inviting prominent Progressive reformers to screenings and then soliciting their support. That same year, D. W. Griffith released *Birth of a Nation.* Although the racist film proved immediately controversial, it fulfilled the fondest wishes of the uplift movement when it received the blessing of President Woodrow Wilson himself (the film liberally quoted from Wilson's *History of the American People* [1902]). After Wilson, a southerner, saw the film at a White House screening, he was alleged to proclaim, "It is like writing history with lightning, and my only regret is that it is all so terribly true!" The lesser-known Cecil B. DeMille, as we have seen, contributed to the uplift movement by adapting Belasco plays for the screen. His contribution to the uplift movement took a giant leap in 1915 when he adapted *Carmen,* securing the services of the celebrated Metropolitan Opera soprano Geraldine Farrar to play the title role. Lasky arranged for "the Brahmin elite" to attend the premiere of *Carmen* at Boston's Symphony Hall, where a full orchestra accompanied the film. Even the *Opera Magazine* took interest, predicting long lines of well-heeled customers.[3] DeMille, who claimed to believe in films with a message, also believed that "to preach is to invite disaster," and he increasingly emphasized middle-class modes of entertainment rather than middle-class modes of reform.[4] This emphasis would serve him well as the film industry shifted strategies. We can ascertain the difference between a Weber product and a DeMille product even at this early point by comparing two films: Lois Weber's *Shoes* (Universal, 1915) and DeMille's *The Golden Chance* (Lasky, 1915).

In the first frames of *Shoes* the face of the film's heroine, Eva Meyer, dissolves into the cover of a 1914 book entitled *A New Conscience and an Ancient Evil*, a treatise on prostitution written by esteemed reformer Jane Addams. The next few frames set up the context of the film by allowing the audience to glimpse a few pages. They reveal the story of a working girl who "sold herself for a new pair of shoes." This device not only validates the film as sociologically accurate but suggests that this specific story is true. The character of Eva Meyer is an upstanding daughter of a hardworking mother and an apparently shiftless father, and she is the eldest of four siblings. She stands for long hours as a clerk in a five-and-dime store and must give all of her meager pay to her mother to keep the family going. Suffering in her dilapidated shoes, Meyer desires a new pair of boots she spies in a store window, but her mother cannot part with the money needed to allow her daughter to buy necessities. Eva tries to make her shoes last, but her cardboard soles prove inadequate in a rainstorm, and her wet feet bring on a cold. Eva does know of a way to get her boots, and more. Her coworker Lil flashes jewelry given to her by male friends, clearly in exchange for sexual favors, and encourages Eva to accept an invitation from a male flirt, Charlie. Eva modestly refuses the attentions of Charlie, but her desire grows as she spies fine boots on a well-dressed girl during a lunch break and feels compelled to hide her own in shame. Although some middle-class viewers might view a poor girl spending earmarked family funds to emulate her betters as selfish, Eva has been thoroughly identified as a modest and hardworking victim of an indolent father. Ultimately, Eva goes to a cabaret with Charlie and returns home in ruin and shame with her new boots, only to find that her father has finally found work.[5]

The publicity surrounding *Shoes* contextualized Weber and the making of the film within the guise of social work. Shelley Stamp notes that when Weber chose the actress who played Eva (Mary MacLaren) from the list of hopefuls standing in line at Universal, she allegedly used the voice of a reformer, asking, "Are you looking for work?" rather than "Are you an actress?"; the latter would have reflected her needs as a director. Although the story itself was inspired by Jane Addams, Weber claimed that Eva's characterization was drawn from her own early missionary work "in the slums of New York and on Blackwell's Island."[6] The publicity of *Shoes* presented the film within the thus-far highly successful context of a Lois Weber social problem film, which firmly grounded its claim for middle-class legitimacy on Weber's own image as a maternalist reformer.

DeMille's *The Golden Chance* also concerns a poor woman who desires things and who is also constrained by an indolent male provider—her alcoholic husband. Mary Denby (Cleo Ridgley), the heroine of the film, is introduced as a judge's daughter who regretfully eloped with an urban "gentleman" and is now an unlikely tenant in a slum. When her lout of a husband, Steve, tells her that just because she was well bred doesn't mean she cannot work, she pulls herself together and finds a job as a seamstress in a wealthy home. Her eyes linger on the beautiful furnishings as she is led to the seamstress's workroom. In a twist of fate a young woman, described by her employer as "the prettiest girl in the world," is unable to show up for a date that has been arranged for a millionaire businessman, Roger (Wallace Reid), who is dining at their home to seal a business relationship. Her employer remembers that Mary is still in the house and convinces the seamstress to don a formal gown and pose as the intended date. Mary agrees, dresses in the finery she admired only a short time ago, and charms the millionaire. As Mary changes back into her own clothes at the end of the evening, DeMille emphasizes her desire with a medium shot of Mary comparing the fancy slipper she was wearing with her own ragged shoe. When Mary is asked to repeat her performance as "the prettiest girl," she and Roger fall in love, but Mary has qualms about her husband. After a series of twists, Roger discovers Mary's true identity, and her husband, Steve, is accidentally killed. Although the film ends with Roger telling Mary that her husband is dead, and Mary ambiguously looking away, the path is now clear for Mary to become Roger's wife. Although Steve Denby is the quintessential alcoholic husband portrayed by the temperance movement, there is little moral uplift in this Cinderella fantasy. It became even more like the fairy tale when DeMille remade the film as *Forbidden Fruit* (Famous Players–Lasky, 1921). The qualms disappeared, Mary was assisted in her transformation by a retinue of maids and boxes of dresses delivered from a chic shop, and the millionaire hero ends the film by placing a slipper on her foot.[7]

By mediating his message with a greater focus on entertainment, DeMille distanced himself from overt social problem films and thus placed himself in the vanguard of a strategic shift within the industry. According to Lee Grieveson the American cinema reached a turning point in 1915. The waffling of the National Board of Censorship over whether it should approve white slavery films revealed an irreconcilable fissure between the progressive social reform adherents, who tended to support the films based on their educational value, and those who wanted to eliminate controversial films and redefine movies as "harmless entertainment."[8] As the white slavery cycle controversy raged, the bill to censor moving pictures at the federal level was introduced into the Senate and the House. Arguments for the bill focused on the ineffectiveness of the National Board of Censorship, which was in any case too closely allied with the producers it was supposed to police. (The Women's Municipal League resigned from the board in 1911 for this reason.)[9] In defending itself, the National Board of Censorship argued that federal censorship of moving pictures would violate freedom of speech. But the Supreme Court struck down that argument. In its ruling, concerning the banning of D. W. Griffith's *Birth of a Nation* in Ohio, the Supreme Court defined moving pictures as "a business pure and simple, organized and conducted for profit" and thus not protected by the first amendment.[10] The decision noted that mixed audiences viewed films and, significantly, that the exploitative films that "pretended" to be serious social problem films (for example, white slavery films) were particularly reprehensible. Films, therefore, could not be protected under the first amendment because of "their potential to do evil."[11] By validating the right of the Ohio censorship board to deny the film's entrance into the state, the Supreme Court mandated other state and local censorship boards and thus further eroded any power still remaining within the National Board of Censorship.

Grieveson argues that the 1915 *Mutual* decision drew a line in the sand. The decision, which mandated increased state regulation, encouraged the film industry to reject the model of uplift based on middle-class reform, which led to mature and often controversial themes, and instead conceive of moving pictures as "harmless and culturally affirmative."[12] This change was not immediate, but the ground began to shift in a manner unfavorable to the ideal of the reformer-filmmaker.

The reception of Lois Weber's post-1916 films demonstrates this trend. A year after the success of *Where Are My Children?* Weber tackled the birth-control issue again, this time in a film based on the trial of Margaret Sanger. Although the film never actually mentioned her name, it was clear that the heroine of *The Hand That Rocks the Cradle* (Universal, 1917) was indeed Sanger. Weber starred as Mrs. Broome, who is converted to the cause when she observes the plight of impoverished, sickly mothers. Soon Mrs. Broome is arrested for spreading birth-control propaganda in violation of state laws, but she is pardoned by the governor. The film ends without resolution, since the state laws against distributing information on birth control were still in place. In contrast to *Where Are My Children?* the reviews of *The Hand That Rocks the Cradle* were mixed. In particular, they began to question the political use of the screen. Edward Weitzel of *Moving Picture World* generally praised *The Hand That Rocks the Cradle*, arguing that Weber used "such good judgement" that "there is no necessity of any one to receive the slightest shock." But he questioned whether the "family photoplay theater" was the proper place for such a film.[13] The critic for the *Dramatic Mirror* admitted that Weber "has presented a powerful appeal for the legalization of birth control in a film play of compelling sincerity," but he, too, questioned whether the moving picture theater was the proper place for politics and called the film "avowedly and entirely propaganda."[14] *Variety* found no redeeming qualities: "This is a weak effort to shoot over a feature that will get some quick money because of a condition [the Sanger case], rather than as a picture."[15]

Part of this reaction may have been due to the fact that Margaret Sanger made her own semiautobiographical film, *Birth Control*, which was shown to reviewers just weeks before Weber's film was released. Sanger's film, in which she starred, was lauded by the critics but never made it to wide release because of quick efforts to censor it. Ironically, Sanger's point in making the film was to soften her image and thereby help her cause.[16] But Sanger's notoriety was enough to cause problems even before the film was shown. The day before its premiere at New York's Park Theatre, the commissioner of licenses suppressed its release. While the Message Feature Film Corporation began suit for $10,000 in damages, Sanger hastily arranged a private showing for some two hundred of New York's most notable citizens.[17] This private showing was a great success, and *Birth Control* was hailed by film critics as a triumph. Eventually, the film won its release, but it was too late.[18] Theaters were afraid to book the film.[19] As Kay Sloan suggests, Weber and Sanger did not receive equal treatment at the censorship board.[20] Sanger's past affiliations with socialists and the Industrial Workers of the World, her short-lived publication *The Woman Rebel*, and her time in jail for violating the Comstock law branded her a dangerous woman.[21] In contrast, Lois Weber's impeccable middle-class credentials allowed her to use full-frontal nudity in *Hypocrites* and even suggest the subject of abortion onscreen.

In 1917 Carl Laemmle rewarded Weber by financing her own studio, located several miles from Universal City. Lois Weber Productions was a quasi-independent company; while Weber enjoyed a great deal of creative freedom, she still had the safety net of a contract with Universal that allegedly made her the highest paid director in Hollywood.[22] It appeared at first that Weber would continue making the sorts of films that she did best: the middle-class social problem film. Weber told Arthur Denison of *Moving Picture World* that her pictures would highlight "the difference between sentimentality and true sentiment," adding that "after nine years of making motion pictures if I see anything clearly, it is that the frothy, unreal picture is doomed." Weber planned to "make constructive pictures of real ideas which shall have some intimate bearing on the lives of the people who see them." She wanted "flesh-and-blood" characters rather than the hero who "can do no wrong," and she hinted (and would later admit) that she was tired of unrealistic happy endings.[23]

Even Weber's idealized collaborative marriage with Smalley began to show signs of wear. There was an increased focus on Weber as the dominant filmmaker after 1916. Shelley Stamp alludes to a cartoon in the *Los Angeles Times* in which a diminished Smalley stands in the shadows of Weber, who is described as "Lois Weber, Wonderful Lois, her note book always filled with clever ideas."[24] Weber wrote nearly all the scripts in addition to directing, and after 1917 critics and journalists focused increasingly on Weber at the expense of Smalley in acknowledgment of her power behind the camera. This is partly because Weber herself began taking more of the credit. Anthony Slide observes "in hindsight, some credits to Smalley seem extraordinary, indefensible, and perhaps evidence of a bias on the part of contemporary commentators."[25] As late as 1917, when *Motion Picture Stories* published "Turning Out Masterpieces," it was a photo of Smalley in his "study," poring over a script, that led the article. Weber's photo, a similar shot of her laboring in her "workroom," followed on the next page.[26] It seems likely, however, that Weber herself, who held traditional notions of marriage despite her own achievements, may have given Smalley more credit than he was due.[27]

In 1917 and 1918 Lois Weber Productions released a series of films starring Mildred Harris, the current Mrs. Charles Chaplin. About this time an undated "Lois Weber Bulletin," created to promote her company's product, claimed that Weber would avoid "propaganda or preachment" but not moralizing: "a story can be entertaining and carry with it a sound idea without obviously pointing a moral." The first Mildred Harris vehicle was *The Price of a Good Time* (November 1917). The title suggested a moralizing yet titillating story, precisely the blend that made Weber famous. The story was a cross-class melodrama concerning an attempt by the rich son of a department store owner, unhappy with the girl his parents want him to marry, to help one of his father's poor female salesclerks. With his parents out of town, he spends six evenings in whatever manner she wishes. She enjoys his company, going for drives in his car, eating meals at restaurants, and visiting the theater, but she cannot forgive herself when the rich son's brother discovers them and tells the family that they have been disgraced. The salesclerk throws herself in front of her benefactor's car and is killed. A reviewer for the *New York Dramatic Mirror* wondered why the ending had to be so tragic, "for the world is sad enough without one's writing photoplays to make it sadder."[28]

One month earlier, when *The Price of a Good Time* would have been near completion, Weber told Elizabeth Peltret of *Photoplay* that she did not intend to make any more "propaganda pictures." Her reason: the war. "The war is the world's jumping toothache," she explained, "and I want to help the world forget about it for a while."[29] This was not a matter of choice. In the summer of 1917 the National Association of the Motion Picture Industry made it clear that producers were to cooperate with the goals of the wartime Committee on Public Information. Social problem films depicting the weaknesses and iniquities of American society were to be replaced by films celebrating the wholesomeness and cheerfulness of American life. Lois Weber, the industry's preeminent producer of the social problem film, now had to rely on other genres in which she was far less successful.[30] On the other hand, she did not leave behind the idea that the screen was her pulpit. The resulting hybrid was less than inspired.

Her next film, *The Doctor and the Woman*, released in the spring of 1918, concerned star-crossed lovers who end up in each other's arms, and the vindication of a doctor accused of malpractice. This film with a happy ending earned tepid reviews, but interestingly, a *Variety* reviewer suggested that Universal's "popular-priced" distribution practices held Weber back from really "worth while" productions.[31] This was untrue, as Universal sold Lois Weber Productions on a states' rights basis: the highest bidder could place them in first-run theaters. It is true, however, that these films did not measure up to her earlier work. The next two Harris films, *For Husbands Only* (1918) and *Borrowed Clothes* (1918), were lackluster marital farces. The same *Variety* reviewer wondered why "so capable a developer of ideas as Lois Weber" managed to miss her mark regarding the former film and claimed the latter, another cross-class romance (this time with a happy ending), was "designed to appeal to the flathead picturegoers that Universal caters to."[32] Weber then tried her hand at a western, *When a Girl Loves* (1919), and two rural dramas, *Home* (1919) and *Forbidden* (1919). None were great successes.

Weber left Universal by the end of 1918. Still considered a top-flight director, she signed a contract with Louis B. Mayer, who was handling the independent productions of star Anita Stewart for her company, Anita Stewart Productions. One headline drew attention to Weber's "enormous salary" of $3,500 a week. The same story printed Weber's telegram to Mayer asking for his opinion of leading man and story, along with Mayer's reply that he had full confidence in her as a "great director" to make such decisions and telling her to spare no expense.[33] Another article claimed that there were "numerous messages of congratulations to the Louis Mayer forces" for securing Weber's services.[34] The first Anita Stewart film, *A Midnight Romance* (1919), a "mystery romance" that echoed the serial queen genre (she saves the hero from blackmailers before he can discover her identity) was given splendid reviews, and Weber herself received much of the credit.[35] The second, *Mary Regan* (1919), was based on a serialized magazine story. Anita Stewart's acting was uninspired. One reviewer quipped that the film reviewed itself in an intertitle that said, "It's too late to argue now, what we need is action . . ." Weber's attempt to be modern was construed as gratuitous "dirtying up" of intertitles, including one in which a honeymooning wife states, "I'm so tired from the trip, let us have luncheon before we (long pause) unpack." Critics again accused Weber of failing "to get the real punches of the story over on the screen."[36]

Despite these mixed reviews, Weber received a vote of confidence in 1920 from the most powerful company in Hollywood, Famous Players–Lasky, who offered her a contract stipulating $50,000 per picture and a percentage of the profits—a hefty but not astronomical salary for a famous Hollywood director.[37] Under this contract Weber was able to make just the kinds of films she wanted under her own Lois Weber Productions brand name. Furthermore, she was guaranteed first-run bookings by Paramount, the distribution arm of the rapidly growing Famous Players–Lasky empire. With all obstacles removed, Weber selected a company of talented actors, taking for her star-in-training the "blue-eyed, slim, chaste, soulful" Claire Windsor.[38]

Armed with a significant budget and apparently complete creative freedom, Weber acted on her long-delayed desire to return to the social problem genre, or at least to serious drama. Like other postwar filmmakers, Weber turned her attentions toward marriage and domestic life. Her first film for Famous Players–Lasky was *To Please One Woman* (1920), a seven-reel "vamp" drama depicting the selfishly philandering wife of a Wall Street broker. Playing the part of the wife was an actress who called herself Mona

Lisa, described by one critic as a "black-eyed, vamp-lashed, voluptuous" woman "with a big symmetry of figure."[39] Playing opposite her as "the sweetest girl in Seagirt," the town in which the action took place, was the delicate, blonde Windsor.

The good/bad, light/dark contrast between the virtuous girl and the vamp was hardly subtle, and *To Please One Woman* was not an auspicious debut. According to the critic for the *New York Times*:

> The picture might be subtitled "In Imitation of Griffith." Mr. Griffith can take such trite, homiletic stories of small-town virtue and corrupting vampires and by the magic of cinematography sometimes give them life, but Lois Weber, who wrote and directed "To Please One Woman," evidently has not his talent or knack. Occasionally her picture shows flashes of inspiration, which may be evidence that with more responsive material she could make a sufficient number of dramatic moving pictures to compose an exceptional photoplay, and this seems the more likely because of similar evidence in some of her earlier works. So, perhaps the production that fulfills the promises made for each of Miss Weber's pictures is to come.[40]

But even Griffith found himself in disfavor among postwar critics and crowds, who appeared to prefer lavish pictures with few morals to teach.[41] *Moving Picture World*'s section "Selling the Picture to the Public" ignored Weber's role as director of *To Please One Woman*—Weber's name was previously a selling point—and even ignored the film's serious subject and tone, suggesting that exhibitors use humorous cartoons showing "various things men would do to please one woman." "Make 'em Laugh," advised *Moving Picture World*. "It's the Best Way to Sell."[42]

Lois Weber now worked for the same studio as Cecil B. DeMille, who by 1920 had fully (if somewhat reluctantly) embraced his role as the director of popular, trend-setting Jazz Age films portraying sumptuous clothing, furnishing, and loose morals. It was Jesse Lasky, now in the studio's New York office and keeping an eye on the bottom line, who convinced DeMille to make "modern" films. DeMille preferred historical dramas and made his first with Geraldine Farrar as Joan of Arc in the critically acclaimed *Joan the Woman* (1916). The film did not do well at the box office despite the praise of critics and its inherently uplifting theme. As a pro-French film it was well poised on the eve of the United States' entry into World War I, but its portrayal of cruel Catholic clergymen, ultimately sacrificing the heroic Joan, displeased Catholic audiences. Sumiko Higashi argues that this signaled rebellion from "an increasingly pluralistic urban audience, fragmented by class and ethnic differences . . . to accept public history validated by the genteel middle class," but I would posit that it was the Catholic middle class that made the difference here, ushering in a new brand of interest-group pressure on the movies and foreshadowing the future role of the Church in regulating Hollywood.[43] In any case, after *Joan the Woman* Lasky urged DeMille to film a novel of "modern" marriage: *Old Wives for New.* In his ghostwritten autobiography DeMille bids "a last good-by to integrity and art" as he embarks on the series of films that made him one of the top postwar directors in Hollywood.[44]

*Old Wives for New* (1918) concerns the breakup of a middle-aged married couple. The fashion-conscious, fastidious husband, disgusted by his overweight, slovenly wife, suggests a divorce, to which his wife agrees. After the divorce the man falls in love with a younger woman, and his former wife remakes herself with the help of fashionable gowns and modern cosmetics, eventually finding her own upscale beau. The former husband and wife both remarry. The film inspired angry letters from wives, resulting in a sequel, *Don't Change Your Husband* (1919). In this film the wife (played by Gloria Swanson) is the fastidious one. She runs off with a flirt at a party after she can no longer stand her husband's bad habits. Now the husband undergoes a makeover, donning fashionable clothes, and wins back his wife. The final film in the trilogy was *Why Change Your Wife?* (1920), which DeMille once again made under the duress of Lasky's office. An unknown Gloria Swanson plays Beth, a woman of conservative taste and reformist tendencies, who represents the middle-class reformer, now considered boring and unfashionable. Her husband, Robert, wants her to listen to his fox-trot records and wear a negligee he bought for her after an attractive gold digger, Sally, modeled it for him in a store. When Beth protests against the negligee and all it represents (resisting commodification), her husband rejects her. Beth goes to her aunt and declares that she is going to "give her life to charity," but when she overhears gossiping women remark that it was her dowdiness that drove her husband away, she decides to make herself over with fashionable clothes, accessories, and makeup. While on honeymoon with Sally, Beth's former husband spies an attractive woman in a fashionable bathing suit, and he is shocked to see that it is his former wife. In a series of plot twists Sally's vindictive ways are revealed, and Robert wins back Beth, who now listens to his fox-trot records and wears negligees.[45]

Film-as-fashion-show was not new; serials had served as fashion shows before the war. But the idea that one must commodify oneself to win and keep love in modern society was different and quite "modern." (As one critic said of *Old Wives for New*, "beauty parlors couldn't have a better ad than this.")[46] And it was not just the fashions; it was the décor as well. Moving pictures now self-consciously illustrated the latest Parisian fashions and interior design for the benefit of the well-heeled classes who might be able to imitate the trends but also for the working class who were now privy to the latest elite styles. Kathy Peiss has argued that working-class street style and behavior influenced the middle classes at the turn of the century, and now luxurious fashions were promoted by Hollywood for the visual consumption of the masses.[47] Thus the moving pictures, particularly DeMille's Jazz Age cycle, contributed to the homogenization of American taste in the 1920s. Certainly fantasy played a role in the appeal of these films even for the middle class, as the uniformed servants, jewels, modern furnishing, and architecture (particularly the famed bathrooms with luxurious tubs and double sinks) were out of reach for all but the wealthiest Americans. Weber's films were not bereft of high fashions or loose morals, but such excesses were illicit. Gowns and jewels signified the immoral vamp; modesty signified the pure heart. The heretofore desired spectatorial position of the Victorian middle class, the core of the uplift movement, was now dismissed by the postwar generation that desired luxury goods without guilt.

What of the censors? How could the divorce-ridden, spouse-swapping postwar DeMille films represent the "harmless entertainment" desired by the film industry in the wake of the 1914–15 censorship crisis? As we saw in the previous chapter, a third censorship crisis (following 1908–9 and 1914–15) arose in the immediate postwar years, just as DeMille's Jazz Age cycle began. Why would Lasky wish to steer DeMille toward a genre so rife with sexually risqué material? The answer, it would seem, was that it sold. It appears that Lasky gambled, hoping that by playing into consumerist desires, DeMille's films would make large profits despite the threat of censorship. He was right.

Allegedly Adolph Zukor of FPL/Paramount hesitated to release *Old Wives for New*. Earlier Zukor had asserted that Paramount's 1919 product would concern "wholesome dramas, uplifting in character" and films "dealing with the more cheerful aspects of life."[48] The film itself was subject to in-house censorship, detailed in a memo sent to Lasky, which warned, for example, that average wives might be offended by the theme of the slovenly wife.[49] *Old Wives for New* was not trimmed enough (and could not be)

to avoid controversy on its release. *Wid's Daily*, which declared the film "classy," also warned that "most of the good folks in your community who have to 'remember their position in society' will probably complain and fuss." The reviewer claimed that "a sequence of café and bed room incidents" was "handled with finesse as far as class and true to life details were concerned but oh boy!—it was rough!" Under the heading "character of story" (following "interiors," "exteriors," "detail," etc.) *Wid's* asserts it "Will shock family trade." Interestingly, in the critic's "box office analysis to the exhibitor" he added that "Cecil de Mille's name won't get you any money but mentioning that this is made by the producer of *Joan the Woman* will help," thus revealing that DeMille had not yet made his reputation and that referring to an earlier and more traditionally highbrow film would help to legitimate the film.[50] Other critics also found the themes problematic but were impressed by the "artistic" images on the screen. DeMille scholar Sumiko Higashi finds the memo sent to Lasky evidence that FPL/Paramount now aimed for broad appeal rather than just upscale audiences.[51] But could appealing to "modern" (and urban) sensibilities be enough to overcome a censorship movement based in an older, but now fragmenting, middle-class sense of respectability?

As we have seen, this sensibility rested on gendered definitions fostered in the nineteenth and early twentieth centuries. In 1920, the year that women gained the vote, the ideology of separate spheres began to unravel. Women were no longer above politics, and although the idea that women were morally superior had some resonance throughout the rest of the century, the idea that women, as a group, took a moralistic view of life disappeared at the same time as the threat of a woman's voting bloc. Women were as politically diverse as men, and a newer generation of women was interested in personal freedoms. Furthermore, the consumerist culture that emerged with the department store, commercial amusements, and the expansion of ever-more effective advertising exploded in the 1920s. The idea that one could remake oneself through the purchase of the correct clothes and cosmetics was again not a new phenomenon, even on the screen, but the combination of postwar politics and the expansion of the U.S. economy, particularly into foreign markets, created a celebration of the American's ability—and right—to consume goods and enjoy personal freedoms. The complaint that women traded their traditional claim to cultural and political power for self-commodification, as evidenced by Beth's initial resistance to wearing the negligee, was overwhelmed by the promise of abundance and freedom, personified not only by DeMille's women but by a string of popular onscreen flappers to come. With regard to the DeMille cycle, tie-ins to fashionable stores helped to deflect attention away from the themes of extramarital sexuality and toward the transformative power of consumerism. And box-office returns muted censorship claims. *Don't Change Your Husband* was another commercial success, breaking attendance records at several theaters.[52]

Interestingly, DeMille's no-star strategy began to work for him following the success of *Old Wives for New*. Advice to exhibitors of *Don't Change Your Husband*, released less than a year after *Old Wives for New*, said, "Here is a case where it is up to you to advertise the director. There is no star's name to work with, so give Cecil B. DeMille just as much prominence as you customarily would give a favorite player," adding, "Fans are coming to recognize the importance of the director."[53]

THE NAME *Lois Weber* HAD ALREADY REACHED name-brand status well before 1919, but it represented increasingly outdated values that no artistic sense could overrule. She made two more films at Paramount before she was released from her contract: *What's Worth While* (1921) and *Too Wise Wives* (1921). Both pitted the vampish Mona Lisa against the virtuous Windsor, and although both films were beautifully staged, photographed, and enacted, they were also dismissed as trite and old-fashioned, indeed, beyond the reach of a new censorship flurry aimed at crime serials and low comedy. In her lavish sets and costumes, and her focus on modern marriage and infidelity, Weber was clearly aiming for a film in the same vein as DeMille. She did not succeed. *Too Wise Wives* deliberately portrayed a young wife too eager to please her husband, suffocating his attempts to relax at home. The film encourages the viewer to sympathize with the husband's wandering eye and then to view the resulting companionate marriage as the ideal. This was Weber's attempt to be "modern," but it was too contrived and still tried too hard to sell a moral.[54] According to one reviewer the secret to DeMille's postwar success was his very shallowness. His style was not "suited to stories of real people and serious import," and that was just as it should be. A DeMille film was "a magnificent puppet show, legitimately and logically excessive in every way." It was pure entertainment.[55]

With Weber's box-office appeal clearly in decline, FPL/Paramount ended its contract with her. She sold two more films, apparently made while still under contract to Famous Players–Lasky, to an independent distributor. The first, *The Blot* (1921), proved to be an outstanding critical success, which must have given Weber some satisfaction. *The Blot* was an unapologetic social problem film, depicting the sinful underpayment of those who "clothe the mind," college professors, librarians, and clergymen. According to one reviewer *The Blot* was free of "offensive preaching" and should "clean up a tidy sum of money." Unfortunately, it did not.[56]

The last release under Lois Weber Productions, *What Do Men Want?* depicted a story of a philandering husband and a virtuous wife. It garnered better reviews than her first films for Paramount, but it drove one critic to despair. Why, he asked, does Weber devote "the really worth while time of herself and her staff to those simplified sermons . . . Why don't all these sufficiently competent people concern themselves with telling a good, straightforward story and let whatever moral it has take care of itself?" One reviewer cracked wise that Weber's version of a "good woman" would lose her husband "simply because the poor fellow is bored to death."[57]

Weber, whose morally upright films bored modern audiences, stood outside the postwar industry. Her views were quaint, her crusading unwanted. She tried to keep up with the times and even tried to portray her relationship with her husband as a modern companionate marriage (referring to him as her "chum") but to no avail.[58] The very tradition and sincerity that made her the ideal filmmaker before the war contained and diminished her afterward, as both the film industry and society took a different tack. Although she made a few more films, Weber's career as a Hollywood director was over. Her personal life began to fall apart as well. Shortly after her release from FPL/Paramount, Weber's marriage ended, and she experienced what she later referred to as a nervous breakdown.[59]

THERE WAS ONE CORNER OF THE INDUSTRY where traditional notions of gender could still make female filmmakers desirable. When the postwar Hollywood scandals broke, once again the assumption of female moral superiority rendered the presence of women as filmmakers politically useful. In articles that smacked of publicity from the Hays Office, women filmmakers were suddenly lauded in the early 1920s as if they had just sprang upon the scene and had not yet had time to cleanse the industry. *Moving Picture World*, which knew better, claimed that Mrs. A. B. Maescher's *Night Life in Hollywood* (1922), which tells the story of a young man who travels to Los Angeles in search of a "perpetual party" only to find the proverbial girl next door, marked the "New Feminine Influence in Picture-Making."[60] In 1923 E. Leslie Gilliams of *Illustrated World* also claimed (incorrectly)

that women were just beginning to enter the film industry. After noting the "new" participation of women behind the camera, Gilliams wondered if they would bring about the reforms needed to save Hollywood, adding, "That is everybody's hope."[61]

Only one woman reaped the limited benefits of this small crack in a closing door: Dorothy Davenport Reid. Dorothy Davenport was a leading actress at Universal when she married Wallace Reid (then a "director and actor of note") in 1913.[62] For several years she was known as Dorothy Davenport Reid, but in 1921, after returning from an acting hiatus, she changed her billing to Mrs. Wallace Reid. A newspaper account noted that it "will be a great shock to the women who are new-fashioned enough to think that every member of the feminine sex should carve her career under her own name—especially when she has one as well known as Dorothy Davenport." The writer surmises it was a move to quash any rumors that the couple was on the verge of a split.[63] It seems quite likely that Wallace Reid was, by that time, succumbing to the morphine habit that would kill him in 1923. In March 1922 Wallace Reid was committed to the Banksia Place Sanitarium for an addiction to morphine, which was blamed at various times on a back injury he had suffered while making a film, spinal injuries from a train accident, or the influence of someone trying to help him cope with exhaustion.[64] Wallace Reid was by then a top star, having appeared in several Cecil B. DeMille films (*The Golden Chance, Joan the Woman, The Affairs of Anatol*).

Reid died in January 1923 at the age of thirty-one, the same month that Will Hays was appointed film czar. Reid's death was one of the largest of the Hollywood scandals, especially after Dorothy Davenport Reid gave the police the names of "Bohemians" who were responsible for introducing her husband to drugs, thus exposing drug trafficking in the movie colony.[65] Shortly afterward, Davenport Reid attended a conference on narcotics held in Washington, DC, with Hearst newspaper reporter Adela Rogers St. Johns. When she returned to Hollywood, she set about producing an antidrug moving picture. Davenport Reid told interviewers that after her husband's death she had just wanted to be left alone, but others persuaded her to "do something." Thus assuming a reluctant stance and armed with moral purpose and "special dispensation" from Will Hays, Reid began work on *Human Wreckage* with coproducer and financier Thomas Ince.[66] (Interestingly, only six months later, Elinor Kershaw Ince, his wife, asked for and received a salary of $250 a week from Ince's board of directors in acknowledgment for her film editing and story selection services, which she had apparently been doing for free.)[67]

According to a copy of the continuity script, *Human Wreckage* began with scenes of dope peddling from street corner to pool hall to school grounds, where pushers are depicted giving "dope" cigarettes to children. Then "Mrs. Wallace Reid" begins speaking in a medium shot: "I wish to deliver a message to every man and woman of our present day civilization; a warning against a great and growing danger." After a close-up another intertitle from Mrs. Reid states: "I have chosen the medium of the screen because it reaches the four corners of the earth." After directly asking the audience to help her "fight," the story begins. *Human Wreckage* depicts the story of an upscale and exhausted lawyer, Alan McFarland, whose doctor introduces him to morphine at their club. Eventually he finds a new supply from street peddlers, and soon he is blackmailed into representing pushers at a trial. Other addicts are graphically depicted, including a young widow who applies a morphine-laced breast salve before nursing her baby. "The baby must have inherited the 'craving' from me," she woefully explains; "it's the only thing that will quiet him." The climax includes a train crash that kills the chief pusher. Mrs. Reid, who played the lawyer's wife, appears once again as "Mrs. Wallace Reid" at the end of the film, sympathizing with the addicts, whose "sole sensuality was *not* to be in pain," and proclaiming the "dope peddler" to be the real enemy. In a close-up at the end of the film she looks directly into the camera and asks "Won't you help?"[68]

Given Lois Weber's postwar failure, it might seem surprising to learn that *Human Wreckage* did extremely well at the box office. Of course, Wallace Reid's shocking death and rumors of continuing addiction among Hollywood stars fueled curiosity. How would a Hollywood insider, particularly the widow of Wally Reid, depict the dope scourge? But there was another reason for the film's box-office success. As Kevin Brownlow argues, Dorothy Reid was a "showman" at heart, willing to inch toward sensationalism and exploitation.[69] Shocking visions of a smuggler offering spiked chocolate to a girl, a middle-class lawyer shooting up, and a young mother drugging her baby were enhanced by garishly tinted expressionistic settings depicting a bad drug trip.[70] As the reviewer for *Variety* noted, "The young can see here things they should not know."[71] Interpreted by *Photoplay* as "a powerful sermon for increased governmental activity in the suppression of the narcotic trade," *Human Wreckage* was really one of the first exploitation films. Although aspects of the genre had been employed since the earliest films, the combination of "forbidden topic," deliberate scare tactics, and direct audience address ("Won't you help?") became common tropes of the B-movie genre. Ironically, coproducer Thomas Ince died a year later under mysterious circumstances. After suddenly falling ill on William Randolph Hearst's yacht, he died within hours. There were initial reports of a gunshot, but the official cause of death was a fatal case of indigestion.[72]

Using the money made from *Human Wreckage* to finance her own production company (and to set up the Wallace Reid Foundation sanitarium for addicts), Reid quickly made *Broken Laws* (1924) at Ince's studio, a simplistic film about juvenile delinquency and parental responsibility. *Broken Laws* did not do nearly as well at the box office as did *Human Wreckage*. Of course it did not have an authentic Hollywood scandal to publicize it.[73] Reid then made *The Red Kimono* (1926), another in the exploitation genre. *The Red Kimono* told a true but outdated white slavery tale of a rural innocent urged to marry her city suitor, only to find herself enslaved as a prostitute. After seeing her erstwhile husband buying a wedding ring for another, she shoots him but is acquitted of the crime. Reid, who appeared at the beginning and end of the film, looked directly at the camera and claimed it was women's duty to help "countless women" like her protagonist. What is interesting about *The Red Kimono* is the manner in which the film vilifies the middle-class social reformer, once the idealized movie patron. "Mrs. Beverley Fontaine" takes the acquitted Gabrielle home because rescuing a former white slave is fashionable (the story takes place in 1917) and fascinating. She throws a tea to impress her friends with her catch and her own largesse, and to Gabrielle's horror one of the female guests corners her and asks her for sex tips. Becoming bored with her pet project, Mrs. Fontaine disappears on a journey and blithely leaves Gabrielle a note telling her she must find some other place to live. Ultimately, Gabrielle marries Mrs. Fontaine's kind chauffeur and begins life anew.[74]

*Human Wreckage* marked the height of Reid's critical acclaim as a producer. *Broken Laws* received lukewarm reviews, and critics panned *The Red Kimono* mercilessly. Worst of all, Adela Rogers St. Johns, who wrote the screenplay for *The Red Kimono* based on a story she covered years earlier as a crime reporter, had not bothered to change the name of the woman in question. Now happily remarried, the woman sued Reid for all she was worth.[75] Yet Dorothy Davenport Reid was one of the very few women whose filmmaking career continued into the sound era. After *The Earth Woman* (1926) received the same critical treatment as *The Red Kimono*, she returned to acting for a time, but she became producer once again in the late 1920s and 1930s, working for small "poverty row" companies that existed on the margins of the studio system.[76]

THE END OF THE UPLIFT MOVEMENT in the years just after 1915 meant that the maternalist reformer was no longer an ideal filmmaker. This, of course, hurt Lois Weber the most, because she had built her reputation on those gendered grounds. Ironically, even as the mainstream industry deemed the feminist version of uplift too dangerous to continue, the postwar industry was robust enough to risk tempting censors in the form of "modern" sex films of the kind made famous by Cecil B. DeMille. It was a marketing decision, and a correct one as far as the bottom line was concerned. The feminized middle-class respectability that informed nineteenth- and early twentieth-century American culture was rapidly dissipating as the Jazz Age commenced. Both inside and outside the film industry, the middle-class matron lost her position as the arbiter of taste for the masses. For Lois Weber this shift spelled the end of an illustrious career. For feature stars, however, whose power derived from audience demand, the expansion of the industry after 1916 once again afforded the opportunity to take the leap into independent production, this time on a scale that dwarfed the earlier generation of star-producers.

# CHAPTER SIX

# A " 'HER-OWN-COMPANY' EPIDEMIC"

## *Stars as Independent Producers*

After 1916, when Hollywood emerged as the moviemaking center of the world, the most powerful individuals were not the nascent movie moguls but the stars of the screen. In the mid-1910s, top stars Mary Pickford and Charlie Chaplin recognized their worth by demanding, and receiving, exorbitant salaries. Soon more stars demanded gigantic pay hikes. Studios capitulated because proven stars were required. But money was not enough. Beginning with Pickford, actors and actresses demanded control of stories, directors, cast, and even distribution practices within the major studios. In a few cases they got it. But stars and near-stars had another option. Between 1916 and 1923 a second wave of screen actors stampeded to a newly flourishing independent market. Since Edison's Motion Picture Patents Company was defunct, this new independent market consisted of small companies operating outside of, and in competition with, the largest studios. These new star-producers, along with activists and famous names from other fields, created their own companies, where they, or their advisers, could call the shots. The new independent market had the same effect as the old; it put large companies in a double bind: the more popular their stars became, the more likely the stars were to capitalize on that popularity by fleeing to the independent market. Only now the stakes were much higher. Just when the largest Hollywood companies enjoyed a healthy global market and greater respectability on Wall Street, this new wave of renegade stars weakened the established studios by heading for the freedom of the open market in what *Photoplay* dubbed a " 'her-own-company' epidemic."[1]

Women played a prominent role in the majority of these companies, and the challenge posed by these entrepreneurial stars was almost certainly part of the reason larger studios moved toward vertical integration in the 1920s. Granted, a number of vital elements inspired studios to expand into theater ownership and distribution, but the advent of the star-producer, and the need to curtail the independent market, was a factor. By the mid-1920s, actors had little alternative but employment by the handful of major studios, and their contracts were on the studio's terms, not the actor's. But for an extended moment, at least, the "her-own-company" epidemic suggested a different future for Hollywood, a future in which the female stars would play powerful and creative roles behind the screen as well as on it.

According to a contemporary film-producer-turned-historian the tumultuous changes that rocked the film industry after 1916 began when Mary Pickford's mother overheard a few Paramount salesmen talking. "As long as we have Mary on the program," they claimed, "we can wrap everything around her neck." The salesmen referred to the distribution technique of block booking. Distributors like Paramount, which handled features produced by Famous Players–Lasky, sold films by the block: if theater owners wanted a Pickford movie, they had to take the program of films Paramount offered with it. Charlotte Pickford logically concluded that if an entire program of movies depended on Mary's popularity, her daughter deserved more. Adolph Zukor, head of Famous Players, agreed. On January 15, 1915, Mary Pickford signed a new contract; she was to receive double her former salary, or $2,000 a week, and, in a stunning display of her newly recognized power, half of the profits from her films.[2]

Pickford's 1915 contract only hinted at what was to come. Charlie Chaplin, who made $150 a week in 1914, and $1,250 a week in 1915, easily negotiated a $10,000-a-week salary and a $150,000 bonus from Mutual in 1916.[3] Not to be outdone, Pickford's mid-1916 contract called for $10,000 a week plus a $300,000 signing bonus, a 50 percent share of the profits from her films, and the creation of the Pickford Film Corporation. This contract not only made Mary Pickford the highest-paid star in Hollywood, but it also made her a quasi-independent film producer, with the power to choose her own stories, directors, and cast.[4]

Pickford and Chaplin were the most powerful individuals in Hollywood because by the mid-1910s industry insiders believed that stardom could not be manufactured. To everyone's surprise, stage stars generally failed to impress movie audiences. Many were famed for their vocal qualities, and others, past their prime, were unable to replicate their youthful stage appearance on the screen. Baffled film producers concluded that it was the audience who ultimately made the star.[5] This understanding inflated the value of screen favorites, and by 1916 proven movie stars knew that they had producers over a barrel. With Pickford and Chaplin as their models, actors with a public image began demanding vast increases in salaries. By 1917 players accustomed to making $150 or $250 a week were making $1,000 to $1,500 a week. "Within an hour after a star received the promise of a raise," remembered Benjamin Hampton, "the secret was known on every stage and in every dressing room." By the next day "a dozen or more cases had to be settled, or the players walked out to find employment with a competitor."[6] A disgusted *Photoplay* writer editorialized that otherwise intelligent studio managers had "fallen for actors' graft like children."[7]

Stars began using their leverage to reshape the industry to their advantage. Their first target was block booking. Stars complained that they received the weakest scripts while their films were used to sell the product of the entire company. Audiences, they feared, would soon associate their names with poor products. Mary Pickford once again set the precedent by forcing Famous Players–Lasky to market her product separately as a more expensive "star series" in 1916. Higher prices meant that more money could be spent on better productions for the star, which, potentially, would generate an even higher salary. Distributors panicked at the star series idea. Who would want to buy their blocks of films if the most popular stars were missing from the program? W. W. Hodkinson, president of Paramount Pictures Corporation, which distributed the Famous Players–Lasky films, refused to cooperate with the star series idea. The head of Famous Players–Lasky, Adolph Zukor, swiftly acquired a 50 percent share of Paramount stock and forced Hodkinson to resign. Within a few years Famous Players–Lasky movies would simply be known as Paramount films.[8]

Famous Players–Lasky's acquisition of Paramount achieved two ends. First, it was more efficient for a studio to distribute its own films. Second, Zukor was right: audiences were willing to pay more to see their favorite stars. But although the higher-priced star series brought in more income, the revenue was still not enough to keep pace with rising costs. Star salaries were but one component. After the wild success of *Birth of a Nation* (1915) and other elaborate feature films, studios competed with each other on the basis of stars *and* enhanced production values. To attract audiences, stories had to be meatier, sets more polished, costumes more authentic, directors more professional, camera operators more creative, and even character actors and extras had to demonstrate real talent. To ensure that stars remained favorites, and to encourage the further popularity of near-stars, moviemakers harnessed story, makeup, sets, costume, and lighting to make the most of the leading players. Distribution, advertising, and exhibition practices focused on the star of the film, while publicists planted stories in fan magazines.[9] These elements came at a price, and together with star salaries, the cost of making the typical movie more than quadrupled after 1915.[10] A pre-1915 two-reeler could be made for a few thousand dollars. By 1917 the average feature film cost between $20,000 and $40,000, and special features cost far more.[11] Within a few more years production costs would jump again.

In the past many manufacturers ground out cheap serials and short films to fund expensive feature productions. But this was not enough. By 1916 major studios like Famous Players–Lasky, Universal, Vitagraph, and Triangle began streamlining production methods to save money. The collaborative filmmaking style that dominated the prenickelodeon era was increasingly replaced by the central-producer system.[12] Under this system efficiencies were reaped when a central producer, studio head, or production head oversaw all of the studio's ongoing productions. Directors lost oversight of scripts, casts, props, locations, costumes, and other production details, which were delegated to specialized departments. Although there were exceptions, directors now received fleshed-out scripts from the studio's continuity writers with instructions for every scene. Studios established rigid schedules for shooting days and required that all expenses, props, and locations be authorized in advance.[13] Stars, accustomed to a degree of creative input, also suffered. At most studios actors and actresses could no longer collaborate or even haggle with directors since the directors themselves were under orders. Now stars had to appeal to the central producer—a more formidable proposition—if they did not like the story assigned to them. The rules changed even for top stars and directors. In 1916 Mary Pickford wisely protected her power to choose story and director by making it part of her contract.[14]

As one might imagine, the stars were not shy about voicing their disapproval over the imposition of what they often referred to as factory methods. The new efficiency methods were, like block booking, believed to cause poor quality. At Universal, where even the studio tour was cancelled in the name of efficiency, a "big shakeup" erupted in the summer of 1916. Among those rumored to be leaving were directors Lois Weber, Phillips Smalley, Otis Turner, Henry McRae, and Hobart Bosworth, plus actors Robert Leonard, Matt Moore, Jane Gail, Mary Fuller, Stella Razeto, and her husband director E. J. Le Saint. In the end only Grace Cunard and Francis

Ford stormed out in protest, but thinly veiled revolts by actors and directors over the erosion of their control were taking place elsewhere.[15]

As studio relations worsened, the market for independent productions expanded. Between 1916 and 1923, independent film distributors included Universal, Pathé, W. Hodkinson, F. B. Warren, William L. Sherry Service, Superpictures, Warner's Features, and First National.[16] In addition, the states' rights market, which sold the right to distribute a film by territory, took a chance on nearly any independent product. As Alfred A. Cohn of *Photoplay* remarked, "It's a mighty poor film that cannot see the light of the projection room via the states rights route." But states' rights distribution was by no means just for poor quality films. On the contrary, established studios sold exceptional films through the states' rights system because the potential returns were much higher than through standard distribution channels. Cohn claimed that "in one instance a five-reel film which cost to produce less than $10,000, was sub-rented in one group of states for the sum of $175,000 merely on the publicity of its New York showing."[17] Although the price of filmmaking was daunting, the open market was a powerful lure.

Lewis J. Selznick (father of David O.) was the first independent producer to exploit the fortuitous intersection of an open market and disgruntled stars. Selznick stunned the industry when he created a company for film star Clara Kimball Young in 1916. Young, a former Vitagraph star, reached the pinnacle of stardom with the World Film Corporation, formed by Chicago mail-order king Arthur Spiegel, theatrical impresario William A. Brady, and Selznick in 1915.[18] When World Film allegedly fired Selznick for excessive self-promotion in 1916, he retaliated by taking Young with him, creating the Clara Kimball Young Film Corporation.[19] At this time only Mary Pickford had her own production company, and industry leaders were not pleased at the suggestion of a trend.[20] If Clara Kimball Young had her own company, every star would want one. Insiders comforted themselves by asking, "How many Youngs or Pickfords are there in America?"[21] As it turned out, there were many. With his wide connections and his own line of film exchanges, Selznick quickly set up independent companies for actresses Kitty Gordon, Norma Talmadge, Alla Nazimova, and director Herbert Brenon.[22] It is difficult to know how the established producers felt about this challenge, but the writer of one editorial described the new female star-producer as "the bane of the industry":

> At present, four more producing companies headed by women [in addition to Pickford and Young] are actually grinding out plays; one stage star of picture

repute is forming her own company, and a lovely minx with whom the country in general has become acquainted only in the last year has a company with a famous director.

The motion picture star-system now imminent is as preposterous, anarchistic and insidious an evil as has ever been introduced into dramatic art in America. The power of combination and co-operation, in the arts and in business, is the premier discovery of this era. These alleged artists would drag film-making back to its days of solitary, suspicious feudal inefficiency.[23]

Within months, the second star-producer movement was in full swing. Joseph Schenck, former booking manager for Marcus Loew's theatrical enterprises, amusement park developer, and an old friend of Selznick's, became a film producer when Selznick encouraged him to promote Vitagraph player Norma Talmadge. Schenck married Talmadge and then created the independent Norma Talmadge Film Co. in 1917. He also created a company for her sister, Constance, another former Vitagraph player. The Talmadge sisters exercised some influence over story material, and they received a percentage of their films' profits, but Schenck enjoyed near total creative control. The Talmadges did not mind, and Schenck proved himself an excellent producer. Schenck formed two companies for male stars as well: Roscoe "Fatty" Arbuckle's Comique Film Company and Buster Keaton Comedies (another family affair, since Keaton was married to Natalie Talmadge, another sister).[24] But overall, companies formed for female stars dominated the movement.

The successful exhibitor and distributor Louis B. Mayer imitated Selznick when he created Anita Stewart Productions, Inc., in 1917. Two years earlier Mayer had broken into the production branch of the industry at Metro. With only one lackluster project behind him, the serial *The Great Secret*, he convinced Stewart, a popular "high-class" Vitagraph actress, to take a chance on independent production under his guidance.[25] Mayer offered Stewart three times the money she was making each week at Vitagraph and a company in her own name. Stewart, who was dissatisfied with her directors at Vitagraph, was granted control over stories, cast, and directors. In the summer of 1918 Stewart began making *Virtuous Wives*, the first of fifteen films produced by Anita Stewart productions.[26]

By the middle of 1917 a "'her-own-company' epidemic" raged among actresses.[27] The independent companies created by Selznick, Schenck, and Mayer were not created by the stars themselves, but they proved to stars who wanted to create their own independent companies that it could be done. Small firms formed around the star-producer appeared (and disappeared) with regularity during this period. Deposed Paramount president W. W. Hodkinson formed a company in 1917 to offer independent filmmakers both production financing and distribution.[28] Robertson-Cole, a banking and export company, entered the film business in 1918, "acting as a banker and exclusive agent for manufacturers of high-grade pictures."[29] In 1921 the Great Northern Finance Corporation offered funding for production, distribution, and exhibition.[30] With financing, studio rental space, and ready distribution outlets the freedoms and possibilities of the independent market were irresistible. Stars who formed, or who were planning to form, their own companies included Olga Petrova (1917), Marie Dressler (1917), Bessie Barriscale (1917), Lois Meredith (1917), Marie Doro (1918), Gail Kane (1918), Louise Glaum (1918), Virginia Pearson (1918), Theda Bara (1919), Alma Rubens (1919), Leah Baird (1920), Madame Mureal (1920), Ethel Clayton (1920), Irene Castle (1920), Justine Johnstone (1921), Juanita Hansen (1921), Mae Marsh (1921), Vivian Martin (1921), Dorothy Gish (1922), and Priscilla Dean (1923).[31] By the early 1920s a number of male stars, such as Sessue Hayakawa, Tom Mix, J. Walter Kerrigan, and Jack Pickford, also formed their own companies, but women dominated the trend.[32]

FACED WITH MASSIVE DEFECTIONS, studios quickly dismantled the more odious aspects of their efficiency efforts. Companies like Famous Players–Lasky and Metro returned to the earlier mode of production for their most popular stars and directors, creating quasi-independent companies that functioned in essentially the same way as the earlier director units. In these companies stars typically exercised some choice in the matter of story, cast, and director, and they usually earned a percentage of the profits from their films. At Famous Players–Lasky new quasi-independent companies were headed by (among others) Marguerite Clark, Geraldine Farrar, and Douglas Fairbanks, as well as by directors D. W. Griffith and Cecil B. DeMille.[33] In 1917 the two-year-old Metro company announced that it was creating units headed by Edith Storey, Viola Dana, and Emmy Wehlen. In 1918 actress Peggy Hyland, a former Vitagraph star, was given her own company at the Fox Film Corporation.[34] Although little is known about most of these companies, it appears that stars exercised at least some degree of creative control within the companies formed under their names, either by choosing their own stories, casts, and directors or by delegating these decisions to male producers and managers of their choice.

With the expansion of the independent market, the screen was theoretically wide open. Rank amateurs were being squeezed out because of the increased cost of filmmaking, but anyone able to draw financial support could still make it to the screen.[35] As in the early 1910s, women who were not screen stars figured prominently in post-1916 filmmaking ventures. Two women joined a growing effort to reach ethnic audiences by creating companies geared toward Asian Americans. Mrs. E. L. Greer headed the Fujiyama Feature Film Company, which was formed in 1916 to produce films in Japan for U.S. release, and Marion E. Wong became president of the Mandarin Film Company, established in Oakland in 1917. Both companies, however, were short-lived.[36]

America's entry into World War I inspired other women filmmakers. In 1916 Agnes Egan Cobb, first noted in 1913 as the only female sales manager in the industry, produced *America Preparing*, which depicted the training of U.S. troops. For unknown reasons it was her only venture into production. By 1921 Cobb was head of the New York–based Motion Picture Enterprises, apparently a sales company, listing branch offices in Chicago, Los Angeles, Yokohama, London, and Montreal.[37] Rita Jolivet, an actress who survived the sinking of the *Lusitania*, turned her experience into a WWI propaganda film with her husband. *Lest We Forget* (1918) featured a re-creation of the *Lusitania* disaster. Jolivet explained: "If I could make every woman understand how much her services are needed if only to save a teaspoonful of flour a day," her purpose in making the film would be fulfilled.[38] The members of the Stage Women's War Relief organization, all prominent stage actresses, funded a motion picture theater for soldiers at a large New York hospital and produced twelve films. Stage stars volunteered to appear in the company's films, one of which was written by its president, Rachel Crothers. Film veteran Eugene Spitz supervised production at the Estee studios in New York City, and they were released on the Universal program.[39]

One interesting example from this era is the Helen Keller Film Corporation, which made the autobiographical *Deliverance* (1919).[40] Film importer and producer George Kleine financed the film, but the origin of the Helen Keller film company itself is unclear. According to the *New York Times*, which took great interest in the film, Keller "actually supervised the production of the entire picture, often causing changes to be made in the arrangements for certain bits of acting and setting." "It is my life," she told the *Times*, "and it must be shown just as I have lived it." George Foster Platt directed Keller by explaining what action he wanted and then cued her by a system of stamps on the floor.[41] *Deliverance* was called one of the "triumphs of the moving picture," but despite critical acclaim, the film was a financial failure. The *Times'* gentle criticism—that "in places it is overburdened with moralizing, and its optimism is sometimes spread too thickly"—hinted at the cause.[42] Years of exhibiting *Deliverance* to schools and colleges could not make up the difference. By 1928 George Kleine claimed that he was left "about $64,000 in the hole."[43]

Stage entertainers also leapt into independent film production. Metro created a production unit for a Hawaiian dancer and café proprietor known as "Madame Doraldina" in 1919.[44] Doraldina made at least two features for Metro in 1920, the Hawaiian-themed *Passion Fruit* and *The Woman Untamed*, the latter awkwardly described as a film in which the actress "lands among the cannibals but she has strange influences over them and this leads to a strange romance."[45] The next year she announced that she was "seeking a suitable vehicle for her first stellar production under her own banner," but no further productions appeared.[46] Doraldina's self-promotion paled next to that of vaudeville star Eva Tanguay, who formed a company in 1916 after "waiting vainly for some moneyed film magnate to meet her price of $10,000 a week." The popular Tanguay, who reputedly spent more than any other vaudevillian on publicity, thus produced her first film, *Energetic Eva*, with her own money. Little is known about the film or how it was received.[47] When Tanguay announced that she was returning to vaudeville, *Photoplay* observed that the waters of independent film production "must have been chilly."[48] Tanguay returned to independent production once more in 1917 in *The Wild Girl*. This time, however, her company was under the aegis of Lewis J. Selznick.[49]

Although the market was fluid, and financing and distribution deals were available, making pictures as a star-producer was much more difficult than collecting a weekly salary from an established studio. Many of the stars who left the comfort of the studios regretted their decision. Others truly preferred control irrespective of the consequences and pushed ahead despite the obstacles, illustrating the strong desire of many stars to control their own work and shape their own films.

THE CAREER OF NELL SHIPMAN, one of the few star-producers who left behind a detailed account of her activities, illustrates the formidable problems facing the star-producer. In 1919 Shipman, a minor Vitagraph star and an experienced scenario writer, partnered with nature writer James Oliver

Curwood. When Curwood wanted to make his own sequel to the 1915 Vitagraph production of *God's Country and the Woman*, a film based on one of his novels, he wanted Shipman to star in the film, since she had played the starring role in the original. Shipman also brought to the new company her skills as an experienced scenarist. *Back to God's Country* (1919) proved a box-office success despite the fact that Curwood detested Shipman's treatment of his novel. According to Shipman the author was angry over the powerful role given to the heroine; the hero of the original short story was a dog. Curwood vowed never to work with Shipman again, but Shipman was now happily typecast as his popular "Girl from God's Country," a plucky, outdoorsy, serial-type action heroine. Furthermore, Shipman was convinced that she could successfully produce her own Curwood-style outdoor films. She established Nell Shipman Productions in 1920 and installed herself as producer, writer, editor, and star.[50] Her partner, codirector, and live-in companion was Bert Van Tuyle, the dashing but hard-drinking production manager she fell for on the set of *God's Country and the Woman*.[51]

Shipman's first independent film was *Something New* (1920). Although Shipman wrote a scenario involving a New Woman–style daredevil heroine, it was a thinly disguised advertisement for the Maxwell automobile and not really what she originally had in mind for a Nell Shipman production. But flush with cash after this extended commercial, Shipman started *The Girl from God's Country* (1922), exploiting her taste for the Curwood-type story and her proven popularity in the role. According to Kay Armatage, Shipman and Van Tuyle formed a board of directors in Spokane, Washington, and the budget of $250,000 was to be raised through stock sales.[52] Shipman played a dual role as the white daughter of an airplane manufacturer and as the "half-breed" daughter of a native of the North, shot in double exposure. Despite the wilderness theme, the film contained scenes of luxury that required expensive sets and costumes, airplanes, an expanding zoo of wild animals, and a well-paid crew. Ultimately it cost twice its estimated budget. Van Tuyle and Shipman spent four months editing the film (apparently twice as long as the estimated time for the entire production). The finished film, at twelve reels long, premiered at Clune's Broadway Theater in September 1921. A reviewer noted it would have made a fine serial and paid particular attention to Shipman's trademark work with wild animals.[53]

Shipman discovered that even for independents, financial entanglements meant giving up some control. After the extremely long feature premiered, Shipman caught a showing and discovered that the distributor had excised three reels of footage without her permission—a direct violation of her contract. Whether it helped the film is impossible to say; the reviewer for *Variety*, who may have seen the original twelve-reeler, advised Shipman to "stick to acting in the future." In any case a furious Shipman claimed to have bought full-page notices in various trade publications asking exhibitors to boycott the "slaughtered" film. (Armatage blames the overly long, over-budget film on Van Tuyle's "delusions of grandeur," but Shipman's efforts to keep the film intact suggest that she shared his vision.) For these actions, Shipman claimed, she was drummed out of certain powerful sectors of Hollywood. Ultimately, when Shipman and Van Tuyle refused to edit the film themselves, the company canceled its contract and took control of the film and its distribution. Shipman and Van Tuyle made no money from the film.[54]

Their next film, *The Grub Stake* (1922), was a convoluted melodrama with elements of white slavery in a Klondike setting. The financial backing for this project was precarious. Shipman and Van Tuyle sold their California home and automobile and put their household goods in storage. They sold $180,000 in shares to three hundred businessmen, all of whom wanted to know how their money was being spent. This became a serious problem when Shipman ran out of funds before shooting was finished. Too intimidated to ask for more money, Shipman sent her cast home minus two week's pay after the film was finished. The disgruntled actors lay in wait, planning to seize the negative when Shipman arrived in Hollywood to edit the film. Shipman managed to get the finished film to New York, where negotiations between producers and distributors took place. A novice, Shipman fell for the poker faces of the trade show audience and sold distribution rights for *The Grub Stake* to the first bidder. She soon realized she had been taken. The price was far too low. The only profit she made was $4,500 for her personal appearances in connection with the film.[55] Financially diminished, but still hopeful, Shipman built a rustic studio at Priest Lake, Idaho. There she hoped to film more outdoor spectacles, but the rigors of the Idaho wilds were more than she had estimated. Her last two films were made while her company was almost completely broke. When finished, the films were quickly "given" to Selznick to distribute. Nell Shipman Productions collapsed in 1924, and Shipman returned to Hollywood to work as a scenario writer.[56]

Clara Kimball Young was a far greater star than Nell Shipman, and unlike Shipman she enjoyed a period of great success as an independent producer. But she also made films with male partners, and here, too, it is tempting to at least partly blame her personal ties for her ultimate failure. In 1916, when Lewis J. Selznick left the World Film Corporation and created the Clara Kimball Young Film Corporation, he appointed Young vice president and treasurer, while he assumed the office of president.[57] Young, like many actresses, "helped revise and reconstruct dozens of stories" earlier in her career, and judging from her later statements it appears that she had strong production ideas.[58] The Clara Kimball Young Film Corporation released four successful seven-reel features: *The Common Law* (1916), *The Foolish Virgin* (1916), *The Easiest Way* (1917), and *The Price She Paid* (1917). In the summer of 1917 Young sued Selznick for fraud, claiming that he denied her a "voice" in her own company and hid her share of the profits through a "manipulation of corporations" when he set up Lewis J. Selznick Enterprises, Inc., as distributor. She only received her actress's salary of $1,000 a week.[59] Because Young's contract did not expire until September 1921, Selznick countersued for breach of contract. Selznick stated that Young received 499,000 shares of stock in the company, and furthermore, she was now taking the advice of Harry Garson, whom she "induced" her company to elect to its board of directors and with whom she was now planning to produce films.[60]

Just two months after Young sued Selznick, *Moving Picture World* reported that Young had "finally realized her ambition to be the head of her own producing company," with Harry Garson as business manager. Young established an office in New York City, and according to *Moving Picture World*, she chose her own stories, casts, and directors.[61] A few weeks later Margaret I. McDonald of *Moving Picture World* interviewed a "businesslike" Young, who waxed forth on interiors, exteriors, and the transition from stage to screen but said nothing about her new company.[62] The powerful Adolph Zukor created the C. K. Y. Film Corporation to distribute her product, and Young leased studios to make ten feature films, some of the most critically acclaimed of her career, including *The Road through the Dark* (1918) and *Cheating Cheaters* (1919). Garson announced in 1918 that Young would build "her own studio" in Pasadena, but their plans changed: Garson and Young sued C. K. Y. Film Corporation for "flagrant violations of the terms" of their contract in January 1919. Ironically, that company was under the supervision of Lewis J. Selznick, who distributed her films through Select Pictures Corporation. Select owned all of its stock of the C. K. Y. Film Corporation outright. It appeared to be another "manipulation of corporations." As he did earlier, Selznick countersued for breach of contract. In June of 1919 Selznick won a settlement in which Young was to pay the C. K. Y. Film Corporation $25,000 for each of her next ten pictures. A week later Garson created a new producing firm, the Fine Arts Film Corporation, exclusively for Young, who now worked out of the Harry Garson Studios in Los Angeles. Her films were distributed by another new firm, Equity Pictures Corporation. Under this configuration she made her most famous surviving film, *Eyes of Youth* (1919), which included a young Rudolph Valentino.

The period 1916 to 1920 was the peak of Clara Kimball Young's career as an actress and an independent producer and perhaps as the dominant filmmaker partner. When a U.S. district court judge found that Young and Garson were illegally keeping the profits of her films, rather than distributing to Selznick his $25,000 per picture, it was Young who received $114,000 of the profits, to Garson's $75,000.[63] Despite her legal problems, Young kept making films and renewed her contract with Equity for a year in December 1920. *Moving Picture World* reported that she was still choosing her stories and cast, as well as taking an interest in "photographing, color, and developing." But she no longer chose the director, as Harry Garson himself directed her films, including *Hush* (1920), *Midchannel* (1920), and *The Soul of Rafael* (1920).[64]

Clara Kimball Young's career reached its pinnacle during the war years. As late as 1921 thousands entered a contest to draw her famous eyes, the mayor of Houston presented her with the keys to the city, and a ten-week publicity tour in that summer created headlines and hundreds of columns of print. But these activities were due to the abilities of Equity's publicist, Milton Crandall.[65] Young suffered under Garson's poor direction, and their finances were in shambles. According to historian Henry R. Davis, at one point Selznick promised to forgive Young all her debts if she got rid of Garson, and Adolph Zukor remarked that he would pay her $7,000 a week and 25 percent of the profits from her films if she would do the same.[66] By October 1921 Young and Garson failed to repay $15,000 in promissory notes, prompting yet another lawsuit. Young made three more silent films, but Garson ultimately lost his production company, and her popularity waned. By the mid-1920s she worked onstage in touring companies and in vaudeville, and for a time the sharp-witted Young lived with her aunt in the Algonquin Hotel at the height of the Round Table.[67]

INDEPENDENT PRODUCTION WAS DIFFICULT, especially without powerful allies. Fortunately, the independent movement received an enormous boost in 1917 when a group of angry theater owners created First National, a distributor-exhibitor combination that sought high-class independent product for its lucrative first-run screens.

Zukor instigated the creation of First National when he began using strong-arm tactics in 1916 and 1917. He tried to force theaters to book only Famous Players—Lasky/Paramount product by threatening to "prefer" theaters that did so, hinting that exhibitors who refused would no longer receive the popular Paramount product. This put exhibitors in a tight spot. Although their first instinct was to boycott FPL/Paramount for Zukor's aggressive arrogance, his studio still had the biggest stars in the movies. If competitors got Zukor's product, how could exhibitors keep their patrons? And even if they kept their patrons, how would they find enough features to make up for the loss of Paramount films?[68]

The owners of several important first-run theaters realized that if they created a distribution system among themselves and recruited top-notch independent producers, they could afford to shut out the obnoxious Zukor and keep their screens "free." On February 1, 1917, twenty-three exhibitors with a total of 117 theaters formed First National Exhibitor's Circuit, Inc. The franchise members agreed to show all the films First National would provide (not enough to monopolize the screen), and they agreed to organize their own exchanges to sell or rent the pictures to other exhibitors, or subfranchises, in their territories. Exhibitors were free to rent films from any other exchanges, but now they secured a reliable source of non-Paramount product.[69] In effect exhibitors fought Zukor on his own terms: as he integrated forward with the acquisition of the distribution company Paramount in 1916, becoming a producer-distributor, First National exhibitors integrated backward, becoming distributor-exhibitors. As of that time, no single company had attempted to do all three: produce, distribute, and exhibit.

It must be remembered that First National was established, above all, to protect the exhibitors' freedom to choose what to put on their screens. But to would-be independent producers, at least half of whom were female stars, First National was a godsend. By 1918 First National snapped up the two most popular stars in Hollywood: Charlie Chaplin and Mary Pickford.

First National attracted the two top stars in Hollywood by granting them comfortable budgets, allowing them complete artistic freedom, and ensuring that their products would be distributed to first-run theaters. Chaplin, who made two-reelers, still the standard format for comedies, received $125,000 to produce each film, while Pickford received $250,000 up front for each of her features. After 30 percent of the profits were subtracted to cover distribution costs and all other costs were accounted for, First National split the profits equally with its producers. With some two hundred first-run theaters controlled by First National, both Chaplin and Pickford expected to make about $1 million a year from this arrangement.[70] In addition, Louis B. Mayer contracted with First National to release Anita Stewart Productions, and in December of 1918 Norma Talmadge Productions Co. signed a two-year contract with First National. Joseph Schenck bragged that the lucrative deal would fulfill "an ambition I have cherished for more than a year" to "pay the price demanded for big stories" and make "bigger pictures."[71] With the rise of First National the independent producer stood on equal footing with the old-line producers. Because most of the strongest independent producers, Chaplin excepted, were those promoting female stars, First National appeared to assure the future not only of the independent movement but of an industry in which some women enjoyed as much—if not more—power than men.

As soon as First National flexed its muscle, the established studios, particularly FPL/Paramount, moved in to destroy it and the power of the stars.[72] Many industry insiders realized that even First National resented the way stars' salary demands inflated film costs. And many realized that a merger between First National and FPL/Paramount would cripple the independent movement and with it the leverage held by movie stars. Even in its infancy First National was not trusted by the producers who fed it. Paranoia deepened when conversations allegedly overheard at First National's 1919 exhibitor's convention indicated that a proposed merger was in the works, one that would allow studios to "tell the stars just where to get off in the matter of salary."[73] According to Charlie Chaplin, when he, Pickford, and Douglas Fairbanks (now Pickford's husband) heard the rumors, they hired a female detective to check the facts. While dining with the spy, an unnamed "executive of an important producing company" bragged that "he and his associates were forming a forty-million-dollar merger of all the producing companies" and that "they intended putting the industry on a proper business basis, instead of having it run by a bunch of crazy actors getting astronomical salaries."[74] Shortly thereafter, Pickford, Fairbanks, Chaplin, William S. Hart, and D. W. Griffith organized United Artists, a company created to finance and distribute their independently made films. Their signed statement claimed that they were taking this step to end the coercive practice of block booking and to "protect the great motion picture public from threatening combinations and trusts that would force upon them mediocre productions and machine-made entertainment."[75]

The big merger never took place. Instead, Zukor now found himself facing not only First National but United Artists, a small company with powerful star appeal. By 1920–21 the biggest box-office draws in the movies worked independently. At United Artists the most popular names on the screen worked for themselves. At First National stars working for their own independent production companies included Norma and Constance Talmadge, Katherine MacDonald, Mabel Normand, Miriam Cooper, Mr. and Mrs. Carter De Haven, Marguerite Clark, Florence Vidor, Anita Stewart, Pola Negri, Hope Hampton, Colleen Moore, Louise Glaum, and Annette Kellerman. First National also attracted the best directors: Thomas Ince, Mack Sennett, Marshall Neilan, Jane Murfin and Larry Trimble, Hobart Bosworth, King Vidor, J. L. Frothingham, Allan Dwan, Maurice Tourneur, and Allen Holubar, who worked in partnership with his wife, star Dorothy Phillips.[76]

All the while, however, Zukor worked on an alternative plan. He would test the underlying validity of the star system itself. Jesse Lasky was one step ahead of him. Lasky spearheaded a view of the director as the pivot of production and exceptional directors (like Cecil B. DeMille) as worthy of top billing, instead of the star. A 1915 letter from Lasky to Samuel Goldwyn claimed, "You know the public go to see a Griffith production, not because it may have a star in the cast, but because Griffith's name on it stands for so much. It seems to me that the time has come for us to do the same with Cecil's name."[77] In 1918 DeMille got his chance to make "all-star" (really no star) films at FPL/Paramount. Zukor built Famous Players by attracting three-quarters of the industry's top stars, but if DeMille was successful, the results would take power back from the stars and return it to the studios.[78] DeMille's "all-star" productions, which included his postwar marital farces such as *Old Wives for New* (1918), were enormously successful. Spending only $40,000 to $70,000 a film, he was able to earn $350,000 to $380,000, or about the same as a star vehicle costing many times more to produce. Following his lead, other notable FPL/Paramount directors, including D. W. Griffith (before his move to United Artists) and Lois Weber, were entrusted to make films without stars. Ultimately it was FPL/Paramount director George Loane Tucker who gained sudden notoriety for the extraordinary success of his no-star film, *The Miracle Man* (1919). The film cost only $120,000 to produce and earned an astounding $3,000,000. The relatively unknown Betty Compson, who played the female lead, was paid only $125 a week for her work on the film.[79] In 1920 Famous Players–Lasky announced the abolition of its star system.[80] This was hyperbole; a major component of the "all-star" strategy was the judicious development of selected players by the studio. The "all-star" film was based on the premise that the best directors on the lot could *create* the stars of the future—and pay them cheaply while they did so.[81] After the success of *The Miracle Man* Betty Compson became a star, and by 1920 she had her own independent production company, but the writing was on the wall.[82]

The success of *The Miracle Man* placed the power enjoyed by Hollywood stars in jeopardy. A movie did not need an established star to reap huge returns at the box office. Moreover, the studio, as much as the public, held the power to confer stardom by placing actors in the right vehicles and under the right directors. After 1920 the quasi-independent production companies created for stars within the major studios began to disappear. Stars still received more freedom and rewards, but the heyday of the star-producer inside the major studios was over.[83] The independent market still existed outside the major studios, but that market was in jeopardy as well. If major studios produced, distributed, and exhibited films in their own theaters, where could independent films go? Some studios, like Vitagraph, had their own showcase theaters years earlier, but a chain of theaters controlled by a major producer could shut out independent producers. Several theater chains owned by several major studios could shut down the independent market altogether.

Historians agree that Zukor's decision in 1919 to begin buying and controlling first-run theaters was inspired by his battle with First National. Wall Street underwriters Kuhn, Loeb, and Co. arranged a loan for Zukor of $10 million to integrate forward into exhibition. With this money he not only built Paramount-controlled theaters, but he built them—or threatened to build them—in the same neighborhoods as First National franchises.[84] The only way for First National to effectively fight Famous Players–Lasky/Paramount was to become vertically integrated itself. Although First National had its own producers, this was not easy. Since First National's raison d'être was to fight the powers of integrated firms like FPL/Paramount, it was deliberately decentralized.[85]

Nevertheless, in 1919 First National centralized control, renaming itself Associated First National Pictures. The controlling owners placed company stock in a voting trust to guard against takeovers, and the firm's distribution exchanges were placed under the purview of the expanded head office.[86] In 1922 First National brought its quasi-independent production companies to the same Burbank studio.[87] After the move to Burbank these companies remained intact, at least in name. Once there, however, First National slowly began to impose the central-producer system, the dreaded "factory methods," thus undermining its independence. From the point of view of its exhibitors, exchange managers, and producers, First National was begin-

 "A BUSINESS PURE & SIMPLE"

ning to look a lot like Paramount. Some were upset, but others undoubtedly saw that the entire industry was becoming centralized. Many of the founding members of First National were exhibitors, after all, who had become powerful by transforming their individual theater businesses into regional chains.[88] Although few recognized it at the time, this was the end of the World War I—era independent movement and the end of the line for female-headed production companies.

The fate of Corinne Griffith productions illustrates these developments. In 1923 the Vitagraph star left her studio (and her husband) to form Corinne Griffith Productions, Inc., with Edward Small and Charles R. Rogers acting as business managers. Griffith's personal contract with Corinne Griffith Productions, Inc., was standard: as the star-producer she enjoyed approval of story, director, and male lead. But Griffith's contract contained a significant new clause. In the event of a creative dispute First National held the right to make final decisions. This clause was much more explicit in the contract drawn up between Small and Rogers and First National. First National had the "power to make all contracts and incur all expenditures, and disburse all funds," and most important, First National retained the right to supervise production.[89] Several weeks into the making of *Black Oxen*, her first film for First National, Griffith realized that she possessed almost no creative control.[90] Unaware of the details of the contract between her business managers and First National, Griffith sent a letter to Small and Rogers accusing them of giving First National "personal supervision" over all her films without her permission. "Excepting for the right to approve stories," she claimed, "none of you have anything whatever to say about the production of pictures." Claiming breach of contract, Griffith stopped working.[91] Three weeks later Griffith sent a terse apology. She realized that neither she nor her business partners retained the degree of creative control enjoyed by star-producers just a few years earlier. In her apology, Griffith agreed that First National "will do the submitting of stories, directors and actor to perform in the male leading parts to me," and she agreed to make her films on the same lot as other First National units.[92]

Corinne Griffith Productions lasted, at least in name, until 1925, but her company clearly did not give her the power enjoyed by earlier star-producers. Griffith was conceded some creative control but within strict boundaries that would not interfere with efficient production practices. In her 1925 contract she was allowed to choose one of four stories, one of three directors, and her own leading man, as long as he did not demand a salary higher than he had received in previous films. All of these choices were to be made within stringent time limits. Worst of all, First National denied Griffith the greatest financial asset of the star-producer, a share of the profits. On straight salary now, Corinne Griffith exercised no more power than any other star at a major studio, despite having her "own" production company.[93]

Anita Stewart's affiliation with Louis B. Mayer and First National did not pan out as she had hoped either. Granted her own company and allegedly given creative control, Stewart was coerced into accepting mediocre scripts that could not be fixed by high production values or top-rated directors (including Lois Weber). After her contract with Mayer expired in 1922, she left to join William Randolph Hearst's Cosmopolitan studio, where she made three films and then faded from view. She later wondered if she would have been better off staying with Vitagraph.[94]

First National, the banner of the independent movement, became another producer-distributor-exhibitor combination. At the production level it imposed economies of scale by centralizing filmmaking activities and limiting the creative freedom of its employees. It had its own distribution system, saving the expense of a "middleman," and its own chain of first-run theaters, just like Famous Players–Lasky/Paramount.

Even the source of the star's power, audience demand, appeared to dissipate after the war. "The only plays that have been a great success recently," said Jesse Lasky in 1920, "have been those that have had a big, popular theme and have been well cast and directed."[95] For the next two years the trend continued. Many insiders blamed the postwar Hollywood scandals. Stars once envied for "living like gods and goddesses in their gorgeous mansions, with their swimming pools, costly limousines, and luxurious clothes" looked like obnoxiously spoiled children.[96] Others thought that movie audiences were showing signs of "maturity." The end result was the same: it appeared that postwar audiences did not always place the stars' names above all other considerations when choosing a film.[97]

By 1922 the trend was confirmed. *Moving Picture World* reported a "world-wide" survey conducted by producer Thomas Ince that established "beyond dispute the new tendency on the part of the public" to "accept big pictures without outstanding stars."[98] In an apparent effort to begin fresh, Famous Players–Lasky formed the Paramount School at its Astoria studio, where twenty-four young "winners" of a national search assembled to learn acting, makeup, costume, dancing, driving, and other screen crafts. At the same time, the studio announced that Paramount actors were "expected to play any part assigned to them," since "ability to do good work can be demonstrated as well in a small bit as in a leading role."[99]

PERHAPS THE MOST IMPORTANT FACTOR in the apparent decline of Americans' love affair with movie stars came from the rise of the "picture palace" after World War I. Now there was something even more spectacular than the star to encourage patrons to go to the movies. Modeled on elaborate opera houses and legitimate theaters, the new picture palaces offered luxury to the masses. Marbled bathrooms, crystal chandeliers, and velvet draperies pleased the eye, and a coterie of uniformed ushers made sure that each patron was seated in comfort. As the 1920s dawned, these picture palaces multiplied and became even grander, many of them designed to place patrons in an exotic environment—an Egyptian temple (Grauman's Egyptian, 1922), an Italian garden (Houston's Majestic, 1923), or Oriental splendor (Grauman's Chinese, 1927).[100] The picture palaces themselves now became the main attraction. Half the evening's entertainment consisted of slickly produced live acts featuring locally or even nationally famous musicians, singers, dancers, and actors.[101] Patrons came not just for the movies but for the whole experience. By 1925, when Chicago chain Balaban and Katz merged with FPL/Paramount to create the huge Publix Paramount chain, its slogan was "you don't need to know what's playing at a Publix House. It's bound to be the best show in town."[102]

As new producer-distributor-exhibitor combinations appeared, they no longer focused on capturing stars but on building huge new picture palaces. Exhibitors integrated backward to obtain a steady source of films for their screens, while producers concentrated on buying or securing access to first-run theaters. Marcus Loew, owner of the largest chain of luxury theaters in New York, bought the ailing Metro in early 1920 to secure a source of films, and movie producer Samuel Goldwyn had gained an interest in about thirty theaters by 1921. In 1924 Loew's, Inc., bought out the struggling Goldwyn Pictures, acquiring its huge Culver City studio, and it bought Louis B. Mayer's small but efficient independent operation. Loew named his large new production company Metro-Goldwyn-Mayer; thus MGM was formed to supply films for Loew's theaters rather than vice-versa.[103] By the time the dust settled in the late 1920s, there were five "majors": Paramount, MGM/Loew's, Warner Brothers (which bought First National), 20th Century–Fox, and RKO, all of which had their own chains of first-run theaters. United Artists was an exception. With its high-quality product and star appeal, United Artists was able to supply not only its own theaters but other first-run theaters as well. The relatively theaterless Universal, however, became only a minor production studio, whose product was aimed at the subsequent-run neighborhood theaters.[104]

The huge producer-distributor-exhibitor combinations with their luxurious picture palaces eliminated the individual star-producer on three counts. First, the picture palaces created a different kind of demand—that for the show, not just the star. Second, by roughly splitting the market among themselves, the new majors cooperated more than they competed.[105] Movie stars were still valuable, and were well compensated, but the outrageous competitive bidding for their services ended. Third, and most important, the control of first-run theaters enjoyed by the integrated majors made it harder for independent productions to reach first-run screens. Without lucrative first-run exhibition, production financing was nearly impossible to secure.[106]

The majors still needed some films made by independent producers to fill their screens, however. In 1925 and 1926, 248 of 696 new releases were independently distributed through the states' rights method, but as the scale of the industry grew, the smaller independents were marginalized.[107] Indeed, the new majors argued that they were doing the industry a service by eliminating its shakier elements, and many investors agreed.[108] And with stardom no longer the only means of winning at the box office, most surviving star-producers were lumped in with other "fly-by-night" companies.[109] By 1925 even Joseph Schenck of United Artists, the company begun by star-producers, said, "It would be a good thing if the so-called 'independents,' who are raising so much agitation about being put out of business, actually were out of business."[110]

By 1925 investors were more interested in whether the director was a "success at keeping within time and cost schedule" and in the distribution arrangements than they were in whether or not a picture boasted a star. Small independents, including star-producers, began to vanish, while investors and major studios applauded the "maturing" of the industry.[111] By the mid-to-late 1920s, even states' rights distributors, traditionally the most open to any kind of producer, began favoring the larger "poverty row" companies like Columbia and Monogram over the individual producer.[112]

THERE WAS ONE PROMISING PLACE where the ambitious star-producer might go in the 1920s, but it was by invitation only. From the beginning the partners of United Artists—Pickford, Chaplin, Fairbanks, and Griffith (who left for Paramount in 1924)—recruited the biggest names in Hollywood to join them. United Artists offered partial financing and distribution, and by 1926 it had its own small chain of first-run theaters and distribution arrangements with the major studios.[113]

The original partners envisioned UA as a company run by independent stars and directors who worked for themselves, but they initially attracted producers. Joseph Schenck transferred Norma and Constance Talmadge from First National to United Artists in 1924, and a few years later he convinced Buster Keaton to move to UA. Producer Samuel Goldwyn joined UA in 1925 but only wanted to release films he financed elsewhere. Two authentic star-producers did join the ranks of United Artists in the 1920s, however, and both were women.[114]

Alla Nazimova, a stage phenomenon who shot to film fame in 1916 in the pacifist *War Brides*, joined United Artists in 1922. At this time UA was just emerging from a rocky adolescence, in which it had difficulty breaking into first-run theaters and attracting investors. In fact, UA had tried to recruit Nazimova, a powerful star, much earlier. Nazimova, who made $13,000 a week at Metro and headed Nazimova Productions, was in the same league as Pickford and Chaplin, with a reputation as a legitimate "artiste" to boot. Nazimova hesitated to assume the risks of independent production at first, particularly since UA producers were required to invest much of their own money in their productions. But she thought her films for Metro (1918–21) were lackluster, and in 1922 she took a chance with UA, believing that she could do better on her own. As a UA partner Nazimova released a film version of her stage triumph, Ibsen's *A Doll's House* (1922), and then quickly launched her most grandiose personal vision, *Salome* (1923), a film adaptation of Oscar Wilde's 1893 play.[115]

*Salome* is remembered today as a campy and over-the-top indulgence; in 1923 it was radically highbrow. Nazimova's friend Natacha Rambova (wife of Rudolph Valentino) designed the fantastical and sensual Aubrey Beardsley—inspired costumes and settings, and Nazimova, as the lithe Salome, appeared in a brief shift and a bubble headdress. In Wilde's play Salome is King Herod's stepdaughter who desires the imprisoned John the Baptist. When King Herod sees her in the dance of the seven veils, he promises to give her whatever she desires. Salome asks for the head of John the Baptist, which she kisses after his decapitation. As Patricia White notes, Nazimova's *Salome* was rife with homosexual as well as heterosexual overtones (she was known for her lesbian affairs).[116] Hiram Abrams, president of

United Artists, reportedly objected to releasing *Salome*, but Mary Pickford and Douglas Fairbanks "declared themselves for it in every way."[117] A producer, probably Abrams, summed up the shaky position of most UA producers when he said, " 'She's putting up all the dough herself. If they flop [*A Doll's House* and *Salome*], it's her hard luck.' "[118] Indeed. Audiences were put off by *Salome*'s pretentiousness and stayed home in droves. Nazimova lost everything. Although she continued to act, primarily on the stage, she never recovered financially.[119]

Gloria Swanson was a huge star when she joined UA in 1925. Ironically, Cecil B. DeMille's "all-star" (no-star) postwar marital farces had turned Swanson into the biggest star of the decade. To keep her, Paramount offered to raise her salary from $6,500 a week to $18,000 a week, and then a flat $1 million a year, but Swanson believed that Paramount's strategy would soon burn out her stardom. In her mind the potential rewards of becoming an independent producer were worth the risk: "I hated making four pictures a year. At UA I could make one or two a year . . . I hated making silly formula pictures to please the studio and the public. At UA I could choose my own stories. I hated contracts and red tape and tight leashes. At UA I would be my own boss and the equal of the founding artists."[120] Swanson's 1925 UA contract stipulated six films, $100,000 of preferred stock, and the creation of the Swanson Producing Corporation.[121]

Swanson soon found that she "could never expect to do anything simply again." Renting a studio, hiring a director, and finding her cast were all more difficult than she had anticipated. Her first film, *The Love of Sunya* (1927), took nine months to make instead of six weeks, and her next, *Sadie Thompson* (1928), ran into tremendous turbulence from the Hays Office, which insisted—to Swanson's bitter anger—on corresponding with UA manager Joseph Schenck instead of Swanson herself. "They refuse to recognize me as a producer," she complained to Schenck. "They expect you to handle me like a silly, temperamental star."[122]

*The Love of Sunya* opened the $10 million new Roxy Theater, the world's biggest picture palace, and the critically acclaimed *Sadie Thompson*—said to be the best film of Swanson's career—made over $850,000 at the box office.[123] But by then Swanson was happy to turn over financial and production matters to Boston banker, film producer, and soon-to-be lover Joseph P. Kennedy: "I was perfectly delighted to be asked to stop doing what I know I had never done well, anyway." Her first UA project with Kennedy at the helm was the $800,000 disaster *Queen Kelly* (1928–29). Bombastic director Erich von Stroheim, known for his attention to detail and excessive retakes, pushed the film way over budget and his "artistic" touches made it impossible for the film to pass muster with the censors. Swanson finally walked off the set, but it was too late to save the film. Schenck continued to support Swanson by giving her a contract for two more films in 1931. Swanson did not have to risk any of her own money in these films—no doubt to her relief—but she was put on straight salary. For her last UA film Swanson tried her hand at production again, this time in Britain to make *Perfect Understanding* (1933). The production was a mess, and Swanson was dangerously in debt when it was finally completed; nevertheless, the film made a small profit. Swanson was cured of producing for good, but her career went into decline.[124]

By THE MID-1920S, STARS WERE STILL WELL PAID, as they were still a necessity, but they had lost most of their clout behind the camera.[125] At MGM, for example, stars "had precious little control over their individual careers or their pictures," according to Thomas Schatz. Only two stars dared challenge Louis B. Mayer and production chief Irving Thalberg in the 1920s—Lillian Gish and Greta Garbo. After becoming a star in late 1926, Garbo demanded a raise from $600 to $5,000 a week. When Mayer refused, she left for Sweden, and after seven months Mayer acceded. Gish's dispute, over creative control, did not end so happily. Gish made two "brooding dramas" in the mid-1920s: *The Scarlet Letter* (1926) and *The Wind* (1928), both directed by Victor Seastrom. Well-made films, they were too dark for Mayer, who ordered happier endings. Although both star and director protested, it was to no avail. Gish and Seastrom never worked for Mayer again.[126]

Gish lost her battle, as did most stars, because stars were no longer the currency of Hollywood—first-run theaters were. Rather than catering to the needs of the star, now stars fulfilled the need of the studios, which was to fill the auditoriums of their theater chains. To do so efficiently, stars were increasingly typecast in the "silly formula pictures" that Swanson hated. Scenario departments rarely even bothered to show interesting material to stars anymore, since most stars had lost the power to choose their own stories.[127]

The hallmark of the reduced status of the star was the famous seven-year contract, which became standard by the end of the silent era. After the imposition of this contract, as Cathy Klaprat argues, stars became well-paid "indentured employees, placed in a subservient position." Stars were contractually tied to the studios, and with a weakened independent field there was little temptation to leave. The studio, however, could choose to keep or drop the star each year. Morality clauses dating back to the postwar scandals gave studios the right to immediately cancel the contract of any stars that incited "public hatred, contempt, scorn, or ridicule," and with regard to actresses a weight limit was sometimes included as well. When a star's option was picked up for another year, he or she would get only the increase already stipulated in his or her contract. No star could renegotiate a contract midyear to "capitalize on a sudden surge of popularity."[128]

By the mid-1920s, stars were stripped of the means to gain leverage behind the camera. They could not shape their own image because the studio had exclusive rights to a star's services, name, and likeness. They could not bargain for more creative control by threatening to leave for another studio. And they could no longer demand their "market price" (the rationale for the huge raises of the 1910s) without being in breach of contract. Because it was through acting, and especially stardom, that most women gained a measure of power in the film studio, the decline of the star-producer closed a critical door for women in Hollywood. No star would ever become as powerful as Mary Pickford, and never again would stars dictate en masse strategic decisions regarding the production, distribution, or exhibition of movies. Although occasional female stars, like Greta Garbo, would so captivate the public's imagination that they once again enjoyed a measure of creative power, they were the exception rather than the rule. By the mid-1920s, stars were employees, plain and simple, albeit with generous paychecks.

Of course, not all women in the American film industry were stars or even actresses. Yet the same developments that served to contain the creative power of stardom—the central-producer system, vertical integration, the rise of the picture palace, and the seven-year contract—had similar effects on other women in the industry. As filmmaking got "out of the class of a game and more in the class of a business," to quote one contemporary observer, women working exclusively behind the camera also found their situation changing distinctly for the worse.[129]

 "A Business Pure & Simple"

Lucille discovers the
torture chambers...

Serial heroine and coproducer Grace Cunard in a still from *The Broken Coin* (1915).
*Photo courtesy of the British Film Institute.*

Lobby advertisement for *The Broken Coin* (1915).
*Photo courtesy of the British Film Institute.*

Serial heroine Ruth Roland, poised with crew on the set of an unidentified film.
*Photo courtesy of the Library of Congress.*

Undated portrait of serial heroine and producer Rose "Helen" Gibson.
*Photo courtesy of the British Film Institute.*

Cecil B. DeMille (with megaphone) on the set of an unidentified film with screenwriter Jeanie Macpherson (right). *Photo courtesy of the Academy of Motion Picture Arts and Sciences.*

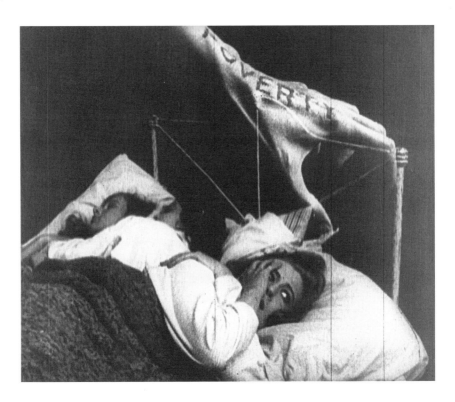

The hand of poverty threatens Eva Meyer in a frame enlargement from Lois Weber's *Shoes* (1915). *Courtesy of the Library of Congress.*

Advertisement for Lois Weber's critical success *The Blot* (1921).
*Courtesy of Madeline Matz.*

Lois Weber with megaphone on the set of *The Angel of Broadway* (1927),
an unsuccessful "comeback" film made several years after the decline of her career
in the early 1920s. *Photo courtesy of the British Film Institute.*

Undated portrait of producer Dorothy Davenport (Mrs. Wallace) Reid.
*Photo courtesy of the British Film Institute.*

Still of a woman shooting morphine, from Dorothy Davenport Reid's
*Human Wreckage* (1923). *Photo courtesy of the British Film Institute.*

Undated photo of Mary Pickford peering into a camera.
*Courtesy of the British Film Institute.*

Marion Fairfax Productions, created by the screenwriter (center). Fairfax produced one film, *The Lying Truth* (1921), "a story of the newspaper world."
*Photo courtesy of the British Film Institute.*

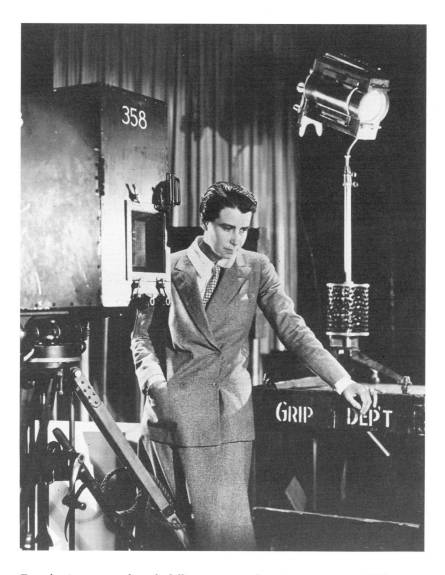

Dorothy Arzner was the only full-time woman director in mainstream Hollywood from the late 1920s to the early 1940s.
*Photo courtesy of the Academy of Motion Picture Arts and Sciences.*

# CHAPTER SEVEN

# "Doing a 'Man's Work'"

## The Rise of the Studio System and the Remasculinization of Filmmaking

In July of 1923 *Photoplay* profiled Grace Haskins, "Girl Producer." At the age of twenty-two Haskins earned her moniker by writing, directing, and producing her first film, *Just like a Woman*. In the space of five years Haskins moved from working in a Hollywood hotel, to answering fan mail, to "talking herself into a job in the cutting room," and then becoming a continuity writer. All the while her ultimate aim was to direct, but "she knew enough of the game to know that no producer was ever going to give her her chance. Not for a long, long time, anyway." Not one to give up, Haskins turned to "several moneyed men," "dusted off" a scenario she was working on, and secured a deal with independent distributor W. W. Hodkinson. With check in hand Haskins collected a company of actors, found a ready-made set, and began making her film. Suddenly "people who had assured her that she couldn't, possibly, hope for success, began to take an unwelcome interest in the proceedings." Damaging rumors surfaced, but Haskins persevered. *Just like a Woman* reached the screen on March 18, 1923.[1]

Haskins intended "to keep right on producing," but like most small independent filmmakers in 1923, she vanished. Unlike many male directors, however, Haskins did not resurface in any of the major, or even the minor, studios. But this scenario in itself could not account for the disappearance of female directors in the 1920s, for the new "majors" hired plenty of experienced filmmakers to grind out the features and programmers that fed their theater chains.[2] Rather, something changed in the very definition of a filmmaker. Although industry writers praised women directors for their "deft touches" and "finesse" as late as 1921, by 1928, the year that the sound film triumphed over the silent film, there was only one working female director, Dorothy Arzner, a filmmaker noted for her masculine persona and approach. By 1928 filmmaking in Hollywood was unquestionably "man's work."[3] Although Haskins could not have guessed it, she was one of the last female filmmakers of her era.[4]

IN MANY WAYS HOLLYWOOD EPITOMIZED modern heterosociability. Mixed-sex groups could be found after a day's shooting at late-night "watering holes" like the Green Room, where in 1913 one might find Jack Holt, Warren Kerrigan, Bob Leonard, Ella Hall, Francis Ford, Grace Cunard, Otis Turner, and Cleo Madison. And writers, split evenly between men and women, formed the heterosocial Screen Writer's Guild and "its social arm, The Writer's Club," in 1920. It boasted a membership roster that was 25 percent female and a clubhouse with a pool and spaces for dances, billiards, and cards.[5] But by and large professionalization segregated the geography of Hollywood by sex. As we have seen, the industry's first trade associations modeled themselves on fraternal societies. In 1915, when the Los Angeles equivalent of the Screen Club moved into "one of the most popular clubhouses on the West Coast," men who belonged to the club could choose from an English bar, a billiard or a pool room, a lounging room, and a dining hall known as "the Stein Room," which held a "weekly good-fellowship dinner, with toasts, vaudeville acts, and music." "Ladies night" occurred one evening a month, when the club held a tango dinner.[6] By 1917, as filmmakers migrated to Hollywood, the Los Angeles Athletic Club surpassed the Screen Club in popularity. According to *Photoplay* the Athletic Club was "the capital of the screen rialto—the Lambs, Players, and Friars rolled into one."[7]

Such organizations were not new. During the last third of the nineteenth century, while middle-class work was still almost exclusively male, clubs and lodges rose to supplement taverns and restaurants as places for masculine recreation and business dealings. But by the early twentieth century, as women entered the world of middle-class work, these all-male organizations took on added significance. Since work itself was "no longer a male club," the spittoon migrated to fraternal organizations. Forced to be genteel at home and at work, middle-class men retired to their clubs, where they might drink, smoke, swear, and revitalize themselves by indulging in the rituals of manhood. But the sexual exclusivity of the Screen Club and similar organizations was not just to insulate men from the encroachment of women in the workplace. The project of professionalization *required*

masculinization. Men needed common ground to put aside their differences and band together for the good of their field. "The exclusion of women," argues Anthony Rotundo, "linked the bitterest of rivals in the solidarity of male professions."[8] Not surprisingly, then, when the Motion Picture Directors' Association (MPDA) was founded in February of 1915, it was described as "a fraternal order." As such, it claimed that "its rituals render impossible the idea of coercion and eliminates any element of partiality or unfairness."[9] It is true that in 1916 the MPDA admitted Lois Weber as an honorary member, but the "rules of the organization" had to be "set aside for this purpose," and just to be certain, the directors added that "no other of the gentler sex will be admitted to membership."[10] Although the MPDA defied its own order in 1923, adding director Ida May Park as an honorary member, it also added actress Lottie Pickford (Mary's sister), theatrical impresario Daniel Frohman, and playwright Augustus Thomas, indicating that the definition of honorary membership meant recognition for good work; it did not confer professional rank as a film director. Thus near the height of their numerical strength, female directors were excluded from the MPDA, which policed its professional neutrality with the rituals of a male lodge.[11]

The rise of these clubs and organizations denied women filmmakers valuable contacts. The list of Athletic Club members in 1917, for example, was a male who's who of Hollywood: Charlie Chaplin, producers Oliver Morosco and William Selig, director Al Christie, and noted actors Ford Sterling, Fred Mace, Tyrone Power, and William and Dustin Farnam. Ironically, a member of the Athletic Club looking for investors to back a new film project might find himself at a gathering of the "The Uplifters," which no longer referred to reformers but to an internal organization of film industry magnates, "democratic millionaires, jurists, doctors, and real estate impresarios." "Every so often they get together in one of the period dining rooms of the club," claimed *Photoplay*, "and lift their voices in song and their arms in—well, we might call it homage, to the spirit of good fellowship."[12] Good fellowship that was by definition off-limits to women.

DESPITE THE INHERENTLY MASCULINE PROJECT of professionalization, women still found opportunities behind the camera as a result of simple pragmatism. Thanks to "doubling in brass," actresses were often well-versed in various duties behind the camera, and it was practical and cost-effective to use all able-bodied employees to their fullest capacity. Thus, at first the reorganization of film production along more efficient lines did not exclude women from behind the camera.

The nascent efficiency movement was not innately masculine. We know that scientific manager Wilbert Melville began his career at Alice Guy Blaché's Solax studio in 1911. In 1913 he visited the Western Lubin plant (which was laid out like "a well-planned kitchen") and introduced to the industry the principle of cost accounting and a new kind of scenario department, where "the scripts are prepared for the directors in such shape that they can be produced as written." "Sooner or later," claimed *Moving Picture World*, "other motion picture manufacturers throughout the world must follow" these innovations "if they wish to meet competition and survive."[13]

According to Janet Staiger, the new sort of script, known as a continuity script, laid the foundation for the central-producer system. It was this system, which first arose in the mid-1910s, that led to the dissolution of the impromptu collaborative style of filmmaking that allowed so many women to gain experience behind the camera. The continuity script included not only dialogue and stage directions for every scene but all the necessary information regarding cinematography, direction, lighting, sets, titling, and costuming. Covering every craft necessary to make a movie, the continuity facilitated the breakdown of filmmaking tasks into discrete crafts and heightened the ability of a central producer to keep a close eye on the progress of all the movies being made on the lot.[14] With continuity scripts for every production, the central producer could allocate all the resources of the studio efficiently, plan budgets well in advance, and monitor each stage of the production process for cost and efficiency.[15]

At first the rise of the continuity script boosted the status of women on the lot. Continuities emerged from the scenario department, and writing was the least sex-typed of all studio crafts. According to Lizzie Francke half the scenarios produced in the silent era were written by women, and female scenario department heads were common.[16] Indeed, many women writers in the 1910s literally defined the craft. Catherine Carr's *The Art of Photoplay Writing* (1914), Marguerite Bertsch's *How to Write for Moving Pictures* (1917), and Anita Loos and John Emerson's *How to Write Photoplays* (1920) were all handbooks on the special requirements of writing for the movies.[17]

Thus, as the continuity script was developed, women were well-positioned to become these new technical experts.[18] Continuity writers became the "brains" of the production process. Creating continuities based on budget, available properties, and available personnel, the continuity writer

"A BUSINESS PURE & SIMPLE"

assumed some of the previous duties of the film producer.[19] The director now worked hand in hand with the continuity writer on the development of the story before a single scene was shot. Indeed, in some studios the director was not allowed to make any changes once the final continuity was approved. He or she had to shoot the film as written.[20]

The role of the continuity script within the central-producer system gave writers, many of them female, a new creative authority behind the camera. Many writers took advantage of their new authority by becoming directors themselves.[21] As continuity writer Marguerite Bertsch remarked in 1916 when she was asked how it felt to direct for the first time: "You know I never wrote a picture that I did not mentally direct. Every situation was as clear in my mind as though the film was already photographed."[22] From the point of view of continuity writers the shift from "mentally directing" to physically directing was a natural one. Initially, then, the dissolution of the collaborative system of filmmaking that characterized the nickelodeon era and its replacement by the central-producer system served to elevate, rather than eliminate, the creative role of some women in the early film industry.

Even women who were not continuity writers appeared to benefit from the early implementation of efficiency methods. Regular scenario writers, those who originated the stories but not the continuities, began to direct more frequently in the mid-1910s as the script in general increased in importance. Writer-actress Nell Shipman's initiation as a Universal director is illustrative. According to Shipman's autobiography, when the director and leading lady ran off together while she was working on location at Lake Tahoe in 1914, "the Universal star [Jack Kerrigan] said I must take up the megaphone. 'You wrote this mish-mash,' he said, 'so you can direct it.' "[23] At about the same time, Universal actress-writer Jeanie Macpherson was said to have won her chance to direct by "pestering" Laemmle when a film she wrote, *The Tarantula*, had to be reshot after the print was accidentally destroyed and the original director was no longer available.[24]

In addition to Nell Shipman and Jeanie Macpherson, Universal writers Ruth Ann Baldwin, Ida May Park, and E. Magnus Ingleton became directors between 1916 and 1917.[25] At least four women directed at Vitagraph after 1916: Marguerite Bertsch, Lucille McVey Drew, Lillian Chester, and Paula Blackton.[26] At other studios, too—large and small—women with writing experience became directors after the introduction of the continuity system, whether they wrote continuities or scenarios. Scenarist Julia Crawford Ivers directed *The Majesty of the Law* (Bosworth, 1915), *The Call of the Cumberland* (Pallas, 1916), *The Son of Erin* (Pallas, 1916), and *The White Flower*

(FPL/Paramount, 1923).[27] Writer Frances Marion directed *Just around the Corner* (Cosmopolitan, 1921) and a Mary Pickford vehicle, *The Love Light* (United Artists, 1921).[28] In the mid-1920s scenarist Lillian Ducey directed *Enemies of Children* (Fisher Productions, 1924); writer Dot Farley directed comedy shorts for Mack Sennett; scenario editor Miriam Meredith worked as an assistant director at the Thomas H. Ince studios; and Elizabeth Pickett, a Wellesley graduate, wrote and directed shorts for Fox, becoming West Coast supervisor for the Fox Variety series.[29]

The implementation of efficiency measures under the nascent central-producer system did not initially prevent actresses from becoming directors either. Economy encouraged Carl Laemmle of Universal to allow actresses to begin directing in the mid-1910s. Ultimately, Universal employed more female directors than all other studios combined. The first actress to become a director while at Universal was Cleo Madison, who, according to *Photoplay*, cajoled studio manager Isadore Bernstein into allowing her to direct in 1915. Madison had worked in production for the legitimate stage prior to joining the film industry. She directed a few two-reelers for Universal, and in 1916 she directed two five-reel features.[30] In 1917 three more actresses became Universal directors: Ruth Stonehouse, Lule Warrenton, and Elsie Jane Wilson.[31]

ONE REASON UNIVERSAL HIRED SO MANY WOMEN to direct was because its pictures were relatively low-cost, low-risk ventures.[32] "By the late teens," observed Richard Koszarski, "Universal was known as a giant factory where work was easily attainable but working conditions (especially salary) remained substandard."[33] But despite the growing tendency to entrust women with only low-budget films, in the context of an open market Universal's female directors gained enough experience to join the growing legion of independent filmmakers. Most had been employed by Universal as writers or actresses for years but left within months after becoming directors to try their luck as independents. Ruth Stonehouse, for example, directed for less than a year at Universal before she signed a contract with the Overland Film Co. to produce six features a year for states' rights release.[34] Like most new independents these companies typically failed. Cleo Madison, who picked up the megaphone to much fanfare in 1915, could not get her own company off the ground in 1917.[35] Lule Warrenton had a single critical success under her own company in 1917 but then returned to the Universal fold and gave up directing entirely.[36] Ida May Park was more successful. After directing for three years at Universal, she created Ida May Park Productions in 1920. She released two features, including the well-reviewed *Butterfly Man* (1920), but her company quickly faded.[37]

Other women directors joined the independent movement as well. Margery Wilson, a Griffith alumnus, directed short comedies and one feature, *That Something* (1921), for Margery Wilson Productions but then lost her company to her lender.[38] Vera McCord directed one film, the poorly received *Good-Bad Wife* (1921), for states' rights release before she too lost her company in a dispute with her lender.[39] In 1921 writer Marion Fairfax wrote and directed *The Lying Truth*, "a story of the newspaper world." Although trade papers announced that she was preparing another film, *The Lying Truth* was her only production.[40] A company formed by Lillian and George Randolph Chester, writer and director, also came and went in 1921.[41] The Cathrine Curtis Corporation, first announced in 1919, appeared to operate along similar lines. Curtis, a "society girl" who starred in one film, described herself as a "screen interpreter." Her company, which ran an impressive advertisement listing officers, a board of directors, and counsel, appeared to act as a literary agency as well. Curtis's first and only production, *The Sky Pilot* (1921), was an expensive "super-special" starring Colleen Moore and directed by King Vidor for First National. After *The Sky Pilot*, which was a success, the company vanished.[42] All of these companies were probably victims of the 1921 recession that forced even the largest studios to suspend production.

Some companies created by women writers after 1916 were not production companies per se but freelance writing services. The Eve Unsell Photoplay Staff, Inc., offered "everything from script to screen": continuities, synopses, "opinions and revisions," subtitling and editing, and representation of authors and publishers. Although literary agents abounded by 1921 (many of them women), Unsell hoped to bring filmmakers and authors to a new level of mutual understanding. When officially unveiled, Unsell's company already held contracts with Famous Players–Lasky and star Katharine McDonald's First National company. Also an apparent victim of the 1921 recession, Unsell's company disappeared by June of that year.[43]

THE INITIAL IMPLEMENTATION OF EFFICIENCY MEASURES, then, most evident in the use of the continuity script, opened up new opportunities for women to direct and produce movies in the 1910s. Indeed, the feminization of filmmaking may well have seemed imminent to some observers, for the number of women directors more than doubled between 1915 and 1919. But even as women were becoming directors and producers in greater numbers than ever before, efficiency measures took on new meaning as film costs skyrocketed after 1916. Although business was booming and well-heeled patrons were attending the movies, studios were experiencing some difficulty generating enough internal financing to cover the budgets of their most expensive stars and feature productions. What the industry needed to continue to grow were reliable sources of outside capital. Thus far nearly all banking interests scorned film producers as fly-by-night operators. The one exception was the Bank of Italy, a California bank founded by two immigrants, A. P. and Attilio (the "Doc") Giannini, that looked favorably on small, struggling businessmen. The Gianninis made several successful loans to exhibitors during the nickelodeon era and by the 1910s had created a system by which they loaned money to fund production of a film but held the negative in a vault until the loan was repaid.[44] Other bankers were frightened off, however, when the two ventures that attracted outside capital in the mid-1910s, Triangle and the World Film Corporation, both collapsed. This "initial lack of recognition by the financial elite," argues Janet Wasko, "drove the movie leaders even harder to build a legitimate industry, in order to be accepted in the financial world."[45]

As the film industry began remaking itself in earnest to attract Wall Street investors, the presence of women in its ranks came under greater scrutiny. Women in the American film industry had thus far enjoyed more latitude and leverage than women in any other industry, including the stage. But women in powerful and visible positions were *not* the norm for most industries, particularly the financial industry.[46] As the film industry began to look at itself through the eyes of the financial community, the theatrical legacy that encouraged the participation of women behind the camera seemed as archaic, and perhaps embarrassing, as the haphazard production methods of the nickelodeon era.

As the theatrical origins of the film industry gave way to the cinema-specific efficiencies of the central-producer system, the transition exposed a fundamental conflict surrounding the female filmmaker, a conflict that had existed since the nickelodeon era: the day-to-day job of a director or producer was to instruct and correct, indeed, to "boss" women and men face-to-face. In any other industry this would easily define the position as masculine. Even a few women who were offered directing opportunities cited the work as too masculine. Ida May Park initially refused to direct features in 1918 "because directing seemed so utterly unsuitable to a woman."[47] When D. W.

Griffith asked Lillian Gish in 1919 to write and direct her sister Dorothy's next comedy, she was "dumbfounded." The *Ladies' Home Journal* agreed, noting that "the Lillian Gish temperament is hardly the kind one would expect to see in executive command."[48]

Even in the allied world of theater, where women had worked as producers and managers for decades, women sometimes hesitated to assume what they believed to be men's work. Musical comedy star Emma Carus professed reluctance when thrust into the job of stage producer in 1911. "I had many hard tussles with refractory working crews," she informed *Green Book* readers, "who did not relish the idea of a woman 'bossing' them," and she found it expedient to adopt "a regime of strict, though not severe, discipline" for her own company after encountering disrespect from some chorus girls.[49] Women in the film industry who pursued filmmaking careers adopted a similar approach. Louis Reeves Harrison asserted in his 1912 profile of Alice Guy Blaché that "she handles the interweaving of movements like a military leader might the maneuvers of an army [yet] she accomplishes gently what a man would attempt by stinging sarcasm."[50] Even during the uplift movement some commentators described women filmmakers as transgressing proper gender boundaries or as women who somehow adopted aspects of masculinity, suggesting that success was not possible without masculine traits. In praise of Weber one source noted that she "has the masculine force combined with feminine sympathies and intuition which seem the peculiarly combined gifts of women of genius."[51] L. H. Johnson of *Photoplay*, visiting Lois Weber on the set of *Hypocrites*, described Weber as a "demon-ess" who "works like a man," turning out films with "super-masculine virility and 'punch.'" Yet she was attractively dressed in "a silk shirt-waist and a smart skirt and chic tan boots" as she issued commands to her "chief subject and vassal, a perspiring camera man, crank[ing] as though Old Nick, instead of [a] pretty woman, were a yard behind him."[52]

ALTHOUGH THE FILM INDUSTRY DID TAKE A NEW TACK AFTER 1916, this transition did not take place overnight but rather over several years, the very years when the numbers of women directors and producers grew most rapidly. Even as the film industry appeared to hold out enormous opportunities for women behind the camera, and even as women took those opportunities and made their own, they were increasingly handicapped by gender. This can be most clearly seen in their treatment by the trade press, which neatly divided women filmmakers into two camps: the New Woman (such as serial-heroine filmmakers) or the artist (such as Lois Weber). Both of these generalizations marginalized and weakened the position of women in the film industry as it began to refashion itself into a serious and modern business.

The first category, the New Woman filmmaker, appeared to liberate women from the constraints of traditional gender roles. But women directors who created New Woman—style characters onscreen, and those who asserted their right to direct films on the grounds of equality with men, were treated as novelties by studio publicity departments and the press. No studio was more adept at promoting and exploiting the novelty of the New Woman than Universal, the undisputed leader in the arena of advertising and exploitation. Throughout the 1910s Universal's publicity department boasted of its modern women. When "Miss Robins" of the accounting department "experienced the novel sensation of riding in an up-to-date flying machine" while on vacation, it was duly reported in the *Universal Weekly*. So, too, was the fact that Ella Hall, a "dainty little actress," was "an expert driver" and that serial star Marie Walcamp rode a motorcycle.[53] Of special interest was the Universal female baseball team, formed in late 1915 by Ida Schnall, former captain and organizer of the New York Female Giants.[54] And at least two directors, Cleo Madison and Elsie Jane Wilson, were depicted in the masculine attire already associated with film directing: tall boots or puttees and riding breeches.[55] Most outstanding, and most often cited by historians, was the apparently feminist politics of Universal City. In 1913, shortly after the incorporation of Universal City, the studio's publicity department declared that it was "the only municipality in the world that possesses an entire outfit of women officials," including a female mayor, Lois Weber, and a female police chief, actress Laura Oakley.[56]

Some historians cite this curious fact as evidence of Laemmle's unusually liberal attitude toward women. But Universal City promoted itself as a land of make-believe. Like a modern-day theme park it was a place where "work is play and play is work" and where for a dollar tourists could take "a new 'rubberneck' autobus" to see the town where " 'movie' actresses control politics."[57] While there is evidence that Universal's female council took its duties seriously, Universal's "suffragettes" were immediately co-opted by the publicity department.[58] In February of 1914 comedy director Al Christie combined filmmaking and promotion for his newest release, *When the Girls Joined the Force*, by parading thirty female members of the Universal police force, each "dressed in regulation cap, blouse, skirts to the knees and silk stockings with a row of buttons down the side," through the business district of Los Angeles. "Traffic became congested and business was at a standstill," reported the *Universal Weekly:* "It was the silk hose." Christie, who secured permission for the parade and the shooting, deemed it "necessary to show a real police station, a real city hall, a real parade, and a real crowd of interested men." As amused onlookers watched, Lois Weber, as mayor of Universal, appointed actress Stella Adams the city's new chief of police.[59] Was this feminism or just spectacle? It was true that the New York suffrage parades of 1911 and especially 1912, with its twenty thousand participants, provided part of the inspiration, but so did the 1912 edition of the Ziegfeld Follies, in which chorus girls dressed as "Lady Policemen," "Lady Soldiers," and "Lady Voters" marched onstage to the patriotic beat of John Philip Sousa and declared that their emblem was "the Chicken," slang for a sexually attractive and possibly available chorus girl. As Susan A. Glenn notes, such spectacles as the Follies deflated the fears that suffragists were coarse Amazons, but at the same time, "they also worked to ridicule the goal of women's political efficacy."[60]

Universal's double-edged publicity undercut the seriousness of its female directors. Universal directors were allowed to appear "feminist" enough to spark interest but not enough to offend. This became apparent when the outspoken Cleo Madison created a rupture in Universal's publicity machine in 1916. Madison told William H. Henry of *Photoplay* that she wrestled her directorship from Laemmle by refusing every director assigned to her. She claimed, "I have seen men with less brains than I have getting away with it, so I knew I could direct if they'd give me the opportunity." In Henry's estimation Madison was "so smart and businesslike that she makes most of the male population of Universal City look like debutantes." Henry asserted that if he ever saw her again onscreen, he would only be able to think of the actress as a "cool, calculating business machine."[61] So, too, might the legions of fans who read *Photoplay.*

Several months later, the studio apparently tried to repair the damage in an article written by a Universal publicist entitled "The Dual Personality of Cleo Madison," in which Madison came across as a competent but unmistakably feminine director. "She is both a professional woman, and a domestic one," claimed the writer, "and it is impossible for her to decide, at times, which is the real taste and which is the cultivated one." The magazine detailed her life in "one of the most charming bungalows in Hollywood, where she lived in apparent domestic bliss with her mother and "invalid sister, to whom she is devoted and for whom she makes her home as beautiful and attractive as a home can be."[62] Madison was quoted in this article as stating that "every play in which women appear needs the feminine touch," adding,

"Lois Weber's productions are phenomenally successful, partly because her woman creations are true to the spirit of womanhood."[63]

Other Universal directors received the same treatment. In 1918 Frances Denton of *Photoplay* was sent to Universal to find out "whether doing a 'man's work' would necessarily make a woman unfeminine." Watching Ida May Park and Elsie Jane Wilson at work, the former in a "dainty pink and white blouse" and the latter in silk gloves, her conclusion was "Unfeminine? Hardly!"[64] The petite Ruth Stonehouse was particularly degraded by Universal's publicity department. In 1917 Universal's *Moving Picture Weekly* claimed in an article entitled "Such a Little Director" that actors were finding it difficult to treat Stonehouse with "the added respect due to her dignity." Next to a photo of an actor peeking flirtatiously at Stonehouse from around a camera, the writer notes that the star of the film "seems to have forgotten that [Stonehouse] is not just a little girl, as he plays 'Ring-around-the-camera' with her."[65] How much Universal's women directors colluded with the publicity department is impossible to know. However, a writer for the *New York Star* treated Gene Gauntier similarly in 1914, when she was still working out of her own studio on 54th Street in New York. On the set of *A Maid of '76*, actress-producer Gauntier necessarily worked while in her Revolutionary-era costume. The writer noted that the rehearsal of a particular scene "was daintily interrupted now by a pretty colonial maid in old-fashioned Dresden silk, hooped, bepuffed, and lace trimmed, and whose coquettish white curls capered from beneath a pink silk bonnet." When she looked through the camera, Gauntier cut "as piquant a figure as any one ever saw doing a businesslike act in this world." Noting a wandering cat on the set, the writer described Gauntier as "scamper[ing] like a pussy down the aisle to the scene of controversy about whether so-and-so or such-and-such would be quite the proper thing under the circumstances." Juxtaposing Gauntier's appearance with her masculine command, the writer described her "feminine mind" as having been "trained" to take quick and decisive action.[66]

Women filmmakers who were touted as artists, and those, like Weber, who argued that women brought special talents to the screen, were also constrained by gender. It is true that Weber created her own persona as the serious "domestic directress."[67] Weber argued that she liked to direct because "a woman, more or less intuitively, brings out many of the emotions that are rarely expressed on the screen."[68] But Universal's publicity department also used implicitly gendered language to laud Weber, praising her, for example, for her "remarkable insight into character."[69] And when Weber was assigned to direct the famous dancer Anna Pavlova in *The Dumb Girl of Portici* (1915), it was reported that since only a director of "supreme artistry" would do, Weber was the natural choice. This artistry was implied by stating that only "a woman would understand a woman."[70]

As long as "feminine" traits were considered important, women filmmakers were tempted to use gendered arguments to bolster their positions in the film industry. The belief that women and men were essentially different was still dominant in American culture in the 1910s, so these arguments carried a particular resonance. But as the definition of filmmaking shifted away from the feminine realm of art and toward the masculine realm of industry, not just these women, but *all* women in the industry, lost ground.

Lois Weber provided the archetype of the feminine director by assuming a maternal persona. When the somewhat cynical Frances Marion interviewed for a job in 1914, she was surprised by Weber's approach: Weber offered to protect and guide Marion under her "broad wing."[71] When Weber finally got her own studio in 1918, she deliberately made it look as much like a home as possible, distancing her studio and herself from "that business air which pervades studios generally." When arriving at Weber's studio for an interview, one writer thought he had the wrong address. Only "a very modest little sign" indicated that the large house with fruit trees and flowers was in fact Lois Weber Productions. Even her production methods were unorthodox; Weber shot her scenes in sequence (to allow for better character development) long after other directors took all the shots requiring a particular background at once to save money. Although Weber said in 1917 that she was not an "idealist," adding that all filmmakers were in business to make money, her methods sacrificed efficiency for art. "Efficiency?" Weber exclaimed in 1918. "Oh, how I hate that word!"[72]

As a highly successful director Weber was excused from the economies of film production. Other women were not. But Weber's fame and her outspokenness on the issue of filmmaking and gender reinforced the view that *all* women filmmakers were intrinsically different from men in their approach. Many women filmmakers were already being relegated to genres that would later be called "women's films." When Frances Denton of *Photoplay* visited the Universal lot in 1918, she found Ida May Park working on "a melodrama," and Elsie Jane Wilson directing some "sob stuff."[73] Universal directors Ruth Stonehouse, Lule Warrenton, and Elsie Jane Wilson all made films centering on children.[74] Lillian Gish's first and only picture as a director, *Remodeling Her Husband* (1920), was described as "a woman's picture. A woman wrote it, a woman stars in it, a woman was its director. And women will enjoy it most."[75] While many women, like Weber, Gish, and Warrenton, appeared to genuinely prefer "feminine" genres, the overall effect was to increase the tendency to define all women as suitable filmmakers *only* when the subject was germane to women.

THE MARGINALIZATION OF THE WOMAN FILMMAKER as either a feminist novelty or a feminized artiste appeared to have had only a limited impact on the activities of women filmmakers in the 1910s. But the ideological consequences of this gendered view became clear in the context of the rapid-fire changes that occurred in the film industry during and after World War I. World War I changed the film industry in two major ways: first, the American film industry finally drew positive attention from government and big business, and second, the decimation of European filmmakers created room for the industry to extend its dominance to the far corners of the world. With these shifts the industry ended its quest for cultural legitimacy and became a bona fide big business. By 1923, when Grace Haskins completed her picture, Hollywood no longer particularly needed nor desired female directors or producers.

High-ranking politicians, like high-ranking bankers, kept their distance from the moving picture industry until a few years before the United States entered the war. In 1915 the film industry gained a major boost with Woodrow Wilson's alleged exclamation after a White House screening of *Birth of a Nation*: "It is like writing history with lightning." By the time that the United States finally declared war in 1917, Wilson marshaled the forces of the film industry as he did other major industries. The major contribution to be made, of course, was in propaganda, both at home and abroad, to be accomplished through the Creel Committee on Public Information. Adolph Zukor and Marcus Loew served the war effort alongside Wall Street tycoon Bernard Baruch, now head of the War Industries Board, and food administrator Herbert Hoover. Meanwhile, Charlie Chaplin, Mary Pickford, and Douglas Fairbanks popularized liberty bonds through live appearances and even short films. The icing on the cake came in 1918, when the government declared filmmaking "an essential industry," so vital to winning the war that it had to keep operating even in the face of shortages. Movies were vital not only to keeping up morale at home but to export American values and goods abroad. "Trade Follows Film" was the new Hollywood mantra.[76]

On top of this sea change in the industry's reputation came staggering new levels of cost and scale. A failure could no longer be so easily forgiven. To meet the pressures of filmmaking at this new level, studio managers began to impose the central-producer system even more stringently. The first to do so was Irving Thalberg, a young man hired to assist Carl Laemmle at Universal in 1918. Thalberg imposed strict compliance, "wherein shooting scripts, production schedules, and detailed budgets were seen as requisites." Directors who demonstrated particular talent as writers were allowed to continue to both write and direct, but for most Universal directors, responsibility for scenarios and continuities rested in the scenario department, freeing them to concentrate on shooting the film.[77] Although the use of the continuity script had earlier opened up opportunities for women in the early central-producer system, the new "businesslike" approach to filmmaking taken by Thalberg imposed a stricter sexual division of labor and enhanced the power of the central producer. After Thalberg became manager at Universal, only one woman director was hired, and she was a special case.[78] Florence Turner, financially destitute after her British company was destroyed by war, was hired to direct a series of short comedies in 1919. Given the minor genre, this was a gesture of kindness.[79]

Thalberg's methods became standard as the studio system coalesced after World War I. As we already know, the rise of the studio system, with its hold on first-run theaters, made it nearly impossible for independent companies, many of them headed by women, to survive in the 1920s. The studio system also curtailed the leverage enjoyed by stars, a critical avenue to power for women, through oligarchical control of the industry and the seven-year contract. There was only one route left for women to continue to participate as filmmakers: as employees of the majors. But the rise in the scale and scope of the film industry just after World War I finalized the masculinization of filmmaking by implementing a strict division of labor, closing this avenue as well. As the studio system emerged, the shift from a theatrically informed model of production and progress to one informed by American industrial norms was completed. Although movies were still a creative product, under the studio system of the 1920s the movie industry became, above all else, a Big Business.

Three events occurring between 1921 and 1923 ushered in the studio system: (1) a brief but devastating recession, (2) Wall Street investment and participation, and (3) the subsequent rise of huge new producer-distributor-exhibitor combinations—the new "majors"—that owned or controlled their own theaters. The recession of 1921 shook the complacency of studios that did well during the war years and during the immediate postwar boom. It was during this brief halcyon period before the recession that independent companies proliferated and the budgets of the most extravagant feature films reached an astonishing $150,000 to $350,000. When the recession hit, the country as a whole suffered from deflation, but the moving picture industry was already facing the consequences of bloated expenses and overproduction. Rental prices fell drastically while budget-conscious patrons stayed home. Larger studios could afford to suspend production, but most of the new independent companies were closed permanently.[80]

The financial health of the larger studios, though wobbling in 1921, rapidly improved as Wall Street investors cast a new eye on the movie business. Otto Kahn, of Kuhn, Loeb, and Company, already showed his favorable take on the movies with a cameo in Reliance's upscale serial *Our Mutual Girl* (1913). A German immigrant and Wall Street financier with a taste for night life, Kahn commissioned a study of the entire industry before granting $10 million in preferred stock to Famous Players–Lasky in 1919 to allow that company to buy its own theaters. From Wall Street's perspective real estate is a good investment—it becomes collateral for future loans.[81] The imprimatur of Kuhn, Loeb, and Company attracted competing Wall Street investment firms such as Goldman, Sachs, who now also desired a toehold in the burgeoning film industry. But Wall Street cast a wary eye on the still-haphazard—and financially dangerous—methods of film production.[82] Demanding stars, profligate directors and producers, and "unbusinesslike methods" fell away under the scrutiny of Wall Street advisers hired not only by Famous Players–Lasky, who won a berth on the Stock Exchange, but by companies that hoped to do so. "There is no room for such items in a report to stockholders," said one banker.[83]

Although the industry itself had long attempted to rationalize production by hiring outside efficiency experts and imposing a modicum of centralized control, the unique culture of moviemaking always mitigated against complete success.[84] Individual directors still retained a great deal of creative control, and quasi-independent director units survived despite the implementation of a central producer. But investment bankers were particularly inspired to see what they could do to rationalize production, and to this end they sent representatives to Hollywood to ensure efficient production and the safety of their investments. Along with its $10 million stock issue, Kuhn, Loeb, and Company imposed rigid scientific management economies on Famous Players. One cost-saving measure had Famous Players personnel dropping friendly salutations like "Regards" from telegrams, transforming requests into rude edicts.[85] All employees were given numbers and color-coded badges (the latter rendered comically ineffective by color-draining klieg lights).[86] According to Lewis Jacobs, the Wall Street "producer-supervisor" assumed "more and more power" in the 1920s, "making the director, stars, and other movie workers mere pawns in production, of which he assumed full charge."[87] It seems likely that these Wall Street producer-supervisors brought with them their own masculine work culture and traditional ideas regarding women and business. Uplift, for example, fell by the wayside. "This is certainly not a campaign to make the world safe for high-brow pictures," said Kahn; "any such effects one way or the other will be entirely accidental."[88] In 1927, on the occasion of its twentieth anniversary, *Moving Picture World* condemned "the absolute domination of the financier" and the "peculiar, steel-mill efficiency in a business that thrives through art."[89]

It was insider Irving Thalberg who became the architect of a regimented production mode that characterized moviemaking under the studio system when he joined MGM in 1923. One-half to three-quarters of MGM's revenues were generated by its first-run theaters. The job of the production supervisor was to make sure that the studio kept the company's first-run screens humming with enough product. To that end Thalberg centralized production on a new scale, implementing a mode of production characterized by "meticulous scheduling and script development, close collaboration with the various department heads to ensure efficiency and to maintain production values, and careful supervision of each picture."[90] In the mid-1920s Thalberg had five male supervisors under him, each assigned projects well before the director or other creative personnel. They closely developed the project, and once in production, "monitored shooting, keeping an eye on budget and schedule as well as the day-to-day activities on the set." Directors were assigned just before production began but were able to contribute to the final script and to the first cut.[91] Similar changes were occurring in other studios. Like other large industries, studios needed to fulfill contracts and maximize overhead through steady production. Under these conditions the formula picture reigned supreme. Minor directors were told to imitate the style of prominent directors, and prominent directors were "asked to repeat successes."[92]

Under the new studio system, moviemaking was increasingly regimented. Stories emerged from the scenario department. After the central producer approved the continuity script and set the budget, a director, cameraman, crew, and cast were assigned; and properties, sets, and costumes were made. When the set was ready, the director shot the film according to the continuity. A new worker, the "script-girl," later called the continuity clerk, worked by his side. Her job was to take copious notes to make sure that every player and every prop was in precisely the same spot in the event of any retakes, which might be shot by a different director.[93] Once the film was "in the can," it went through the laboratory and then to the editing department. By using the continuity script and the slate numbers as a guide, the cutter could assemble a rough cut. From there an editor would make the final cut.[94] The studio then released the film in its predetermined slot on the company's program.

The boundaries between film crafts solidified under the studio system, and as they did, each craft became sex-typed. Some positions experienced little change. Art directors were male.[95] Costume designers were mostly female. And screenwriting remained opened to women throughout the 1920s. But directing, producing, and editing became masculinized. The average director was now a "glorified foreman," chiefly valued for his administrative ability rather than his artistic leanings. He was no longer the sole creator but the "representative of a creative team; he is the man in authority, the field commander who accepts responsibility."[96] According to Cecil B. DeMille the director in the 1920s was an administrator who "never sleeps": "Because if he superintends a staff of brilliant and infallible scenario writers, temperamental stars and untemperamental actors, helpless extra people, nut cameramen, artistic artists, impractical technical directors, excitable designers, varied electricians and carpenters, strange title writers, the financial department and the check signers; if he endeavors ultimately to please the exhibitors, the critics, the censors, the exchangemen, and the public, it's a perfect cinch he won't have time to sleep."[97] Directors need to be "dominating," a quality DeMille believed to be "rare in men and almost absent in women."[98] The masculine image of the film director that characterized associational life in the Screen Club and the MPDA emerged full-blown. Cecil B. DeMille and Erich von Stroheim, in particular, honed this image in the 1920s. Both were among the chosen few who were partly excused from the economies of the studio system on the basis of their past success, and both enjoyed a measure of creative control far out of reach for most directors. But despite their exceptionalism, DeMille and von Stroheim became the popular archetypes of the Hollywood director, placing an insurmountable ideological distance between the "domesticated directress" of the 1910s and the masculinized ideal of the 1920s.

DeMille began work on his masculine persona as soon as he arrived in Hollywood in the mid-1910s, where he told his brother William "real men lived." William recalled thinking it strange that his brother, as well as other directors, wore outdoor gear even while filming inside on stages but chalked it up to the frontier mentality of early Hollywood.[99] Indeed, the high-profile, gun-toting, puttee-wearing male directors of the 1920s defined the work of directing movies as highly physical, even as sets moved indoors and microphones replaced megaphones. By mid-decade the primary argument put forward by the major studios to explain the exclusion of female directors centered on the physical demands of the job. Cecil B. DeMille argued that although indoor sets eliminated long hours in the saddle, and although a director had plenty of assistants to "perform much of the trying labor," most women would "crumple from the strain" of eighteen-hour days.[100] In a 1927 article Carl Laemmle, the man who once employed more female directors than any other studio head, concluded, "It costs from fifty thousand dollars to a million or two to make a picture, and I can't afford to bet that much money on uncertain physical strength . . . I would rather risk my money on a man."[101]

The safari attire donned by the Hollywood director was, of course, more than just practical. DeMille consciously created a hypermasculinized image to instill respect and authority. "Commanding absolute loyalty from his staff," according to Kevin Brownlow, "he directed as though chosen by God for this one task."[102] When Gloria Swanson worked with DeMille, she recalled that he entered the set "like Caesar, with a whole retinue of people in his wake."[103] Even DeMille's office was a shrine to masculinity, crammed full of trophy heads, guns, swords, cannons, leathers, and furs.[104] Writer Frances Marion described it as a *"sanctum sanctorum"* inspiring awe among DeMille's underlings.[105]

Observers of DeMille noted his military bearing, but Erich von Stroheim took this common directorial metaphor literally.[106] Von Stroheim, born into a petit-bourgeois family in Vienna, turned an exceedingly brief military career into a full-blown character sketch—the severe Austrian officer—which he played onscreen and off. Von Stroheim became "The Man You Love to Hate," among fans and production managers. After directing the critical and financial success *Blind Husbands* (Universal, 1919), he was labeled a genius in Hollywood, and von Stroheim took the masculinized role of the film director to its most brutal extreme. According to Richard Koszarski, von Stroheim once physically beat an actress to achieve the emotional effect that he desired.[107] Although most of his films were expensive and overblown box-office failures, von Stroheim carried such a presence that producers continued to employ him throughout the 1920s, even as at least one writer called him a "poser supreme." Writing not too long after Thalberg fired him from the set of *Merry-Go-Round* (Universal, 1923), the journalist described von Stroheim to be "as vain as a girl, and as egoistic as a third-class poet." Von Stroheim, however, saw himself as a maligned artist of the Griffith school and embodied a masculinized idea of the director-as-artist. Bombastic and patriarchal, he bullied those who criticized him for delays, cost overruns, and films so absurdly lengthy they could not possibly be exhibited without severe editing. Ultimately, he submitted. As Koszarski observes, "after the first blush of success, he soon learned that the relationship" between director and studio head was that of "employer and employee."[108]

By the 1920s the most important qualities that the director brought to the set were leadership and discipline. Female directors fell outside these parameters, stereotyped as soft, emotional, and intuitive. The arguments put forward in the 1910s—that the movies needed a "woman's touch"—now served to exclude women from becoming directors in the major studios. Even as independent filmmakers, women ran into interference on the grounds of gender. When Margery Wilson rented studio space from Robert Brunton in 1920 to make *That Something*, he warned her that the set would be "bedlam" because as a woman she could not control the cast and crew.[109] As production values swelled after the war, requiring huge casts and dozens of specialized workers, the quiet, refined, and even domestic style of filmmaking associated with women seemed to have no purchase in the new Hollywood.[110] Even Lois Weber joined the chorus. In a 1927 article about directing entitled "The Gate Women Don't Crash," the author introduced Weber as the woman who "retired from motion pictures with about $2,000,000 in cash and property, a nervous breakdown, and the record of being the only woman who had been able consistently to stand the gaff of directing." When asked if she would recommend filmmaking as a potential career for girls, Weber issued a stern warning: "If you feel a heaven-sent call, take careful stock of your qualifications," she advised. "If you haven't got a superabundant vitality, a hard mind that can be merciless in shutting off disturbances, and the ability to keep going from sunrise to midnight, day after day, don't try it. You'll never get away with it."[111]

Directors, largely shorn of the creative input they enjoyed before the studio system, based the masculine definition of directing primarily on the physical demands of the job. But the technical nature of film direction served the same purpose. Although female directors like Alice Guy Blaché and Lois Weber were among the first to employ cutting-edge cinematic techniques, DeMille argued that the technical and mechanical aspects of direction lie "outside a woman's mind." To prove it, during a 1927 interview about female directors he turned to his screenwriter and constant aide Jeanie Macpherson and asked if she understood how he had made the Red Sea part in *The Ten Commandments* (Famous Players–Lasky, 1923). Macpherson, who wrote, directed, and starred in her own weekly two-reelers in 1913, replied with a demure, "No, Mr. DeMille."[112]

ALTHOUGH DIRECTORS GARNERED THE LION'S SHARE OF PUBLICITY, producers became the most powerful individuals on the lot under the studio system. Not surprisingly, production duties were masculinized as well. As budgets reached into the hundreds of thousands of dollars, if not millions, writer-producer Jane Murfin claimed that studios assumed women were simply not smart enough or experienced enough to handle this level of fiscal responsibility: "Men don't expect women to understand the intricacies of business, the cost of production and distribution, the percentage of overhead, locked up capital and liquid assets, and especially the complications of banking transactions. I admit I've sometimes wondered just how clearly the men themselves understood them, and one or two unwisely frank gentlemen have even admitted that they were congenitally hazy about 'earned and unearned profits' and the 'circuit velocity of money,' doubtless due to the parental influence of their mothers."[113] Cecil B. DeMille represented both production and direction when the film-friendly Gianninis of the Bank of Italy made him vice president of one of their branches: the Commercial National Trust and Savings Bank in Los Angeles. The Gianninis "packed the board" of each of their branches with industry insiders—producers, directors, actors, and actresses—but industry insiders were likely drawn from the new major studios, so they likely favored each other over independent filmmakers when extending credit. DeMille illustrated this when he made a $200,000 unsecured loan to Samuel Goldwyn as one of his first acts as a banking official.[114]

A few women did find work as producers in the 1920s. Paramount hired Elinor Glyn, author of the scandalous novel *Three Weeks* and the woman who coined the phrase "IT," or sex appeal, as an all-round production adviser in the 1920s.[115] Jane Murfin wrote and coproduced five films for First National (most of them featuring a dog, Strongheart, predecessor to Rin Tin Tin) between 1921 and 1924, and codirected one film for First National, *Flapper Wives* (1924), listed as "her own production." Murfin did not direct again, but she became RKO's first female production supervisor in 1934.[116]

Much more significant was June Mathis, one of the most influential figures in 1920s Hollywood and the woman who might have set a precedent for female producers under the studio system. Mathis became chief of Metro's scenario department in 1919 after working as a screenwriter for only one year. At the age of twenty-seven Mathis made script selections, adaptations, and continuities for the studio. In 1921 she adapted Ibanez's war novel *The Four Horsemen of the Apocalypse*, insisting that the studio hire Rex Ingram to direct it and cast bit player Rudolph Valentino as the male lead. *The Four Horsemen* became known as one of the best films of the year. Already well known, Mathis became a celebrity.[117] After her triumph with *The Four Horsemen* she was hired by Samuel Goldwyn as editorial director of Goldwyn Pictures. She set studio policy and handled continuities and scenarios. It was during her term that some of the most prominent directors in the movies worked for Goldwyn: King Vidor, Victor Seastrom, Marshall Neilan, and Erich von Stroheim.[118]

Mathis, who enjoyed what the *Los Angeles Times* called the "Most Responsible Job Ever Held by a Woman," survived as a Hollywood producer despite her involvement in two of the greatest production fiascos of the 1920s.[119] The first occurred after she hired von Stroheim to film *Greed* (1924). When he was finally done, von Stroheim handed Mathis a "finished" forty-two-reel film. Mathis cut the film to thirteen reels herself, and further editing was done by her best title writer, but von Stroheim loudly complained that his masterpiece was ruined, and his fans blamed Mathis for tampering with genius.[120] The headaches caused by *Greed* paled next to those stemming from *Ben Hur* (1926), however. Mathis fought for months with the Goldwyn studio over casting and crew for the filming of the $1 million script, which was shot, per her request, in Italy. Mathis won most of her battles, but when she arrived on the set in early 1924, director Charles Brabin refused to permit her to "interfere." To make matters worse, labor disputes and permissions from the Italian government slowed progress on the film. When Goldwyn became part of MGM in 1924, the studio fired Mathis along with the director and star.[121] Despite these difficulties, First National hired Mathis as editorial director, where she "demonstrate[d] her successful supervision of a major studio's entire output." But in 1927, at the age of thirty-five, Mathis died after a seizure.[122] Although female producers survived in less important companies, no woman of her stature emerged within the studio system to take her place.

As direction and production became rapidly masculinized after World War I, the increasingly important craft of film editing was still open to women, at least for a while. With the rise of the continuity script and the central-producer system, the position of "film cutter," like that of other workers, became a specialized element of the postproduction process. Using the continuity script and the slate numbers as a guide, the cutter could assemble a rough cut, and even a final cut, often without the director's personal instruction. By 1922, editing had reached the status of a creative craft, one that was just below that of director in terms of its creative impact on the final product.[123] It was true that inexperienced "boys" were often hired to piece together films according to continuities, but in the 1910s and early 1920s many studios recruited women from the joining room to become "cutter girls." Margaret Booth recalled her move from joiner to cutter in the mid-1910s in a matter-of-fact manner: "Irene Morra was the negative cutter and she took me to help her and showed me how to cut." Viola Lawrence, a former film polisher, learned how to cut film at Vitagraph in 1915. By 1918 the "master cutter" gained recognition as an important creative force behind the finished film. Rose Smith cut Griffith's *Intolerance* with her husband James Smith, and at least a dozen other women were counted among the first editors in Hollywood, among them Anne Bauchens, Blanche Sewell, Anne McKnight, Barbara McLean, Alma MacCrory, Nan Heron, and Anna Spiegel.[124] Editor Adrienne Fazan recalled that in the early 1920s "every studio had a few women editors . . . [A] woman could get started then."[125]

Women who learned to edit in the 1910s and early 1920s enjoyed long careers. Viola Lawrence became head editor at Columbia in 1925, Margaret Booth became MGM's supervising film editor in 1936, and Cecil B. DeMille employed editor Anne Bauchens for more than forty years.[126] According to Douglas Gomery, "all filmmakers from the late 1930's through the late 1960's who worked for MGM had, in the end, to go through Margaret Booth to have the final editing of sound and image approved."[127] But editing was masculinized as well. Viola Lawrence recalled having "all boy assistants" (but for one) in the 1920s and 1930s, as did Adrienne Fazan.[128] Even female editors who began their careers in the 1910s and early 1920s ran into hostility from male editors. Viola Lawrence's husband, Frank, who taught her to cut film in 1915, was "mean" to the female assistant editors he supervised at Paramount in the 1920s. "He just hated them," she claimed. "If any of the girls were cutting—if they did get the chance to cut—he'd put them right back as assistants," but he "broke in a lot of boys." In the early 1930s, editor Adrienne Fazan recalled, "MGM didn't want me to become a feature cutter." Production head Eddie Mannix told her that film editing was "just too tough work for women," who "should go home and cook for their husbands and have babies." (It was Dorothy Arzner, the only woman to survive the purge of female directors in the 1920s, who took Fazan out of the short film department by asking specifically for a female editor.)[129] As editing became recognized as a critical step in the production of what were now often million-dollar films, it, like direction and production, became a masculinized craft.

IN THE 1920S A NEW GENERATION of female studio workers faced an occupationally sex-typed industry. After 1916 the dominant paradigm for the American film industry shifted from the stage, with its egalitarian work culture, to a model based on American business. As this shift took place, opportunities for women behind the camera, from office jobs to the director's chair, grew increasingly gendered. The American film industry was not unique in this regard. There are now enough histories of gender and business to perceive a pattern. As industries grew from being small and decentralized at the beginning of the twentieth century to becoming larger and more "professional," women who had once been welcomed were now defined as unfit. Susan Coultrap-McQuinn found that female writers who flourished in the mid-nineteenth century (and indeed wrote the best-sellers of the era) did so under the paternalistic guidance of "gentlemen publishers" who spurned commercialism. By the early twentieth century, however, a "new idea of the Businessman Publisher" had emerged. The new publishers "were workaholics who emphasized activity, energy, and time orientation. They organized their offices for efficiency and profits." Coultrap-McQuinn asserts that "the increased emphasis on vigor and marketing made authorship seem more than ever to be a male activity."[130] Wendy Gamber found similar changes in the once "female economy" of milliners. Milliners were dependent on credit from wholesalers to secure the fabric, lace, and other materials needed to make their product. In the mid-nineteenth century women in small business actually received credit more easily than their male counterparts because of the belief in the moral superiority of women. By the late nineteenth century, however, as wholesalers began to adopt more "rational" methods, they began to view their female clients as "unbusinesslike." Since ready-made hats infiltrated the market at about the same time, the number of female milliners plummeted; whereas there had been almost 128,000 in 1910, there were fewer than 45,000 in 1930.[131] A female economy also ruled in the beauty business before the turn of the century. Several early twentieth-century entrepreneurs made fortunes when the beauty industry was small and decentralized but suffered when national markets necessitated the replacement of woman-to-woman customer culture with dealer relations. Interestingly, Kathy Peiss found that women in the beauty business "often struggled with husbands or relatives for control of their companies," a finding that bears a striking resemblance to what happened to male-female filmmaking partnerships. In particular, women in the beauty industry, like women in the film industry, were hampered by access to distribution outlets. Instead of theater screens, women were fighting for space on department store shelves, where they were elbowed out by "prestigious male perfumers, considered skilled craftsmen," and the established brands supplied by large wholesale suppliers. Max Factor used the nascent film industry to launch a line of nationally advertised cosmetics in 1928.[132] A few women were able to survive and even begin businesses in the 1920s, but most withdrew. Even the powerful Helena Rubenstein sold her company to Lehman Brothers in 1928. By 1935, observed Catherine Oglesby, "the great majority" of the beauty firms once owned by women "have passed over into the hands of large companies controlled by men who are directors in large holding companies."[133] That same year a *Fortune* magazine writer concluded that "women's place is not the executive's chair": women who succeed in feminized markets are "not professional women . . . Elizabeth Arden is not a potential Henry Ford . . . It is a career in itself, but it is not a career in industry."[134] As Gamber notes, "large-scale enterprise meant male enterprise."[135]

By the mid-1920s, the power of stardom diminished, the independent movement ended, and the gendered studio emerged. Directors and producers were almost exclusively male, and film editing was rapidly becoming masculinized. Of the creative crafts behind the camera, only screenwriting, a job that often paid poorly and was chronically disrespected, remained open to women. By 1928, when the "talkies" triumphed, the reorganization of studio work that followed merely codified the sexual division of labor that had emerged after World War I. The era when actresses and writers easily slipped into the director's chair, when a woman was one of America's most critically acclaimed and successful directors, and when "America's Sweetheart" was one of the most powerful producers in Hollywood was over. Yet the very presence of this generation of women filmmakers demonstrates that the male domination of Hollywood moviemaking was not a foregone conclusion but rather the outcome of a historical struggle that might have had a different ending.

# Getting Away with It

When Lois Weber warned would-be female directors in 1927 that they would "never get away with it," the age of the female filmmaker appeared to be over.[1] Career-advice literature for women stopped suggesting creative filmmaking careers and began urging girls to think about feminized studio work.[2] Women might become manicurists, script girls, or "secretaries to the stars."[3] Constance Talmadge, for example, brightly suggested that "girls" interested in Hollywood become scenario and title writers, publicity writers and film splicers, seamstresses and stenographers. "There are girls who have become directors," she noted, but she spent far more time writing about the "extra girl" who was now running a Hollywood hairdressing salon.[4]

An unmistakable pattern emerged by the first decades of the twentieth century: big business, that is business with large capital requirements and national distribution, was masculine. "The search for order" was in fact a gendered search; rationalization and national distribution were freighted with gendered concepts that proved to be ultimately, but not immediately, unfavorable to women.[5] When industries were small and decentralized, as the film industry was during the nickelodeon era, female entrepreneurs tended to abound. But as industries became larger, centralized, and dependent on outside capital, as did so many at the start of the twentieth century, work once defined as suitable for women was redefined as suitable only for men.

The one woman who "got away with" filmmaking after the silent era was Dorothy Arzner. In the annals of both film history and film theory Arzner looms large as the great exception. In 1927, as the previous generation of female filmmakers disappeared, Arzner made the (now) giant leap from editing to directing with Paramount's *Fashions for Women*. By 1928, with two brief exceptions, Arzner became the only female director in Hollywood to find work during the height of the studio system. Between 1927 and 1943 Arzner directed sixteen feature films: ten for Paramount and six as a freelance director under the aegis of four different studios: RKO, United Artists, MGM, and Columbia.[6] Arzner's very exceptionalism, however, raises an obvious question: why was Arzner able to assume the now-masculinized title of "Hollywood film director" while other women could not?

One vital reason was the fact that Arzner entered the film industry just in time to gain the kind of varied experience behind the camera that characterized the unofficial apprenticeships of nearly all the women in this study. According to her biographer, Judith Mayne, Arzner's entrance into Hollywood was due to a "fortuitous coincidence." The well-to-do Arzner, a Los Angeles native and college student, served as a volunteer in the Los Angeles Emergency Ambulance Corps during World War I. Soon after the war William DeMille, head of the Ambulance Corps, gave Arzner a tour of the Famous Players–Lasky/Paramount studio. Impressed, Arzner decided to leave college for a career as a film director. At perhaps the last moment when directing was still considered an appropriate goal for women, Arzner took a studio position as a script typist in order to learn the business. Her rise through the ranks was extraordinary. Although she was not very good at typing, within three months Arzner acquired a position as a "script girl" on the Alla Nazimova production *Stronger than Death* (1920). Although the precise circumstances are unclear, Arzner soon gained a position as an apprentice under female cutter Nan Heron on *Too Much Johnson* (Famous Players–Lasky, 1919). Her progress was so impressive that she was allowed to finish cutting the film and soon became chief editor at Realart, a Paramount subsidiary. In 1922 Arzner transferred back to Paramount to cut Rudolph Valentino's first starring vehicle, *Blood and Sand*. It was while on this project that she cut her teeth as the director of the bullfight scenes for this film. What really impressed the studio brass was her artful integration of stock footage, thus saving the studio money without sacrificing aesthetic value.[7]

Eventually Arzner made a name for herself as a "resourceful" director, one who made "A" films on "B" budgets. She remained an editor for several more years at Paramount, working primarily for director James Cruze. Impressed by the editing on *Blood and Sand*, Cruze hired Arzner to cut his extremely successful western *The Covered Wagon* (1923). Arzner claimed that Cruze treated her like a "son, without any frills but with a sort of comradely friendship." While working as an editor at Paramount, Arzner also wrote scripts for Columbia. This was in keeping with her efforts to learn all production crafts. It was also in keeping with what was left of the fluid boundaries between screen crafts, which persisted in screenwriting, a craft that was (and is) often taken up as a sideline. Significantly, Arzner wrote the adaptation for Dorothy Davenport Reid's *The Red Kimono* (1925).[8]

In 1927 Columbia, then a struggling poverty row studio, offered Arzner a job writing and directing her own films. She wanted to take the offer, but Paramount tried to convince her to stay, maintaining that she would get a chance to direct when the appropriate opportunity came along. Arzner insisted that unless she was on the set in two weeks as the director of a Paramount "A" production, she would leave. In 1927 Arzner's first directorial effort was released: Paramount's *Fashions for Women*. Reviewers gave Arzner top billing as "Paramount's first woman director," overlooking Lois Weber's string of films for Paramount only seven years earlier.[9]

*Fashions for Women*, by its title alone, signified Paramount's view of an appropriate vehicle for its lone woman director. Arzner's apprenticeship as an editor on James Cruze's masculine westerns and historical dramas was not sufficient to categorize her as simply another new director. After *Fashions for Women* (1927) Arzner was assigned to *Ten Modern Commandments* (1927) and *Get Your Man* (1927), variations on the husband-finding theme; *Manhattan Cocktail* (1928), a film with a Broadway theme and a happy romantic ending; *The Wild Party* (1929), Clara Bow's first talking film, in which Bow played a college coed; and *Sarah and Son* (1930), a mother-and-child melodrama that prefigured the "woman's film," which would emerge as a fullfledged genre during the 1930s and 1940s.[10]

Arzner sought early to prevent such codification, which she correctly believed would marginalize her career. "I was so averse to having any comment made about being a woman director," Arzner said, "that in my first contract I asked that I didn't even have screen credit on the picture, because I wanted to stand up as a director and not have people make allowances that I was a woman." Likewise, Arzner undercut the assumption of sisterhood behind the camera. She often mentioned that she had far more troubles from women than from men and in 1929 told an interviewer that she experienced "a certain antagonism in the women I was directing," which she blamed on the fact "that a woman director was a novelty." In addition Arzner worked mightily to prove her competence; an effort that may have caused her to take few artistic risks. As Mayne suggests, Arzner's ability to bring in well-crafted films on time and under budget went far in securing her continued employment.[11]

Against her wishes, publicists focused on the fact that Arzner was a woman. One subtle way was to peg her as a "star-maker" from the very beginning of her career. Arzner's immediate reputation as a star-maker stemmed from the assumption that a woman filmmaker would instinctively bring out the talents of her stars—woman to woman—in a way that male directors could not.[12] More directly, publicists and journalists attempted to define Arzner in ways that emphasized her femininity. Although Mayne registers surprise at the extent to which Arzner's looks were deemed important, this emphasis is not, as we know, particular to Arzner; female producers and directors throughout the silent era were similarly constructed in the press as traditionally feminine. In the case of Arzner journalists could not feminize and domesticate Arzner through the tried-and-true method of describing clothes (mannish), house (shared with female companion), and husband (none and no prospects). As Mayne notes, Arzner was a lesbian who wore her dark hair cropped short and combed back and sported full brows, no makeup, low shoes, and often wore a tie. She was what was called a "handsome" woman; journalists tended to focus instead on her eyes, skin, and quiet voice—aspects of Arzner's appearance most open to feminization.

The immediate desire to define Arzner as feminine both inside and outside the studio indicates that the impulse to contain powerful women in Hollywood continued after the decline of the female filmmaker in the mid-1920s; in fact, gender boundaries appear to have grown even more rigid. As we have seen, Arzner was immediately labeled a woman's director. But with solidified gender boundaries and the prompt containment of the remaining woman filmmaker, an allowance could be made in the case of Arzner. Indeed, it seems the unique combination of what Mayne refers to as Arzner's butch style, her economy as a filmmaker, and her competence—the latter learned at the very end of the industry's openness to women filmmakers—allowed her to enter the fully masculinized ranks of American directors. Arzner may not have truly been one of the boys, but her refusal to voluntarily draw attention to the fact that she was a women minimized the ideological challenge she presented to the masculine definition of movie directing.[13] In other words, Arzner got away with being a female director because Hollywood filmmaking could remain a man's world, even with Dorothy Arzner in it.

No REVIVAL OF WOMEN DIRECTORS took place in the immediate wake of Arzner's departure. Once in place, the gendering of studio crafts was codified in the industry's work culture and associational life. The decay of the studio system after World War II suggested promise—Virginia Van Upp became executive director for Columbia in 1945, and Ida Lupino joined a new wave of independent actor-director-producers in 1949—but not until the 1970s did women begin to reappear as directors in any number, and only in the 1980s and 1990s did women direct more than 1 percent of mainstream Hollywood product.[14] Even after the studio system fell apart in the late 1940s and 1950s, film direction and production retained the masculine aura that congealed so swiftly in the early 1920s. Although lauded in the press, female filmmakers remain a minority. A century after the integration of women as filmmakers, the promise held out by the early American cinema is still unfulfilled.

# Notes

INTRODUCTION. MAKING MOVIES AND INCORPORATING GENDER

*Epigraph 1.* Robert Grau, "Woman's Conquest in Filmdom," *Motion Picture Supplement,* Sept. 1915, 41.

*Epigraph 2.* Mlle. Chic, "The Dual Personality of Cleo Madison," *Moving Picture Weekly,* July 1, 1916, 24.

1. Although the term *actress* is no longer in current usage, I use it rather than the gender-neutral *actor* in this study to clarify that the work in question was performed by a woman. Clara Beranger, "Feminine Sphere in the Field of Movies," *Moving Picture World* [MPW], Aug. 2, 1919, 662; Mlle. Chic, "The Dual Personality of Cleo Madison," *Moving Picture Weekly,* July 1, 1916, 24; Henry MacMahon, "Women Directors of Plays and Pictures," *LHJ,* Dec. 1920, 140; E. Leslie Gilliams, "Will Woman's Leadership Change the Movies?" *Illustrated World,* Feb. 1923, 38, 860, 956; Wendy Holliday, "Hollywood's Modern Women: Screenwriting, Work Culture, and Feminism, 1910–1940" (PhD diss., New York University, 1995); Hugh C. Weir, "Behind the Scenes with Lois Weber," *Moving Picture Weekly,* July 3, 1915, 28.

2. Henry MacMahon, "Women Directors of Plays and Pictures," 140.

3. From *Wid's Year Book* (Hollywood: Wid's Film and Film Folks, 1918, 1919–20, 1920–21, and 1921–22).

4. Amy L. Unterburger, ed., *The St. James Women Filmmakers Encyclopedia: Women on the Other Side of the Camera* (Boston: Visible Ink, 1999), xix–xx, 306–8, 346–49, 352–54, 342–44, 390–91, 250–52; Elissa J. Rashkin, *Women Filmmakers in Mexico: The Country of Which We Dream* (Austin: University of Texas Press, 2001), 35–40.

5. Charles S. Dunning, "The Gate Women Don't Crash: The Story of Lois Weber, Famous Film Director, and Why There Aren't More like Her," *Liberty,* May 14, 1927, 33.

6. A few female directors were active between the end of the silent era and the 1970s, when a significant number of women began to appear among mainstream directors again. Ida Lupino, who directed films from 1949 to 1966, was by far the most successful. Virginia Van Upp was a mainstream producer (Columbia Pictures) active between 1934 and 1952, but she was the only woman producer working at a major studio at that time. A handful of other female producers were confined to small independents, and most made avant-garde films with limited distribution. See Ally Acker, *Reel Women* (New York: Continuum, 1991).

7. Judith Mayne, *Directed by Dorothy Arzner* (Bloomington: Indiana University Press, 1994); Alison McMahan, *Alice Guy Blaché: Lost Visionary of the Cinema* (New York: Continuum, 2002); Kay Armatage, *The Girl from God's Country: Nell Shipman and the Silent Cinema* (Toronto: University of Toronto Press, 2003). Shelley Stamp's biography on Lois Weber is forthcoming.

8. Figures based on *Film Daily Yearbook of Motion Pictures* (1951), 90; repr. in Richard Koszarski, *An Evening's Entertainment: The Age of the Silent Feature Picture, 1915–1928* (Berkeley: University of California Press, 1994), 26.

9. Janet Staiger, "Authorship Approaches," in *Authorship and Film*, ed. David A. Gerstner and Janet Staiger (New York: Routledge, 2003), 29.

10. *Wid's Film Yearbook* (1919–20), n.p.

11. Amelie Hastie, "Circuits of Memory and History: The Memories of Alice Guy Blaché," in *A Feminist Reader in Early Cinema*, ed. Jennifer M. Bean and Diane Negra (Durham, NC: Duke University Press, 2002), 29–59; Jane M. Gaines, "Of Cabbages and Authors," in ibid., 103.

12. Nancy F. Cott, *The Grounding of Modern Feminism* (New Haven, CT: Yale University Press, 1987), 16–22.


PROLOGUE. "THE GREATEST ELECTRICAL NOVELTY IN THE WORLD"

1. *New York Mail and Express*, April 24, 1896, 12, as quoted in Charles Musser, *The Emergence of Cinema: The American Screen to 1907* (Berkeley: University of California Press, 1990), 116.

2. Advertisement in the *New York Clipper*, quoted in Musser, *Emergence of Cinema*, 124.

3. Reproduced in Musser, *Emergence of Cinema*, 126.

4. Charles Edward Hastings, "Cinematic Beginnings," MPW, March 26, 1927, 289–99.

5. Musser, *Emergence of Cinema*, 20–32, 40–41.

6. Ibid., 48–53.

7. Ruth Oldenziel, *Making Technology Masculine: Men, Women, and Modern Machines in America, 1870–1945* (Amsterdam: Amsterdam University Press, 1999), 10.

8. Ibid., 157–58.

9. Ruth Oldenziel, "Gender and the Meanings of Technology: Engineering in the U.S., 1880–1945" (PhD diss., Yale University, 1992), 31, 40–42.

10. See Virginia Scharff, *Taking the Wheel: Women and the Coming of the Motor Age* (New York: Free Press, 1991); and Julie Wosk, *Women and the Machine: Representations from the Spinning Wheel to the Electronic Age* (Baltimore: Johns Hopkins University Press, 2001).

11. Oldenziel, *Making Technology Masculine*, 11.

12. Musser, *Emergence of Cinema*, 54–60.

13. Gordon Hendricks, *The Edison Motion Picture Myth* (Berkeley: University of California Press, 1961), 110.

14. Tino Balio, "A Novelty Spawns Small Businesses, 1894–1908," in *The American Film Industry*, ed. Tino Balio, rev. ed. (Madison: University of Wisconsin Press, 1985), 7–10.

15. David Puttnam, with Neil Watson, *Movies and Money* (New York: Knopf, 1998), 21, 40.

16. Benjamin B. Hampton, *A History of the American Film Industry from Its Beginnings to 1931* (New York: Dover, 1970), 12, 21.

17. Fred F. Balshofer and Arthur C. Miller, *One Reel a Week* (Berkeley: University of California Press, 1967), 13–15.

18. Puttnam and Watson, *Movies and Money*, 40–41, 51.

19. Barbara Drygulski Wright, Myra Marx Ferree, Gail O. Mellow, Linda H. Lewis, Mara-Luz Daza Sampler, Robert Asher, and Kathleen Claspell, eds., *Women, Work, and Technology* (Ann Arbor: University of Michigan Press, 1987), 16–17.

20. Wendy Gamber, *The Female Economy: The Millinery and Dressmaking Trades, 1860–1930* (Urbana: University of Illinois Press, 1997), 30.

21. Susan Coultrap-McQuinn, *Doing Literary Business: American Writers in the Nineteenth Century* (Chapel Hill: University of North Carolina Press, 1990), 36–46.

22. See, e.g., G. W. Bitzer, *Billy Bitzer: His Story* (New York: Farrar, Straus, and Giroux, 1973); Balshofer and Miller, *One Reel a Week*; and oral histories in Kevin Brownlow, *The Parade's Gone By* (Berkeley: University of California Press, 1968).

23. Janet Wasko, *Movies and Money: Financing the American Film Industry* (Norwood, NJ: Ablex, 1982), xix.

24. Hampton, *History of the American Film Industry*, 22–24; Musser, *Emergence of Cinema*, 168–69, 177, 290, 484.

25. Brownlow, *The Parade's Gone By*, 216–17.

26. Albert E. Smith, "The Beginning," 170, from the Albert E. Smith Collection, Special Collections, UCLA.

27. Iris Barry, *D. W. Griffith: American Film Master* (Garden City, NY: Doubleday, 1965), 9.

28. For the "cameraman system" argument see Janet Staiger, "The Hollywood Mode of Production to 1930," in *The Classical Hollywood Cinema: Film Style and Mode of Production to 1960*, by David Bordwell, Janet Staiger, and Kristin Thompson (New York: Columbia University Press, 1985), 116–17; for the "collaborative system" see Charles Musser, "Pre-Classical American Cinema: Its Changing Modes of Production," in *Silent Film*, ed. Richard Abel (New Brunswick, NJ: Rutgers University Press, 1996), 85–108.

29. In 1994 *American Cinematographer* published a membership list of the American Society of Cinematographers from 1919 to 1994. Active members: 4 potentially female of 571 total (Terry K. Meade, Brianna Murphy, Ariel L. Varges, Zoli Vidor); associate members: 2 female of 272 total (Sharon O'Brien and Stacy O'Brien); honorary members: 1 female of 25 total (Barbara Prevedel); see *American Cinematographer* 75, no. 8 (Aug. 1994): 69–90.

30. Quoted in Judy Wajcman, "Reflections on Gender and Technology Studies: In What State Is the Art?" *Social Studies of Science* 30, no. 3 (June 2000): 451.

31. Catherine Weed Barnes, "Woman's Work: A Woman to Women," *American Amateur Photographer* 2, no. 5 (May 1890): 185–88; Emile Colston, "Female Employment in Photography," *Photographic News* 32, no. 1547 (April 27, 1888): 266; C. Jane Gover, *The Positive Image: Women Photographers in Turn of the Century America* (Albany: State University of New York Press, 1988), 5–7, 14–17; "Photography

as a Field of Employment," *Wilson's Photographic Magazine*, Aug. 20, 1892, 31–32; "Women Who Press the Button," *NYT*, Oct. 1, 1893, 18; Frances Benjamin Johnston, "The Foremost Women Photographers of America," parts 1–7, *LHJ*, May 1901–Jan. 1902; "Photography as a Field of Employment," 11.

32. Catherine Weed Barnes, "Why Ladies Should Be Admitted to Membership in Photographic Societies," *American Amateur Photographer* 1, no. 6 (Dec. 1889): 233–34.

33. Gover, *The Positive Image*, 5–7, 14–17.

34. *American Cinematographer* 5, no. 1 (April 1924): 23.

35. See Ava Baron, ed., *Work Engendered: Toward a New History of American Labor* (Ithaca, NY: Cornell University Press, 1991); Peter G. Filene, *Him/Her/Self: Sex Roles in Modern America*, 2nd ed. (Baltimore: Johns Hopkins University Press, 1986), 140–42; E. Anthony Rotundo, *American Manhood: Transformation of Masculinity from the Revolution to the Modern Era* (New York: Basic Books, 1993), 248–55.

36. Theodore Roosevelt, *The Strenuous Life: Essays and Addresses* (New York: Century, 1900), www.bartleby.com/58/1.html.

37. "The Cameraman," *American Cinematographer* 1, no. 1 (Nov. 1920): 1; repr. in *American Cinematographer* 50, no. 1 (Jan. 1969): 49.

38. Filene, *Him/Her/Self*, 71; also see Bitzer, *Billy Bitzer*; Balshofer and Miller, *One Reel a Week*; and oral histories in Brownlow, *The Parade's Gone By*.

39. Staiger, "Hollywood Mode of Production to 1930," 116–17.

40. See Lauren Rabinovitz, *For the Love of Pleasure: Women, Movies, and Culture in Turn-of-the-Century Chicago* (New Brunswick, NJ: Rutgers University Press, 1998), 7–8; also see Giuliana Bruno, *Streetwalking on a Ruined Map: Cultural Theory and the City Films of Elvira Notari* (Princeton, NJ: Princeton University Press, 1993).

41. See Laura Mulvey, "Visual Pleasure and Narrative Cinema," *Screen* 16, no. 3 (autumn 1975), repr. in *Feminism and Film Theory*, ed. Constance Penley (New York: Routledge, 1988), 57–68.

42. *American Cinematographer* 50, no. 1 (Jan. 1969): 50.

43. Gover, *The Positive Image*.

44. Anthony Slide, *Early Women Directors* (New York: A. S. Barnes, 1977), 10; Katherine Synon, "Francelita Billington: Who Can Play Both Ends of a Camera against the Middle," *Photoplay*, Dec. 1914, 58–60; "All Ready! Now the Villain Enters! Camera!" *Photoplay*, Nov. 1915, 91; "This Is the New Fall Style in Camera 'Men,'" *Photoplay*, Oct. 1916, 103.

45. "The History of the ASC," www.theasc.com/clubhouse/inside/aschist_index.htm (2004); "A.S.C. Occupies New Offices," *American Cinematographer* 5, no. 7 (Oct. 1924): 4–5; "Camera Men's Static Club a Flourishing Body," *MPN*, April 3, 1915, 172.

46. Oldenziel, *Making Technology Masculine*; Margaret W. Rossiter, *Women Scientists in America: Struggles and Strategies to 1940* (Baltimore: Johns Hopkins University Press, 1982); Barbara F. Reskin and Polly A. Phipps, "Women in Male-

Dominated Professional and Managerial Occupations," in *Women Working: Theories and Facts in Perspective*, ed. Ann H. Stromberg and Shirley Harkness (Mountain View, CA: Mayfield, 1988); and Barbara F. Reskin and Patricia A. Roose, *Job Queues, Gender Queues: Explaining Women's Inroads into Male Occupations* (Philadelphia: Temple University Press, 1990).

47. Alice Kessler-Harris, *Out to Work: A History of Wage-Earning Women in the United States* (New York: Oxford University Press, 1982), 70–71, 137–39; Julie A. Matthaei, *An Economic History of Women in America: Women's Work, the Sexual Division of Labor, and the Development of Capitalism* (New York: Schocken, 1982), 141–42.

48. C. Frances Jenkins, *Animated Pictures* (Washington, DC: C. Frances Jenkins, 1898), 55–56.

49. Musser, *Emergence of Cinema*, 145.

50. "The Art of Moving Picture Photography," *Scientific American*, April 17, 1897, 248–50.

51. Reese V. Jenkins, *Images and Enterprise: Technology and the American Photographic Industry, 1839–1925* (Baltimore: Johns Hopkins University Press, 1975), 19, 27.

52. Colston, "Female Employment in Photography," 26.

53. Adelaide Steel, "Mrs. Christmas; A Retoucher's Story," *Anthony's Photographic Bulletin*, Dec. 28, 1889, 745–48.

54. Jenkins, *Images and Enterprise*, 112–15.

55. Harriet Bradley, *Men's Work, Women's Work* (Minneapolis: University of Minnesota Press, 1989), 9.

56. S. M. Spedon, *How and When Moving Pictures Are Made by the Vitagraph Company of America* (c. 1912), not paginated, Florence Lawrence Collection, LACMNH; Kristin Thompson, "Film Style and Technology to 1930," in *The Classical Hollywood Cinema: Film Style and Mode of Production to 1960*, by David Bordwell, Janet Staiger, and Kristin Thompson (New York: Columbia University Press, 1985), 278–79.

57. Spedon, *Vitagraph Company of America*.

58. Eugene Dengler, "Pathé Frères' New Jersey Factory," *Nickelodeon*, April 1, 1910, 181–82; Spedon, *Vitagraph Company of America*.

59. *MPW*, Oct. 8, 1910, 817; Spedon, *Vitagraph Company of America*.

60. Spedon, *Vitagraph Company of America*; Dengler, "Pathé Frères," 182.

61. James E. McQuade, "Making 'Selig' Pictures," *Film Index*, Nov. 20, 1909, 4–6; repr. in Kalton C. Lahue, *The Motion Picture Pioneer: The Selig Polyscope Company* (New York: A. S. Barnes, 1973), 61.

62. Eleanor Francis Humphrey, "The Creative Woman in Motion Picture Production" (master's thesis, University of Southern California, 1970), 123.

63. Dengler, "Pathé Frères," 182.

64. "The Crystal Film Company, New York," *MPW*, March 16, 1912, 946–47.

65. Musser, *Emergence of Cinema*, 116, 33, 36.

66. "Coloring Song Slides," *MPW*, Sept. 18, 1909, 375; also see "Slide Colorists Scarce," *MPW*, Oct. 16, 1909, 527.

67. "Trade Notes," *MPW*, Aug. 11, 1907, 406.

68. David S. Hulfish, "Colored Films Today," *Nickelodeon*, Jan. 1909, 15.

69. Henri Destynn, "How Pathé Films Are Colored," *Nickelodeon*, March 1, 1910, 121.

70. Thompson, "Film Style and Technology to 1930," 276, 279.

71. Quote from Bradley, *Men's Work, Women's Work*, 1–2.

72. Wm. H. Jackson, "A Casual Visit to the Vitagraph Plant," *MPW*, Sept. 17, 1910, 624.

73. Thompson, "Film Style and Technology to 1930," 276–79; Spedon, *Vitagraph Company of America*.

74. Dengler, "Pathé Frères," 182.

75. From a list of employees dated Oct. 6 and Oct. 17, 1917, George Kleine Collection, MD-LC.

76. McQuade, "Making 'Selig' Pictures," 61.

77. Musser, *Emergence of Cinema*, 6.

78. Eric Smoodin, "Attitudes of the American Printed Medium toward the Cinema, 1894–1908," unpublished paper, University of California, Los Angeles, 1979, cited in Richard deCordova, *Picture Personalities: The Emergence of the Star System in America* (repr., Champaign: University of Illinois Press, 2001), 46.

### CHAPTER 1. A QUIET INVASION

1. "A Woman Invades the American Moving Picture Field," *MPW*, Nov. 13, 1909, 680; "More European Films for the American Market," *MPW*, Feb. 19, 1910, 261; Lee Grieveson, *Policing Cinema: Movies and Censorship in Early-Twentieth-Century America* (Berkeley: University of California Press, 2004), 100. Lauren Rabinovitz identifies Chicago as the national film distribution center at this time; see Lauren Rabinovitz, *For the Love of Pleasure: Women, Movies, and Culture in Turn-of-the-Century Chicago* (New Brunswick, NJ: Rutgers University Press, 1998, 170.

2. "With the Film Men," *New York Dramatic Mirror*, April 12, 1913, 26.

3. Few women entered the distribution arm of the American film industry; see "A Woman Exchange Manager," *MPW*, Aug. 19, 1911, n.p.; "A Successful Woman Sales Manager," *MPW* [c. 1912], n.p.

4. Janet Staiger, "Blueprints for Feature Films: Hollywood's Continuity Scripts," in *The American Film Industry*, ed. Tino Balio, rev. ed. (Madison: University of Wisconsin Press, 1985), 175; Catherine E. Kerr, "Incorporating the Star: The Intersection of Business and Aesthetic Strategies in Early American Film," *Business History Review* 64 (autumn 1990): 388.

5. Charles Musser, *The Emergence of Cinema* (Berkeley: University of California Press, 1990), 6.

6. Ibid., 312; Robert Allen, "The Movies in Vaudeville: Historical Context of the Movies as Popular Entertainment," in *The American Film Industry*, ed. Tino Balio, rev. ed. (Madison: University of Wisconsin Press, 1985), 72–75.

7. Musser, *Emergence of Cinema*, 314, 320–22, 325, 335; John Izod, *Hollywood and the Box Office, 1895–1986* (New York: Columbia University Press, 1988), 8–9; Miriam Hansen, *Babel and Babylon: Spectatorship in American Silent Film* (Cambridge, MA: Harvard University Press, 1991), 44.

8. Musser, *Emergence of Cinema*, 366.

9. Allen, "The Movies in Vaudeville," 75.

10. Kathryn Fuller-Seeley and Karen Ward Mahar, "Exhibiting Women: Gender, Showmanship, and the Professionalization of Film Exhibition in the United States, 1900–1930," in *Women Film Pioneers Sourcebook*, ed. Jane Gaines (Durham, NC: Duke University Press, forthcoming).

11. Izod, *Hollywood and the Box Office*, 15; H. T. Lewis, "Distributing Motion Pictures," *Harvard Business Review* 7, no. 3 (April 1929): 267; Allen, "The Movies in Vaudeville," 69.

12. "The Nickelodeon," *MPW*, May 4, 1907, 146.

13. Russell Merritt, "Nickelodeon Theaters, 1905–1914: Building an Audience for the Movies," in *The American Film Industry*, ed. Tino Balio, rev. ed. (Madison: University of Wisconsin Press, 1985), 86.

14. For women in the nickelodeon audience see Rabinovitz, *For the Love of Pleasure*; Kathryn Fuller-Seeley, *At the Picture Show: Small-Town Audiences and the Creation of Movie Fan Culture* (Washington, DC: Smithsonian Press, 1996); Judith Mayne, "Immigrants and Spectators," *Wide Angle* 5, no. 2 (1982): 32–41; Elizabeth Ewen, "City Lights: Immigrant Women and the Rise of the Movies," *Signs* 5, no. 3 (spring 1980): S45-S65; Kathy Peiss, *Cheap Amusements: Working Women and Leisure in Turn-of-the-Century New York* (Philadelphia: Temple University Press, 1986); Hansen, *Babel and Babylon*, 62, 104, 145; for general audiences see Robert C. Allen, "Motion Picture Exhibition in Manhattan, 1906–1912: Beyond the Nickelodeon," in *Film before Griffith*, ed. John Fell (Berkeley: University of California Press, 1983), 162–75; Merritt, "Nickelodeon Theaters, 1905–1914," 83–102; Garth S. Jowett, "The First Motion Picture Audiences," in *Film before Griffith*, ed. John Fell (Berkeley: University of California Press, 1983), 196–206; for rural and small-town audiences see George Potamianos, "Hollywood in the Hinterlands: Mass Culture in Two California Communities, 1896–1936" (PhD diss., University of Southern California, 1998).

15. "The Masses Are Being Educated," *MPW*, July 4, 1908, 7.

16. "Why the 'Nickels' are Popular," *MPW*, Oct. 25, 1907, 541; Musser, *Emergence of Cinema*, 425.

17. "The Nickelodeon," *MPW*, May 4, 1907, 146.

18. "Picture Shows Popular in the 'Hub,'" *MPW*, May 16, 1908, 433.

19. Benjamin B. Hampton, *A History of the American Film Industry from Its Beginnings to 1931* (New York: Dover, 1970), 59.

20. Fuller-Seeley and Mahar, "Exhibiting Women."

21. Exceptions were few for the entire silent era. See "Rochester Has Real

Woman Operator," *MPW*, Feb. 1, 1919, 618; Editorial, "Stop Swearing at the Operator," *MPW*, Feb. 1, 1919, 626; for WWI-era recruitment of female projectionists see Cal York, Plays and Players, *Photoplay*, Nov. 1918, 92.

22. F. H. Richardson, *Motion Picture Handbook: A Guide for Managers and Operators of Motion Picture Theaters* (New York: Moving Picture World, c. 1912); Musser, *Emergence of Cinema*, 441–43.

23. Lewis Jacobs, *The Rise of the American Film: A Critical History* (New York: Columbia University Press, 1968), 56–57; Musser, *Emergence of Cinema*, 441–43; "Correspondence School for Operators," *MPW*, April 30, 1910, 693; Eileen Bowser, *The Transformation of Cinema, 1907–1915* (Berkeley: University of California Press, 1994), 12–13.

24. James R. Cameron, *Instruction of Disabled Men in Motion Picture Projection: An Elementary Text Book* (New York: Red Cross Institute for Crippled and Disabled Men, 1919), vi–vii.

25. Robert Osborne Baker, "The International Alliance of Theatrical Stage Employees and Moving Picture Operators of the United States and Canada" (PhD diss., University of Kansas, 1933), 24.

26. For department store clerks see Susan Porter Benson, *Counter Cultures: Saleswomen, Managers, and Customers in American Department Stores, 1890–1940* (Chicago: University of Illinois Press, 1986), 23, 128, 182–85.

27. David S. Hulfish, "Economy in Picture Theater Operation," *Nickelodeon*, Jan. 1, 1910, 15–16; Samuel L. Rothapfel, "Pertinent Facts about Box Offices and the Young Woman Who Sells the Ticket," *MPW*, July 17, 1920, 311; Ina Rae Hark, "The 'Theater Man' and 'The Girl in the Box Office': Gender in the Discourse of Motion Picture Theatre Management," *Film History* 6 (1994): 183.

28. Hulfish, "Economy in Picture Theater Operation," 15–16; for comparative wages see Richard Koszarski, *An Evening's Entertainment: The Age of the Silent Feature Picture, 1915–1928* (Berkeley: University of California Press, 1994), 14; Benson, *Counter Cultures*, 182–85; Elizabeth Beardsley Butler, *Women and the Trades, Pittsburg, 1907–8* (Pittsburgh: University of Pittsburg Press, 1984), 47, 114–21, 155, 163, 212–23, 229, 243, 261, 263, 267, 279–80, 304.

29. Given the number of women in vaudeville, it is no surprise that women were well represented; see "Where the Profits Go—No. 1: The Lady Vocalist," *MPW*, April 30, 1910, 693; Hulfish, "Economy in Picture Theater Operation," 15–16; Merritt, "Nickelodeon Theaters, 1905–1914," 5; Bowser, *Transformation of Cinema*, 15–16; Hans Leigh, "How the Vogue of the Motion Picture May Be Preserved," *MPW*, Aug. 8, 1908, 101; Allen, "Motion Picture Exhibition in Manhattan," 171–73; for women in vaudeville see Elsie Janis, *So Far, So Good!* (New York: E. P. Dutton, 1932); Shirley Staples, *Male-Female Comedy Teams in American Vaudeville, 1836–1932* (Ann Arbor, MI: UMI Research Press, 1981); Charles and Louise Samuels, *Once upon a Stage: The Merry World of Vaudeville* (New York: Dodd, Mead, 1974); Charles W. Stein, ed., *American Vaudeville as Seen by Its Contemporaries*

(New York: Knopf, 1984); and Robert W. Snyder, *The Voice of the City: Vaudeville and Popular Culture in New York* (New York: Oxford University Press, 1989).

30. Robert Sklar, *Movie-Made America: A Social History of American Movies* (New York: Random House, 1975), 30; Lary May, *Screening Out the Past: The Birth of Mass Culture and the Motion Picture Industry* (Chicago: University of Chicago Press, 1980), 44; Merritt, "Nickelodeon Theaters, 1905–1914," 89; Roy Rosenzweig, *Eight Hours for What We Will: Workers and Leisure in an Industrial City* (New York: Cambridge University Press, 1983), 193; Bowser, *Transformation of Cinema*, 10; quote from Hampton, *History of the American Film Industry*, 47.

31. This conclusion is based on readings published in MPW from 1907 to 1911; examples cited are from MPW, April 20, 1907, 102; MPW, June 26, 1909, 875; MPW, Aug. 7, 1909, 196. Eileen Bowser also makes this observation, and as she argues, a statistical study on female nickelodeon ownership needs to be done; see Bowser, *Transformation of Cinema*, 45–47.

32. Fuller-Seeley and Mahar, "Exhibiting Women."

33. Angel Kwolek-Folland, *Incorporating Women: A History of Women and Business in the United States* (New York: Twayne, 1998), 14–15.

34. Christine Stansell, *City of Women: Sex and Class in New York, 1789–1860* (Urbana: University of Illinois Press, 1987), 14–15; for more examples of female proprietors of commercialized leisure see Linda Kerber, *Women of the Republic: Intellect and Ideology in Revolutionary America* (New York: Norton, 1980), 139–55, 154; Suzanne Lebsock, *The Free Women of Petersburg: Status and Culture in a Southern Town, 1784–1860* (New York: Norton, 1984), 176; Julie A. Matthaei, *An Economic History of Women in America: Women's Work, the Sexual Division of Labor, and the Development of Capitalism* (New York: Schocken, 1982), 65–68.

35. Kathleen Ann Morgan, "Of Stars and Standards: Actress-Managers of Philadelphia and New York, 1855–1860" (PhD diss., University of Illinois, Urbana-Champaign, 1988), 7–8; Alfred L. Bernheim, *The Business of the Theatre* (New York: Actor's Equity Association, 1932), 8; Faye E. Dudden, *Women in the American Theatre: Actresses and Audiences, 1790–1870* (New Haven, CT: Yale University Press, 1994): 123–24.

36. Kwolek-Folland, *Incorporating Women*, 57.

37. "The Actorless Theater," *Current Literature*, Nov. 1909, 556; "Recipe," MPW, May 4, 1907, 146.

38. Koszarski, *An Evening's Entertainment*, 54–59; Hansen, *Babel and Babylon*, 43–44; quote from "The Nickelodeon," MPW, May 4, 1907, 146.

39. Leigh, "How the Vogue of the Motion Picture May be Preserved," 101.

40. According to Lewis Jacobs, in 1906 films were changed twice a week, and by 1907 they were changed daily. See Jacobs, *Rise of the American Film*, 5.

41. Gene Gauntier, "Blazing the Trail," TS, 11–12 (Gauntier's "Blazing the Trail" exists in two forms: a typescript, archived at the Museum of Modern Art's Film Study Center, and a series of articles published in the *Woman's Home Compan-*

*ion* from 1928 to 1929. Because I cite from both versions, I will distinguish between them by labeling each citation TS or *WHC.*); Janet Staiger, "Hollywood Mode of Production to 1930," in *The Classical Hollywood Cinema: Film Style and Mode of Production to 1960*, by David Bordwell, Janet Staiger, and Kristin Thompson (New York: Columbia University Press, 1985), 117–19.

42. Harvey Douglass, "Movie Stock Stars of 1907 Got $30 per Week," *Brooklyn Daily Eagle*, Feb. 21, 1933, M2.

43. "'Nickel Actors' in the Making," *MPW*, March 11, 1911, 274; Gerald D. McDonald, "Origin of the Star System," *Films in Review* 4, no. 9 (Nov. 1953): 450; Anthony Slide, "The Evolution of the Film Star," *Films in Review* 25, no. 10 (Dec. 1974): 592.

44. David S. Hulfish, *Cyclopedia of Motion-Picture Work* (Chicago: School of Correspondence, 1911), 108.

45. Mabel Rhea Dennison, "Actresses in Moving Picture Work," *Nickelodeon*, Jan. 1909, 19.

46. Florence Lawrence, "Growing Up in the Movies," *Photoplay*, Nov. 1914, 28, 40; Larry Lee Holland, "Florence Lawrence," *Films in Review* 31, no. 7 (Aug.–Sept. 1980): 383, 387–88.

47. Gauntier, *Blazing the Trail*, TS, 3–8.

48. Net income at Vitagraph grew from $38,860 in 1905, to $279,814 in 1908 (Albert E. Smith Collection, SC-UCLA); Gauntier, "Blazing the Trail," WHC, Jan. 1929, 13; Janet Staiger, "Mass-Produced Photoplays: Economic and Signifying Practices in the First Years of Hollywood," *Wide Angle* 4, no. 3 (Dec. 27, 1980): 16.

49. Hampton, *History of the American Film Industry*, 32; Janet Staiger, "Hollywood Mode of Production to 1930," 117.

50. Edward W. Townsend, "Picture Plays," *Outlook*, Nov. 27, 1909, 704.

51. Harvey Douglass, "Movie Stars Got Start on Flatbush Lots," *Brooklyn Daily Eagle*, Feb. 17, 1933, M2.

52. Stuart Blackton to Albert E. Smith, Feb. 15, 1908, Albert E. Smith Collection, SC-UCLA; J. Stuart Blackton, "The World in Motion (A Complete History of Moviedom)," n.p., Albert E. Smith Collection, SC-UCLA; Jacobs, *Rise of the American Film*, 57. The expression "doubling in brass" was also used by Carlton Miles, dramatic editor of the *Minneapolis Journal*, to describe the dramatic tent show, in which "the prize packages of candy are sold by the juvenile, walking up and down the aisles, wearing his stage costume and makeup . . . [T]he leading man does a comic monologue and the ingenue sings a popular ballad. The actors double in brass, adding their efforts to swell the volume of the orchestra" (*Theatre Arts Monthly* 10 [Oct. 1926]: 685, quoted in Bernheim, *Business of the Theatre*, 98).

53. Quote from Douglass, "Movie Stars Got Start on Flatbush Lots"; unidentified newspaper clipping in scrapbook, Oct. 15, 1933, Florence Lawrence Collection, LACMNH; Anthony Slide, *Early American Cinema* (New York: A. S. Barnes, 1970), 30–31; Gauntier, "Blazing the Trail," TS, 38.

54. Augusta Cary, "An Interview with Miss Blanche Lasky," *Photoplay*, Dec. 1914, 139–40.

55. Frances Marion, *Off with Their Heads!* (New York: Macmillan, 1972), 2, 8–9, 12–13; Dewitt Bodeen, "Frances Marion," *Films in Review* 20, no. 2 (Feb. 1969): 76–77, 83–88.

56. Kevin Brownlow, *The Parade's Gone By* (New York: Knopf, 1968), 275–76.

57. Lillian Gish, interview by Raymond Daum, Dec. 12, 1978, transcript, CUOHC, 11–12.

58. Jacobs, *Rise of the American Film*, 59–61; also see Terry Ramsaye, *A Million and One Nights* (New York: Simon and Schuster, 1926), 443.

59. McDonald "Origin of the Star System," 455.

60. Jack Cohn, "Fourteenth Street," *Film Daily*, Feb. 28, 1926, 57.

61. Gauntier, "Blazing the Trail," TS, 51–54.

62. Gauntier, "Blazing the Trail," WHC, Nov. 1928, 169–70.

63. Staiger, "Hollywood Mode of Production to 1930," 117.

64. Jacobs, *Rise of the American Film*, 59.

65. Linda Arvidson, *When the Movies Were Young* (New York: Dover, 1969), 15; Jacobs, *Rise of the American Film*, 59–60, 95–98; Musser, *Emergence of Cinema*, 492; Staiger, "Hollywood Mode of Production to 1930," 117.

66. Richard Koszarski, "The Years Have Not Been Kind to Lois Weber," *Village Voice*, Nov. 11, 1975, 140; Gauntier, "Blazing the Trail," TS, 20–34.

67. Hulfish, *Cyclopedia of Motion-Picture Work*, 95.

68. Contests for the public tapered off with the rise of the feature film and the professionalization of film writing, but those that remained, which became entirely promotional, also became extremely lucrative. See "St. Louis Girl Wins Thanhouser $10,000 Prize," *MPN*, March 13, 1915, 46; "Fortune for an Idea," *Photoplay*, Feb. 1917, 34; Staiger, "Hollywood Mode of Production to 1930," 118; Tino Balio, "A Novelty Spawns Small Businesses, 1894–1908," in *The American Film Industry*, ed. Tino Balio, rev. ed. (Madison: University of Wisconsin Press, 1985), 22. Whether this is the same Florence Turner that was gaining fame as "The Vitagraph Girl" at the Brooklyn Vitagraph studio is impossible to know; see *MPW*, Oct. 2, 1909, n.p.

69. *MPW*, Nov. 19, 1910, 1167.

70. Ibid. Lewis Jacobs has argued that five dollars to fifteen dollars was good pay; see Jacobs, *Rise of the American Film*, 61.

71. *MPW*, Aug. 5, 1911, 297; *MPW*, Dec. 11, 1911, 1063.

72. Wendy Holliday, "Hollywood's Modern Women: Screenwriting, Work Culture, and Feminism, 1910–1940" (PhD diss., New York University, 1995), 93–96, 107–9, 127–28.

73. Epes Winthrop Sargent, "The Photoplaywright," *MPW*, Aug. 9, 1913, 630.

74. Quoted in Holliday, "Hollywood's Modern Women," 161.

75. Gauntier, "Blazing the Trail," TS, 10–12.

76. Ibid., 20–21; Gauntier, "Blazing the Trail," WHC, Nov. 1928, 170.

77. Gauntier, "Blazing the Trail," TS, 36–37.

78. Gauntier, "Blazing the Trail," WHC, Nov. 1928, 26.

79. Ibid., 32–34.

80. Ibid., 2, 67; ibid., Dec. 1928, 132; Jan. 1929, 13, 56; Feb. 1929, 20.

81. Hulfish, *Cyclopedia of Motion-Picture Work*, 96.

82. Staiger, "Blueprints for Feature Films," 175–76; Gauntier, "Blazing the Trail," WHC, Nov. 1928, 15.

83. Kerr, "Incorporating the Star," 386.

84. "The Solax Company," MPW, Oct. 8, 1910, 812.

85. Interview with Alice Guy Blaché, in *The Lost Garden: The Life and Cinema of Alice Guy Blaché*, dir. Marquise Lepage, 53 minutes, National Film Board of Canada, 1996 (MOMA).

86. Alison McMahan, *Alice Guy Blaché: Lost Visionary of the Cinema* (New York: Continuum, 2002), 22.

87. Alice Guy Blaché, *The Memoirs of Alice Guy Blaché*, ed. Anthony Slide, trans. Roberta Blaché and Simone Blaché (Metuchen, NJ: Scarecrow Press, 1986), 16–24, 31, 40.

88. McMahan, *Alice Guy Blaché*, 98.

89. Ibid., 102, 27, 115–16, 36, 92.

90. Ibid., 55–57.

91. Ibid., 64–66.

92. Guy Blaché, *Memoirs*, 52–55; "Talking Pictures: The Gaumont Chronophone C.," MPW, March 27, 1909, 362; Lepage, *Lost Garden*.

93. Guy Blaché, *Memoirs*, 65; McMahan, *Alice Guy Blaché*, 70–71.

94. MPW, Feb. 8, 1913, 551–52, quoted in Janet Staiger, "Combination and Litigation: Structures of U.S. Film Distribution, 1896–1917," *Cinema Journal* 23, no. 2 (winter 1985): 50, 45–47.

95. Staiger, "Combination and Litigation," 50.

96. MPW, May 21, 1910, 822; Anthony Slide, *The American Film Industry: A Historical Dictionary* (New York: Limelight Editions, 1990), 221–22; Staiger, "Combination and Litigation," 51.

97. *Nickelodeon*, Nov. 15, 1910, 273–74; Musser, *Emergence of Cinema*, 485.

98. Henry Blaché to Messrs. Sussfeld, Lorsch, and Co., July 21, 1909; George Kleine to Henry Lorsch, Dec. 17, 27, 1909; Leon Gaumont to Sussfeld, Lorsch, and Co., Dec. 16, 1909; Leon Gaumont to Kleine, Dec. 21, 1909; Gaumont to Kleine, Jan. 19, 1910; Sussfeld, Lorsch, and Co. to Kleine, Jan. 24, 1910; Kleine to Gaumont, Jan. 27, 1910; George Kleine Papers, MD-LC.

99. H. Sussfeld to George Kleine, Feb. 18, 1910; Sussfeld, Lorsch, and Co. to Kleine, April 9, 1910; Sussfeld, Lorsch, and Co. to Kleine, June 17, 1910; Kleine to Henry Lorsch, June 18, 1910, George Kleine Papers, MD-LC; "The Truth of the Gaumont Rumors," MPW, June 25, 1910, 1094.

100. Guy Blaché, *Memoirs*, 62–63.

101. H. Schreier, Acct. and Auditor, Dec. 31, 1911, report, signed March 2, 1912, Albert E. Smith Collection, SC-UCLA; "Studio Efficiency: Scientific Management as Applied to the Lubin Western Branch by Wilbert Melville," *MPW*, Aug. 9, 1913, 624.

102. "The New Solax Plant," *MPN*, Sept. 21, 1912, 10–11; "New Solax Plant at Fort Lee," *MPN*, Sept. 14, 1912, 1061–62.

103. "The Solax Company," *MPW*, Oct. 8, 1910, 812. Strangely, Alice Guy Blaché did not even mention the dispute between Gaumont and the Patents Company in her memoirs. "Out of Oblivion: Alice Blaché," *Sight and Sound* 40, no. 3 (summer 1971): 153.

104. Advertisement, *MPW*, Oct. 8, 1910, 837; advertisement, *MPW*, Dec. 24, 1910, 1452.

105. "Who's Who in the Film Game," *Motography*, Oct. 12, 1912, 293.

106. Hampton, *History of the American Film Industry*, 46, 73–74.

107. Janet Staiger, "Seeing Stars," in *Stardom: Industry of Desire*, ed. Christine Gledhill (New York: Routledge, 1991), 9–10.

108. "Solax Quality," *MPW*, Feb. 18, 1911, 372.

109. Guy Blaché, *Memoirs*, 65.

110. Louis Reeves Harrison, "Studio Saunterings," *MPW*, June 15, 1912, 1007–10.

111. Hugh Hoffman, "New Solax Plant at Fort Lee," *MPW*, Sept. 14, 1912, 1061.

112. Advertisement, *MPW*, Dec. 10, 1910, 1330; see also *MPW*, Dec. 17, 1910, 1390.

113. This judgment is based on the following films viewed by the author at the MPBRSD-LC: *A House Divided* (c. 1910); *A Greater Love Hath No Man* (c. 1911); *The Girl in the Arm-Chair* (1912); *Canned Harmony* (1912); *The Detective's Dog* (c. 1913); *Brennan of the Moor* (1913); and *Matrimony's Speed Limit* (1913).

114. "The Vogue of Western and Military Dramas," *MPW*, Aug. 5, 1911, 271. One of Solax's very first films, *The Sergeant's Daughter*, featured "scenes of thousands of troops embarking on a man-of-war" (advertisement, *MPW*, Oct. 29, 1910, 1017).

115. Advertisement, *MPW*, April 29, 1911, 929; review of *Across the Mexican Line*, *MPW*, April 29, 1911, 970.

116. McMahan, *Alice Guy Blaché*, 229–31.

117. Ibid., 241.

118. Gaumont Scenario Collection, 1906 folder, file G0010, no. 3811, 1906, translated and quoted in McMahan, *Alice Guy Blaché*, 235–37.

119. McMahan, *Alice Guy Blaché*, 234–35.

Chapter 2. "To Get Some of the 'Good Gravy'" for Themselves

1. *MPW*, March 12, 1910, 365.

2. *St. Louis Times*, n.d., scrapbook, Florence Lawrence Collection, LACMNH.

3. "The IMP Leading Lady," *MPW*, April 2, 1910, 517; Gerald D. McDonald, "Origin of the Star System," *Films in Review* 9, no. 9 (Nov. 1953): 453–54; Janet

Staiger, "Seeing Stars," in *Stardom: Industry of Desire*, ed. Christine Gledhill (New York: Routledge, 1991), 3.

4. Richard deCordova, *Picture Personalities: The Emergence of the Star System in America* (1990; repr., Urbana: University of Illinois Press, 2001).

5. "Observations by Our Man About Town," MPW, May 21, 1910, 825.

6. For Florence Turner see "Picture Personalities," MPW, July 23, 1910, 187; and Lux Graphicus, "On the Screen," MPW, Oct. 8, 1910, 807. For Mary Pickford see MPW profile, Dec. 24, 1910, 1462. For Pearl White see "The New Magdalen," MPW, Nov. 26, 1910, 1240; and "Picture Personalities," MPW, Dec. 3, 1910, 1281.

7. deCordova, *Picture Personalities*, 7.

8. Hugh Hoffman, "Florence Turner Going to England," MPW, March 22, 1913, 1205.

9. Lewis E. Palmer, "The World in Motion," *Survey* 22 (June 5, 1909): 356, cited in Constance Balides, "Cinema under the Sign of Money: Commercialized Leisure, Economies of Abundance, and Pecuniary Madness, 1905–1915," in *American Cinema's Transitional Era : Audiences, Institutions, Practices*, ed. Charlie Keil and Shelley Stamp (Berkeley: University of California Press, 2004), 288.

10. Eileen Bowser, *The Transformation of Cinema, 1907–1915* (Berkeley: University of California Press, 1994), 87, 94, 96–97, 102.

11. Johnson Briscoe, "Photoplays versus Personality," *Photoplay*, March 1914, 39.

12. "Observations by Our Man About Town," MPW, Nov. 20, 1909, 714; Catherine Kerr, "Incorporating the Star: The Intersection of Business and Aesthetic Strategies in Early American Film," *Business History Review* 64 (autumn 1990): 392–97; Richard deCordova, "The Emergence of the Star System in America," in *Stardom: Industry of Desire*, ed. Christine Gledhill (New York: Routledge, 1991), 17–28.

13. Alfred L. Bernheim, *Business of the Theatre* (New York: Actors' Equity Association, 1932), 27.

14. Ibid., 27–28; Benjamin McArthur, *Actors and American Culture, 1880–1920* (Philadelphia: Temple University Press, 1984), 10, 24–25.

15. Bernheim, *Business of the Theatre*, 29–30; McArthur, *Actors and American Culture*, 18.

16. Kathleen Anne Morgan, "Of Stars and Standards: Actress-Managers of Philadelphia and New York, 1855–1860" (PhD diss., University of Illinois, 1988), 7–8; Alice M. Robinson, Vera Mowry Roberts, and Milly S. Barranger, eds., *Notable Women in the American Theatre: A Biographical Dictionary* (New York: Greenwood Press, 1989), 207–8, 230–32, 282–83, 537–39, 925–97. Regarding Mrs. Drew's large influence on a generation of American actors, see A. Frank Stull, "Where Famous Actors Learned Their Art," *Lippincott's Monthly Magazine*, March 1905, 372–79; obituary of Mrs. John Drew, *New York Dramatic Mirror*, Sept. 11, 1897, 15; untitled clipping, *New York Spirit of the Times*, July 24, 1888, Robinson Locke Collection, ser. 2, v. 83, p. 355, NYPL-PA.

17. Bernheim, *Business of the Theatre*, 26–27; Robinson, Roberts, and Barranger, *Notable Women in the American Theatre*, 207–11.

18. Robinson, Roberts, and Barranger, *Notable Women in the American Theatre*, 207–11

19. Weldon B. Durham, ed., *American Theatre Companies, 1888–1930* (New York: Greenwood Press, 1987), 425–27.

20. William H. Henry, "Cleo the Craftswoman," *Photoplay*, Jan. 1916, 110.

21. Anthony Slide, "Evolution of the Film Star," *Films in Review* 25, no. 10 (Dec. 1974): 592; McArthur, *Actors and American Culture*, 4–5.

22. David Nasaw, *Going Out: The Rise and Fall of Public Amusement* (New York: Basic Books, 1993), 36–37.

23. Acton Davies, "What I Don't Know about Vaudeville," *Variety*, Dec. 16, 1905, 1.

24. Charles and Louise Samuels, *Once upon a Stage: The Merry World of Vaudeville* (New York: Dodd, Mead, 1974), 50–52.

25. Robert C. Allen, "The Movies in Vaudeville: Historical Context of the Movies as Popular Entertainment," in *The American Film Industry*, ed. Tino Balio, rev. ed. (Madison: University of Wisconsin Press, 1985), 65, 68–69, 71; Charles W. Stein, ed., *American Vaudeville as Seen by Its Contemporaries* (New York: Knopf, 1984), 4–5.

26. *Variety*, March 31, 1906, 11; *Variety*, May 5, 1906, 5.

27. Susan A. Glenn, *Female Spectacle: The Theatrical Roots of Modern Feminism* (Cambridge, MA: Harvard University Press, 2000), 13.

28. William Gould, "Vaudeville versus Musical Comedy," *Variety*, Dec. 14, 1907, repr. in Stein, *American Vaudeville*, 78.

29. Edwin Milton Royale, "The Vaudeville Theatre," *Scribner's Magazine*, Oct. 1899, 485–89, repr. in Stein, *American Vaudeville*, 26–27.

30. Paul C. Spehr, *The Movies Begin: Making Movies in New Jersey, 1887–1920* (Newark, NJ: Newark Museum, 1977), 148; McArthur, *Actors and American Culture*, 5; scrapbook, Florence Turner Collection, LACMNH; scrapbook, Florence Lawrence Collection, LACMNH.

31. Gauntier, "Blazing the Trail," clipping from *Woman's Home Companion*, in MOMA file, 2–3 (see note 41 of chap. 1 above).

32. Kerr, "Incorporating the Star," 396–98.

33. "Getting together," MPW, Jan. 7, 1911, 37.

34. "Observations by Our Man About Town," MPW, Nov. 20, 1909, 714; see also "Photographs of Moving Picture Actors," MPW, Jan. 15, 1910, 50; and "Vitagraph Notes," MPW, April 2, 1910, 515.

35. Kerr, "Incorporating the Star," 397–99; Gauntier, "Blazing the Trail," TS, 39, 52, 54; McDonald, "Origin of the Star System," 451–55; Bowser, *Transformation of Cinema*, 113. For Mary Pickford see MPW, Dec. 24, 1910, 1462. Pickford's birth name, Gladys Smith, was changed by theatrical producer David Belasco before

she entered the movies; see Scott Eyman, *Mary Pickford, America's Sweetheart* (New York: Donald I. Fine, 1990), 31. Florence Lawrence was "Flo," for example, in both the tragic *Song of the Shirt* (1908), in which she played an impoverished seamstress, and the "Jonesy" series of domestic comedies, such as *Jones Entertains* (1908) and *Her First Biscuits* (1909). All films above viewed by the author at MPBRSD-LC.

36. See deCordova, *Picture Personalities*, 42–43, 38.

37. Staiger, "Seeing Stars," 10–11.

38. Song quoted in McDonald, "Origin of the Star System," 455. For live appearances see "Vitagraph Girl Feted," *MPW*, April 23, 1910, 644; Lux Graphicus, "On the Screen," *MPW*, Oct. 8, 1910, 807: F. H. Richardson, "A Vitagraph Girl Night," *MPW*, Dec. 24, 1910, 1521.

39. Female stars predominated on the stage, according to a chart published in the *New York Mirror Annual* in 1888, which listed seventy-three female stars, or 7.7 percent of a total of 945 actresses, versus sixty-eight male stars, or 4.7 percent of 1,436 actors (repr. in McArthur, *Actors and American Culture*, 13–14). The popularity of female vaudeville stars and their subsequently higher salaries is discussed in Allen, "The Movies in Vaudeville," 65, 68–69, 71; Stein, *American Vaudeville*, 4–5; Samuels, *Once upon a Stage*, 50–52.

40. *MPW*, Nov. 29, 1911, 1223.

41. Slide, "Evolution of the Film Star," 591–94; McDonald, "Origin of the Star System," 449–58.

42. C. H. Claudy, "Too Much Acting," *MPW*, Feb. 11, 1911, 288.

43. Glenn, *Female Spectacle*, 6–7.

44. See, e.g., Miriam Hansen, *Babel and Babylon: Spectatorship in American Silent Film* (Cambridge, MA: Harvard University Press, 1991); and Judith Mayne, *Cinema and Spectatorship* (New York: Routledge, 1993).

45. Janet Staiger, "The Hollywood Mode of Production to 1930," in *The Classical Hollywood Cinema: Film Style and Mode of Production to 1960*, by David Bordwell, Janet Staiger, and Kristin Thompson (New York: Columbia University Press, 1985), 113–41.

46. Charles Musser, "Pre-Classical American Cinema: Its Changing Modes of Production," in *Silent Film*, ed. Richard Abel (New Brunswick, NJ: Rutgers University Press, 1996), 85–108.

47. Staiger, "The Hollywood Mode of Production to 1930," 117, 121–22.

48. Gauntier, "Blazing the Trail," TS, 51–54.

49. Untitled clipping, *New York Spirit of the Times*, July 24, 1888, Robinson Locke Collection, ser. 2, v. 83, p. 355.

50. Janet Wasko, *Movies and Money: Financing the American Film Industry* (Norwood, NJ: Ablex, 1982), 25.

51. "Investing in the Movies, "*Photoplay*, Dec. 1915, 59–60.

52. David Puttnam, with Neil Watson, *Movies and Money* (New York: Knopf, 1998), 61.

53. Lawrence, "Growing Up with the Movies," *Photoplay*, Nov. 1914, 30–31; Larry Lee Holland, "Florence Lawrence," *Films in Review* 31, no. 7 (Aug.-Sept. 1980): 389.

54. "A Moving Picture Jewelry Shop," MPW, Dec. 30, 1911, 1075; Gauntier, "Blazing the Trail," TS, 36.

55. Advertisement, MPW, Nov. 25, 1911, 737.

56. "A Moving Picture Jewelry Shop," MPW, Dec. 30, 1911, 1075.

57. "Gem Company Announces First Releases," MPW, Jan. 6, 1912, 48.

58. See "An Explanation," MPW, Jan. 20, 1912, 177; "A New Gem in the Rex Crown," MPW, Jan. 20, 1912, 191; MPW, Jan. 27, 1912, 269.

59. Quoted in Holland, "Florence Lawrence," 389, 392–93; MPW, Dec. 11, 1909, 866; MPW, Jan. 31, 1910, 134.

60. Harry Salter to Florence Lawrence, Aug. 1, 1910, Florence Lawrence Collection, LACMNH; Cochrane also left IMP for Lubin that same year, so Salter and Cochrane ended up together again.

61. Janet Staiger, "Combination and Litigation: Structures of U.S. Film Distribution, 1896–1917," *Cinema Journal* 23, no. 2 (winter 1985): 53–54.

62. MPW, June 8, 1912, n.p.; MPW, June 15, 1912, 1045; "Independent Factions Organize," MPW, June 1, 1912, 807.

63. "Independent Factions Organize," MPW, June 1, 1912, 807.

64. "The Photoplaywright," MPW, June 15, 1912, 1024.

65. Enclair and IMP usually produced four reels per week, Nestor three, and Powers, Rex, and Bison two or three; see Universal ads, MPW, July 13, 1912, 107; MPW, Aug. 17, 1912, 608; review of *The Mill Buyers*, MPW, Aug. 17, 1912, 676; review of *After All*, MPW, Sept. 7, 1912, 984; "Who's Who in Picturedom," *Buffalo Courier*, Sept. 8, 1912, scrapbook, Florence Lawrence Collection, LACMNH.

66. Review of *In Swift Waters*, MPW, July 20, 1912, 255; review of *The Players*, MPW, July 20, 1912, 205.

67. See Universal advertisements: MPW, July 13, 1912, 107; MPW, July 20, 1912, 204–5; MPW, July 27, 1912, 307–8; MPW, Aug. 17, 1912, 608.

68. Florence Lawrence to Lotta Lawrence, Oct. 25, 1912; Lotta Lawrence to Florence Lawrence, July 9, 1912; Lotta Lawrence to Florence Lawrence, July [?] 1912; Salter to Lawrence, Sept. 6, 1912; Florence Lawrence Collection, LACMNH.

69. Long series of letters between Salter and Lawrence, and Lawrence and her mother, Lotta Lawrence, from Aug. 1912 through Feb. 1913 have been archived in the Florence Lawrence Collection, LACMNH; see also *Photoplay*, Nov. 1912, 105; Holland, "Florence Lawrence," 391, 393.

70. Salter to Lawrence, Sept. 6, 1912, Lawrence Collection, LACMNH.

71. Lotta Lawrence to Florence Lawrence, Aug. 7, 1912; Harry Salter to Florence Lawrence, c. Aug. 1912; Lawrence's leaving her husband is cited in a letter from Salter to his mother, Sept. 27, 1912; paycheck receipts for Florence Lawrence

and Harry Salter, July 7, 1914 through Aug. 15, 1914; Florence Lawrence to Harry Salter, Oct. 12, 1916; Florence Lawrence Collection, LACMNH.

72. Spehr, *The Movies Begin*, 78; MPW, Aug. 9, 1913, 620; MPW, Aug. 16, 1913, 701.

73. Holland, "Florence Lawrence," 391; *New York Star*, April 4, 1914, 21. Lawrence sued Laemmle for damages stemming from her accident; see Florence Lawrence to Carl Laemmle, Oct. 15, 1916; Florence Lawrence to Lotta Lawrence, Oct. 15, 1916; copy of contract between Florence Lawrence and Max D. Josephson, Nov. 1, 1916; advertisement to rent rooms in her farm estate, July 1916; Florence Lawrence Collection, LACMNH.

74. *Billboard*, May 29, 1916, n.p., *New York Morning Telegraph*, June 1, 1916, n.p.; Lotta Lawrence to Florence Lawrence, July 10, Nov. 1, 1917; insurance contract, Nov. 2, 1918, Florence Lawrence Collection, LACMNH; Holland, "Florence Lawrence," 391–92, 394; Sandra Hansen, "The Silver Screen's First Star," *Box Office*, Feb. 1982, n.p.

75. Hoffman, "Florence Turner Going to England," 1225.

76. Elizabeth Peltret, "The Return of Florence Turner," *Motion Picture Classic*, Dec. 1919, n.p., Locke Collection Scrapbook, NYPL-PA.

77. Rachel Low, *History of the British Film, 1906–1914* (London: George Allen and Unwin, 1949), 109–11, 133–37.

78. Hoffman, "Florence Turner Going to England," 1205.

79. Low, *History of the British Film*, 107.

80. *The Rose of Surrey*, pamphlet (London: Hepworth, n.d.), n.p.; advertisements, *Cinematograph Exhibitor's Mail*, July 16, 1913, n.p.; *Bioscope*, July 17, 1913, 194; *Kinematograph and Lantern Weekly*, July 17, 1913, n.p.; *Cinema*, July 30, 1913, 26, all from the Florence Turner Collection, LACMNH; Anthony Slide, with Alan Gevinson, *The Big V: A History of the Vitagraph Company* (Metuchen, NJ: Scarecrow Press, 1987), 40.

81. Advertisements and ephemera from British trade journals, esp. *Bioscope, Cinematograph and Exhibitor's Mail, Kinematograph and Lantern Weekly*, from the Florence Turner Collection, LACMNH; Anthony Slide, *Early American Cinema* (New York: A. S. Barnes, 1970), 33–34.

82. Slide, *Big V*, 41; Lewis Jacobs, *The Rise of the American Film: A Critical History* (New York: Columbia University Press, 1968), 159. Reference to Turner's plan to produce is in a letter from Butcher's Film Service, Ltd., London, to Turner, May 24, 1922, Florence Turner Collection, LACMNH.

83. Durham, *American Theatre Companies*, 425–27; "Cecil Spooner in 'Dancer and the King,'" MPN, Dec. 5, 1914, 26; Anthony Slide, *The American Film Industry: A Historical Dictionary* (New York: Limelight, 1990), 43.

84. "Eleanor Gates in Photoplay Field," MPN, May 30, 1914, 1269.

85. "Women Manage Liberty with Success," MPN, April 3, 1915, 163.

86. Ibid.

87. "The Suffragists Go Filming," MPW, May 25, 1912, 714; review of *Suffrage and the Man*, MPW, June 8, 1912, 962; advertisement, MPW, June 1, 1912, 796.

88. Russell E. Smith, review of *Votes for Women*, *Photoplay*, July 1912, 35–41; Kay Sloan, *The Loud Silents: Origins of the Social Problem Film* (Chicago: University of Illinois Press, 1988), 112.

89. "Recruiting Stations of Vice," MPW, March 12, 1910, 370–71; Sloan, *The Loud Silents*, 112–13; Lary May, *Screening Out the Past: The Birth of Mass Culture and the Motion Picture Industry* (Chicago: University of Chicago Press, 1980), 44; review of *Votes for Women*, MPW, June 1, 1912, 811.

90. W. Stephen Bush, "Eight Million Women Want—?." MPW, Nov. 15, 1913, 741; advertisement, *Moving Picture World*, Nov. 15, 1913, 763.

91. Slide, *American Film Industry*, 35, 69, 268, 338, 401.

92. Ibid., 173–74, 179, 251–52.

93. Ibid., 46, 118, 146, 165.

94. Ibid., 000.

95. Briscoe, "Photoplays versus Personality," 40.

96. Paul H. Davis, "Investing in the Movies," *Photoplay*, June 1916, 137–38. Production costs for the epic *Birth of a Nation* (Biograph, 1915) were approximated at $100,000; see Jack C. Ellis, *A History of Film* (Englewood Cliffs, NJ: Prentice-Hall, 1979), 146.

97. Bowser, *Transformation of Cinema*, 33, 78, 192, 196.

98. "The Moving Picture Play," MPW, Feb. 25, 1911, 418; L. F. Cook, "Of Interest to the Trade," *Nickelodeon*, March 4, 1911, 252; "Three Reel Pictures a Success," MPW, March 25, 1911, 639.

99. Bowser, *Transformation of Cinema*, 192.

100. "Helen Gardner to Have a Company of Her Own," MPW, June 8, 1912, 917; *Vitagraph Life Portrayals* 1, no. 12 (Dec. 17, 1911–Jan. 1, 1912), n.p., Albert E. Smith collection, SC-UCLA; Bowser, *Transformation of Cinema*, 195; Slide, *Early American Cinema*, 36; Jacobs, *Rise of the American Film*, 91.

101. "Helen Gardner to Have a Company of Her Own," 917; "C. L. Fuller with Helen Gardner Co.," MPW, June 22, 1912, 1120; advertisement, MPW, Dec. 4, 1912, 1048; Jean Darnell, "A Day Spent with Helen Gardner," *Photoplay*, Jan. 1914, 72–73.

102. Louis Reeves Harrison, "Helen Gardner's Idealization of Cleopatra," review of *Helen Gardner in Cleopatra*, MPW, Nov. 30, 1912, 859; advertisement for *The Wife of Cain*, MPW, June 14, 1913, 1160–61; advertisement for *A Princess of Bagdad*, MPW, Nov. 22, 1913, 991; advertisement for *A Daughter of Pan*, MPW, Nov. 28, 1913, 1071; Patricia King Hansen, ed., *The American Film Institute Catalog of Motion Pictures Produced in the United States: Feature Films, 1911–1920* (Berkeley: University of California Press, 1988), 105, 148–49, 716, 740, 845 (hereafter cited as Hansen, *AFI Catalog . . . 1911–1920*).

103. Bowser, *Transformation of Cinema*, 192; Harrison, "Helen Gardner's Idealization of Cleopatra," 859; Hansen, *AFI Catalog . . . 1911–1920*, 149.

104. "Marion Leonard Joins Monopol Company," *MPW*, Dec. 7, 1912, 988; Slide, *American Film Industry*, 216; advertisement, *MPW*, March 1, 1913, 855.

105. "Warren, Taylor and Leonard with Warners," *MPW*, Aug. 23, 1913, 851; review of *In the Watches of the Night*, *MPW*, Oct. 18, 1913, 266; advertisement, *MPW*, Nov. 29, 1913, 1070.

106. See review in *MPW*, Dec. 20, 1913, 1425.

107. Hansen, *AFI Catalog . . . 1911–1920*, 638; advertisement for *The Journey's Ending*, *MPW*, Dec. 20, 1913, 1444; *The Light Unseen*, in Hansen, *AFI Catalog . . . 1911–1920*, 519.

108. Marion to Olcott, Sept. 10, 1912, private collection of Robert Britchard.

109. "Gauntier Feature Players," *MPW*, Dec. 21, 1912, 1169; for Warners' method of distribution see "*Redemption*, A 3-Reel Éclair Feature," *MPW*, April 6, 1912, 48; advertisement, *MPW*, April 6, 1912, 13; "Name of F. B. Warren Corporation Has Been Changed to Wid Gunning, Inc.," *MPW*, Dec. 3, 1921, 549.

110. Anthony Slide, *The Cinema and Ireland* (Jefferson, NC: McFarland, 1988), 40–43; Kevin Rockett, Luke Gibbons, and John Hill, *Cinema and Ireland* (Syracuse, NY: Syracuse University Press, 1988), 7–12; "Taking Pictures under Difficulties," *MPW*, Sept. 16, 1911, 784; "Kalem Announces Egyptian Pictures," *MPW*, Sept. 16, 1911, 784; "Kalem Announces Egyptian Pictures," *MPW*, April 20, 1912, 311–12; "Striking Kalem Subject," *MPW*, May 18, 1912, 614. Any films dealing with Christ were controversial; see, e.g., "The Matinee Girl," *New York Dramatic Mirror*, June 16, 1909, 2; Epes Winthrop Sargent, "Handling the Kalem Release," *MPW*, Oct. 19, 1912, 233; W. Stephen Bush, review of *From the Manger to the Cross*, *MPW*, Oct. 26, 1912, 324; "A Leap Forward," *MPW*, Oct. 26, 1912, 330; W. Stephen Bush, "Are Sacred Pictures 'Irreverent'?" *MPW*, Dec. 7, 1912, 957; Rev. E. Boudinot Stockton, "The Pictures in the Pulpit," *MPW*, Dec. 28, 1912, 1284.

111. Advertisement, *MPW*, Dec. 21, 1912, 1148–49; George Blaisdell, At the Sign of the Flaming Arcs, *MPW*, Dec. 21, 1912, 1196; reference to Clark as Gauntier's husband is in Slide, *Cinema and Ireland*, 41.

112. "Gene Gauntier Players at Work," *MPW*, Jan. 18, 1913, 274; Richard Alan Nelson, "Florida, the Forgotten Film Capital," *Journal of the University Film Association* 29, no. 3 (summer 1977): 9.

113. Gauntier asked for appropriate scenarios for three-reel subjects with "thrilling situations and strong plots" ("Miss Gauntier Wants," *MPW*, Jan. 18, 1913, 258).

114. "Gauntier Feature Players," *MPW*, Dec. 21, 1912, 1169; Blaisdell, At the Sign of the Flaming Arcs, *MPW*, Dec. 21, 1912, 1196.

115. H. C. Judson, review of *A Daughter of the Confederacy*, *MPW*, March 1, 1913, 892; advertisement, *MPW*, Feb. 15, 1913, 645; review of *The Girl Spy*, *MPW*, June 5, 1909, 672; Gauntier, "Blazing the Trail," TS, 54.

116. "Gene Gauntier Players Return," *MPW*, May 13, 1913, 926.

117. "Gauntier Players Go to Ireland," *MPW*, Aug. 1913; review, "For Ireland's Sake," Jan. 31, 1914, 526, Scrapbook Collection, NYPL-PA.

118. Columnist George Blaisdell was particularly taken with Gauntier's screen-writing abilities; see review of *In the Power of the Hypnotist*, MPW, Nov. 1, 1913, 499; and Blaisdell, At the Sign of the Flaming Arcs, MPW, Dec. 27, 1913, 1548.

119. Wig-Wag at the Movies, "At the Gene Gauntier Studio," *New York Star*, May 9, 1914, 20; "Gene Gauntier Takes European Vacation," MPW, July 18, 1914, Scrapbook Collection, NYPL-PA.

120. "Miss Gauntier Returns from Europe," MPW, Sept. 5, 1914, 1524; "Gene Gauntier Returns," *New York Star*, Feb. 10, 1914, n.p., Scrapbook Collection, NYPL-PA.

121. "Gene Gauntier with Universal," MPW, Nov. 27, 1914, n.p., Scrapbook Collection, NYPL-PA.

122. Mabel Condon, "Hot Chocolate and Reminiscences at Nine in the Morning," *Photoplay*, June 1915, 69–72, Scrapbook Collection, NYPL-PA.

123. *Motion Picture Magazine*, Dec. 1917, n.p., Scrapbook Collection, NYPL-PA.

124. "Once upon a Time," *Photoplay*, March 1918, n.p., Scrapbook Collection, NYPL-PA.

125. Slide, *American Film Industry*, 313.

126. George Mitchell, "Sidney Olcott," *Films in Review* 5, no. 4 (April 1954): 179.

127. Untitled clipping, *New York Spirit of the Times*, July 24, 1888, Robinson Locke Collection, ser. 2, v. 83, p. 355.

128. Alice Guy Blaché, *Memoirs of Alice Guy Blaché*, ed. Anthony Slide, trans. Robert Blaché and Simone Blaché (Metuchen, NJ: Scarecrow Press, 1986), 73.

129. Grandin was known chiefly for her appearance in the controversial white slavery film *Traffic in Souls* (1913). See "Ethel Grandin," MPW, Dec. 6, 1913, 1127. In 1914 Smallwood's company was simply the Smallwood Film Corporation, renamed Grandin Films, Inc., in 1915; see Slide, *American Film Industry*, 315–16.

130. Deposition, "Narrative of Our Dealings with Raymond C. Smallwood and Ethel Grandin, His Wife"; Kleine to unknown, detailing rumor of injunction, July 31, 1915; New York attorneys to Kleine, Oct. 11, 1915; R. C. Smallwood to Kleine, Dec. 22, 1915; Kleine to Smallwood, Jan. 24, 1916; Kleine to Smallwood, Feb. 10, 1916; Henry Melville to Kleine, March 4, 1916; Kleine to Melville, Feb. 20, 1917; Kleine to Melville, April 20, 1918; Melville to Kleine, April 3, 1919; Kleine to H. H. Breunnerof, April 11, 1919, George Kleine Collection, MD-LC.

131. Anthony Slide, *Early Women Directors* (New York: A. S. Barnes, 1977), v–vi, 23.

132. McMahan, *Alice Guy Blaché*, 77, 121, 172–73.

133. Ibid., 115, 173–84; Guy Blaché, *Memoirs*, 79, 100–103.

134. Guy Blaché, *Memoirs*, 89–98.

135. McMahan, *Alice Guy Blaché*, xxii–xxiii, 203–5, 241.

CHAPTER 3. "SO MUCH MORE NATURAL TO A WOMAN"

1. MPW, Nov. 16, 1907, 593–94.

2. "Notes and Comments," MPW, June 6, 1908, 491.

3. "Kansas City after Immoral Pictures," MPW, July 25, 1908, 63.

4. "Keeping the Pictures Clean," MPW, Aug. 8, 1908, 100.

5. Gene Gauntier, "Blazing the Trail," WHC, Oct. 1928, 8.

6. Robert C. Allen, "The Movies in Vaudeville," in *The American Film Industry*, ed. Tino Balio (Madison: University of Wisconsin Press, 1985), 71, 77–78, 80; Lary May, *Screening Out the Past: The Birth of Mass Culture and the Motion Picture Industry* (Chicago: University of Chicago Press, 1983), 35–40; Charles Musser, *The Emergence of Cinema: The American Screen to 1907* (New York: Charles Scribner's Sons, 1990), 432; Kathy Peiss, *Cheap Amusements: Working Women and Leisure in Turn-of-the-Century New York* (Philadelphia: Temple University Press, 1986), 145–58; Steven J. Ross, *Working-Class Hollywood: Politics, Class, and the Rise of the Movies* (Princeton, NJ: Princeton University Press, 1998).

7. Daniel Czitrom, "The Politics of Performance: Theater Licensing and the Origins of Movie Censorship in New York," in *Movie Censorship and American Culture*, ed. Francis G. Couvares (Washington, DC: Smithsonian Institution Press, 1996), 31.

8. W. Stephen Bush, "The Question of Censorship," MPW, Jan. 9, 1909, 32.

9. Robert Bruce Fisher, "The People's Institute of New York City, 1897–1934: Culture, Progressive Democracy, and the People" (PhD diss., New York University, 1974), 189–208; "Woman's League Investigates," MPW, Feb. 22, 1908, 137; May, *Screening Out the Past*, 54–55; Robert Sklar, *Movie-Made America: A Cultural History of American Movies*, rev. ed. (New York: Vintage Books, 1994), 31.

10. Shelley Stamp, *Movie-Struck Girls: Women and Motion Picture Culture after the Nickelodeon* (Princeton, NJ: Princeton University Press, 2000), 6.

11. Czitrom, "The Politics of Difference," 34.

12. "Woman's League Investigates," MPW, Feb. 22, 1908, 137.

13. Lee Grieveson, *Policing Cinema: Movies and Censorship in Early-Twentieth-Century America* (Berkeley: University of California Press, 2004), 101.

14. Nancy Cott, *Bonds of Womanhood: 'Woman's Sphere' in New England, 1780–1835* (New Haven, CT: Yale University Press, 1977), chap. 2; Barbara Welter, "The Cult of True Womanhood, 1820–1860," *American Quarterly* (summer 1966): 151–74.

15. Paula Baker, "The Domestication of Politics, 1870–1920," *American Historical Review* 89, no. 3 (June 1984): 632.

16. Ibid., 641–42.

17. "Baltimore Censorship Flurry," MPW, July 22, 1911, n.p.

18. Mary P. Ryan, *Women in Public: Between Banners and Ballots, 1825–1880* (Baltimore: Johns Hopkins University Press, 1990), 79–80; Bruce A. McConachie, *Melodramatic Formations: American Theatre and Society, 1820–1870* (Iowa City: University of Iowa Press, 1992), 204; "Dean of Vaudeville Celebrities," *Variety*, March 24, 1906, 5; Allen, "The Movies in Vaudeville," 65, 68–69, 71; Charles W. Stein, *American Vaudeville as Seen by Its Contemporaries* (New York: Da Capo Press, 1984), 4–5.

19. David Nasaw, *Going Out: The Rise and Fall of Public Amusements* (New York: Basic Books, 1993), 26; *Variety*, May 12, 1906, 4.

20. "Play to the Ladies," *Nickelodeon*, Feb. 1909, 33–34.

21. See Stamp, *Movie-Struck Girls*, 197–99.

22. Ibid., 15, 27.

23. For this generational shift see Nancy F. Cott, *The Grounding of Modern Feminism* (New Haven, CT: Yale University Press, 1987).

24. Russell Merritt argued against the assumption that all nickelodeon audiences were working class, asserting that as early as 1914 Boston's moving picture theaters were patronized by middle-class audiences; see Russell Merritt, "Nickelodeon Theaters, 1905–1914: Building an Audience for the Movies," in *The American Film Industry*, ed. Tino Balio, rev. ed. (Madison: University of Wisconsin Press, 1985), 85. Steven J. Ross argues, however, that middle-class audiences did not patronize the movies regularly until the rise of the picture palace after World War I; see Ross, *Working-Class Hollywood*. Meanwhile, other scholars working on rural exhibition are finding a mixed-class pattern of patronage throughout the silent era; see George Potamianos, "Hollywood in the Hinterlands: Mass Culture in Two California Communities, 1896–1936" (PhD diss., University of Southern California, 1998).

25. "The Colonial Ladies Matinee," MPW, Oct. 26, 1907, 540–41; "Baby Show Draws at Nickel Theater," *Nickelodeon*, May 1909, 142; "Souvenirs," MPW, March 4, 1911, 465; Merritt, "Nickelodeon Theaters, 1905–1914," 95–96.

26. See, e.g., "The Way a Live Wire Appeals to the Ladies," *Universal Weekly*, Sept. 6, 1913, n.p.

27. "The Princess of Milwaukee," MPW, Oct. 22, 1910, 928. An interesting and insightful analysis of moving picture theaters in the silent era is offered in May, *Screening Out the Past*, chap. 6.

28. Grieveson, *Policing Cinema*, 69.

29. George J. Anderson, "The Case for Motion Pictures," *Nickelodeon*, Aug. 15, 1910, 97.

30. "Mrs. Clement and Her Work," MPW, Oct. 15, 1910, 859.

31. "The Current Problem and Opportunity of the Moving Picture Show," *MPN*, Jan. 11, 1913, 20, quoted in Eileen Bowser, *The Transformation of Cinema, 1907–1915* (Berkeley: University of California Press, 1994), 123.

32. Bowser, *Transformation of Cinema*, 123–24.

33. "How a Woman Plans to Run a Theatre," *MPN*, Dec. 19, 1914, 37.

34. *Nickelodeon*, Nov. 1909, 134–35; "Vaudeville in Picture Theaters," *Nickelodeon*, Feb. 5, 1910, 86; "Vaudeville Losing Out," *Nickelodeon*, May 1, 1910, 224; Bowser, *Transformation of Cinema*, 19.

35. Prof. Preston, "The Permanent Feature," MPW, May 18, 1912, 610; "Montgomery Likes Miss Russell," MPW, Feb. 24, 1912, 683; "Miss Russell's Lecture Tour," MPW, Dec. 14, 1912, 1066.

36. Not surprisingly, Ms. Bona found work at Mrs. Clement's Boston Theater;

see "A Lecturer and the Pictures," *MPW*, Oct. 1, 1910, 750; "Lora Bona," *MPW*, Nov. 12, 1910, 1113; Bowser, *Transformation of Cinema*, 123.

37. Grieveson, *Policing Cinema*, 37–42.

38. Henry Jenkins, *What Made Pistachio Nuts? Early Sound Comedy and the Vaudeville Aesthetic* (New York: Columbia University Press, 1992), 42–43.

39. Bowser, *Transformation of Cinema*, 40–41, 44.

40. A. H. Giannini, "Financing the Production and Distribution of Motion Pictures," *Annals of the American Academy of Political and Social Science* 128 (Nov. 1926): 46.

41. Louis Reeves Harrison, "A Wondrous Training School," *MPW*, Aug. 16, 1913, 720.

42. William Uricchio and Roberta A. Pearson, *Reframing Culture: The Case of the Vitagraph Quality Films* (Princeton, NJ: Princeton University Press, 1993).

43. Bowser, *Transformation of Cinema*, 87–97.

44. Roberta Pearson, "Cultivated Folks and the Better Classes," *Journal of Popular Film and Television* 15, no. 3 (fall 1987): 122.

45. Richard deCordova, *Picture Personalities: The Emergence of the Star System in America* (1990; repr., Urbana: University of Illinois Press, 2001), 34.

46. Grieveson, *Policing Cinema*, 79.

47. Kay Sloan, *The Loud Silents: Origins of the Social Problem Film* (Chicago: University of Illinois Press, 1988); for films made by unions and radicals see Ross, *Working-Class Hollywood*; Kevin Brownlow, *Behind the Mask of Innocence: Sex, Violence, Prejudice, Crime: Films of Social Conscience in the Silent Era* (Berkeley: University of California Press, 1990), xv–xviii.

48. Grieveson, *Policing Cinema*, 81, 91, 95, 119.

49. "Recruiting Stations of Vice," *MPW*, March 12, 1910, 370–71; Sloan, *The Loud Silents*, 112–13; May, *Screening Out the Past*, 44; review of *Votes for Women*, *MPW*, June 1, 1912, 811.

50. Grieveson, *Policing Cinema*, 174.

51. Ibid., 155–57, 180.

52. Ibid., 166.

53. Stamp, *Movie-Struck Girls*, 42–45, 52–53, 55–60.

54. Ibid., 59–60.

55. W. Stephen Bush, "Gauging the Public Taste," *MPW*, May 11, 1912, 505; W. Stephen Bush, "Factory or Studio?" *MPW*, Sept. 12, 1912, 1153.

56. deCordova, *Picture Personalities*, 38–39.

57. Bowser, *Transformation of Cinema*, 192, 212–15; Tino Balio, ed., *The American Film Industry* (Madison: University of Wisconsin Press, 1985), 111–13.

58. "Well-Known Writers Turning to a New Field—That of Writing Film Scenarios," *MPW*, Jan. 29, 1910, 120.

59. "On Filming a Classic," *Nickelodeon*, Jan. 7, 1911, 4, quoted in Uricchio and Pearson, *Reframing Culture*, 50.

60. Alice Guy Blaché, "Woman's Place in Photoplay Production," *MPW*, July 11, 1914, 195.

61. Louis Reeves Harrison, "Studio Saunterings," *MPW*, June 15, 1912, 1007; advertisement, *MPW*, June 22, 1912, 1093; Louis Reeves Harrison, review of *Fra Diavolo*, *MPW*, June 22, 1912, 1114–15.

62. Alice Guy Blaché, *The Memoirs of Alice Guy Blaché*, ed. Anthony Slide, trans. Roberta Blaché and Simone Blaché (Metuchen, NJ: Scarecrow Press, 1986), 37–38.

63. "Who's Who in the Film Game," *Motography*, Oct. 12, 1912, 293.

64. See, e.g., W. Stephen Bush, review of *The Shadows of the Moulin Rouge*, *MPW*, Jan. 24, 1914, 417; W. Stephen Bush, review of *The Heart of a Painted Woman*, *MPW*, May 1, 1915, 739.

65. Alison McMahan, *Alice Guy Blaché: Lost Visionary of the Cinema* (New York: Continuum, 2002), 174.

66. L. H. Johnson, "'A Lady General in the Motion Picture Army,' Lois Weber-Smalley, Virile Director," *Photoplay*, June 1915, 42; H. H. Van Loan, "Lois the Wizard," *Motion Picture Magazine*, July 1916, 44; Fritzi Remont, "The Lady behind the Lens," *Motion Picture Magazine*, May 1918, 126; Aline Carter, "The Muse of the Reel," *Motion Picture Magazine*, March 1921, 63; McMahan, *Alice Guy Blaché*, 71; Anthony Slide, who has devoted his career to the history of silent film and wrote a biography of Weber, claims that Weber made no reference to Alice Guy Blaché in any published forum. See Anthony Slide, *Lois Weber: The Director Who Lost Her Way in History* (Westport, CT: Greenwood Press, 1996), 44–45.

67. "Phillips Smalley Talks," *MPW*, Jan. 24, 1914, 399.

68. "The First Birthday of Rex" and "What Rex Is Going to Do," *MPW*, Feb. 24, 1912, 670–73; advertisement for "The Rex Motion Picture Masterpiece Co.," *MPW*, March 16, 1913, 925.

69. George Blaisdell, At the Sign of the Flaming Arcs, *MPW*, April 5, 1913, 59.

70. Karen Halttunen, *Confidence Men and Painted Women: A Study of Middle-Class Culture in America, 1830–1870* (New Haven, CT: Yale University Press, 1982).

71. "All Star Cast in Distinctive Rex Drama," *Universal Weekly*, March 21, 1914, 13–16.

72. George Blaisdell, review of *The Jew's Christmas*, *MPW*, Dec. 5, 1913, 1132; "The Story of the 'Jew's Christmas,'" *Universal Weekly*, Dec. 13, 1913, 13, 16; "'Merchant of Venice' Is Supreme Adaptation of Shakespeare," *Universal Weekly*, Feb. 14, 1914, 5; Weber and Smalley also made typical films during their first years with Universal: *The Mask* (1913), a Jekyll and Hyde–type film, and *The Wife's Deceit* (1913), a marital farce, see *Universal Weekly*, Sept. 20, 1913, 5; Dec. 6, 1913, 20; Dec. 13, 1913, 16.

73. "Lois Weber—Mrs. Phillips Smalley," *Universal Weekly*, Oct. 4, 1913, 8–9.

74. Minerva Marvin, "A Versatile Couple," *Photoplay*, April 1914, 84.

75. Cott, *The Grounding of Modern Feminism*, 16–22.

76. Blaisdell, At the Sign of the Flaming Arcs, *MPW*, April 5, 1913, 59.

77. Slide, *Lois Weber*, 18–25; Weber quote from Aline Carter, "The Muse of the Reel," *Motion Picture Magazine*, March 1921, 105.

78. Blaisdell, At the Sign of the Flaming Arcs, *MPW*, April 15, 1913, 59.

79. Slide, *Lois Weber*, 25–28.

80. Shelley Stamp, "Lois Weber and the Celebrity of Matronly Respectability," forthcoming in *Looking Past the Screen: Case Studies in American Film History and Method*, ed. Jon Lewis and Eric Smoodin (Durham, NC: Duke University Press).

81. Slide, *Lois Weber*, 52–53.

82. My discussion of this film was previously presented in a conference paper entitled "Bearing It All: Nudity, Censorship, and Privilege in Lois Weber's *Hypocrites* (1914)," Annual Meeting of the Organization of American Historians, Boston, March 25, 2004.

83. Review of *Hypocrites*, *Variety*, Nov. 7, 1914, n.p.

84. Slide, *Lois Weber*, 72.

85. "Remarkable Record of 'Hypocrites,' " *MPW*, Feb. 13, 1915, n.p.

86. Slide, *Lois Weber*, 72–73.

87. "Remarkable Record of 'Hypocrites,' " n.p.

88. Joy S. Kasson, *Marble Queens and Captives: Women in Nineteenth-Century American Sculpture* (New Haven, CT: Yale University Press, 1990), 46–48.

89. Nicola Beisel, "Morals Versus Art: Censorship, the Politics of Interpretation, and the Victorian Nude," *American Sociological Review* 58 (1993): 150.

90. *Hypocrites*, prod. and dir. Lois Weber, videocassette, 50 min. (1914; Kino Video, 2000).

91. See House Motion Picture Commission, *Hearings before the Committee on Education*, 63rd Cong., 2nd sess., 1914 (Washington: GPO, 1914).

92. "Remarkable Record of 'Hypocrites,' " n.p.

93. Julian M. Solomon, review of *Hypocrites*, *MPW*, Oct. 10, 1914, 45; review of *Hypocrites*, *Variety*, Nov. 7, 1914, n.p.

94. Review of *Hypocrites*, *Variety*, Nov. 7, 1914, n.p.

95. " 'Hypocrites' Enjoys Successful Broadway Run," *MPN*, Feb. 6, 1915, 28.

96. " 'Hypocrites' Plays to Big Houses in New York," *MPN*, Feb. 20, 1915, 40.

97. "Hypocrites," continuity script, Paramount Collection, AMPAS.

98. Arthur Denison, "A Dream in Realization," *MPW*, July 21, 1917, 417–18.

99. "The Greatest Woman Director," *Moving Picture Stories*, July 7, 1916, 26.

100. Slide, *Lois Weber*, 54–55.

101. Clifford H. Pangburn, review of *Scandal*, *MPN*, June 19, 1915, 69.

102. Review of *Hop, the Devil's Brew*, *Variety*, Feb. 4, 1916, n.p.

103. Review of *Shoes*, *MPW*, June 24, 1916, 2257; Peter Mime, review of *Shoes*, *MPN*, June 24, 1916, 3927; review of *Shoes*, *Moving Picture Weekly*, June 24, 1916, 12–13, 34; Larry Lee Holland, "Shoes," in *Magill's Survey of Cinema—Silent Films*, ed. Frank N. Magill (Englewood Cliffs, NJ: Salem Press, 1982), 1:975–77.

104. Peter Mime, review of *Saving the Family Name*, MPN, Sept. 9, 1916, 1576; Claire Marchand, review of *Saving the Family Name*, Photoplay, Nov. 1916, 91–98, 168.

105. Peter Mime, review of *Idle Wives*, MPN, Oct. 7, 1916, 2231; "Shadow Stage," Photoplay, Dec. 1916, 84; review of *Idle Wives*, Variety, Sept. 22, 1916, n.p.

106. "The Greatest Woman Director," *Moving Picture Stories*, July 7, 1916, 20, 27; Slide, *Lois Weber*, 101.

107. "Background to the filming of 'The Dumb Girl of Portici,'" Kevin Brownlow, private collection.

108. "Anna Pavlova Finally Decides to Appear on the Screen," MPW, June 19, 1915, 52.

109. Slide, *Lois Weber*, 98–101.

110. "Capital Punishment Film Play's Theme," review of *The People vs. John Doe*, NYT, Dec. 11, 1916, 7; review of *The People vs. John Doe*, Variety, Dec. 15, 1916, n.p.

111. Linda Gordon, *Women's Body, Women's Right: A Social History of Birth Control in America* (New York: Penguin, 1976), 228–29.

112. Shelley Stamp, "Taking Precautions, or Regulating Early Birth Control Films," *A Feminist Reader in Early Cinema*, ed. Jennifer M. Bean and Diane Negra (Durham, NC: Duke University Press, 2002), 270–97.

113. *Where Are My Children?* partial print, viewed at MPBRSD-LC; Brownlow, *Behind the Mask of Innocence*, 50–55.

114. "Review Board Again Sees 'Children' Picture," MPW, May 20, 1916, 1321.

115. Stamp, "Taking Precautions," 284–88.

116. "Pennsylvania Turns Down 'Where Are My Children?'" MPN, Oct. 7, 1916, 2206; also see "Lewis Makes War on Film," NYT, June 18, 1916, 3.

117. Advertisement, "Where Are My Children?" MPW, June 3, 1916, 1615; "'Children' Film in Boston," Variety Weekly, July 7, 1916, 22.

118. "The Screen Club Is a Fact," MPW, Sept. 21, 1912, 1163; "Screen Club a Winner," MPW, Sept. 28, 1912, 1283; "Screen Club Opens Its Door," MPW, Nov. 23, 1912, 778; MPW, Oct. 12, 1912, 128; Doings in Los Angeles, MPW, Nov. 30, 1912, 870; Doings in Los Angeles, MPW, Dec. 12, 1912, 1175; Doings in Los Angeles, MPW, March 15, 1913, 1090.

119. *The Screen Club*, 1st annual ball, pamphlet, April 10, 1913, Terrace Garden, New York, Screen Club general file, AMPAS.

120. "Screen Club Is a Fact," 1163.

121. *Screen Club Annual Ball*, pamphlet, 1914, Screen Club general file, AMPAS; *The Screen Club*, 1st annual ball, pamphlet, AMPAS. Women were allowed into the clubhouse but only as guests and only during limited hours.

122. "Doings in Los Angeles," MPW, Jan. 11, 1913, 142.

123. "Photoplay Author's League One Year Old," MPN, April 3, 1915, 171.

## Chapter 4. The "Girls Who Play"

1. Creighton Hamilton, "Girls Who Play with Death," *Picture Play*, May 1916, n.p., Helen Holmes clipping file, Billy Rose Theatre Collection, NYPL-PA.

2. A female detective appears in Famous Players' *An Hour before Dawn* (Oct. 1913); Zane Grey's *Riders of the Purple Sage* (1918) featured a female cattle rustler; Edith Storey starred as gunfighter "Colonel Billy" in *As the Sun Went Down* (1919); cross-dressing Agnes Vernon shot the hat off of William Desmond in *Bare-Fisted Gallagher* (1919); Mutual represented female soldiers in *Miss Jackie of the Navy* (Nov. 1916) and *Miss Jackie of the Army* (Dec. 1917). These examples pale in comparison to the dozens of reels featuring intrepid, gender-defying serial heroines that sustained the weekly program. The kind of public flirtation exhibited in many Keystone comedies was also a rarity in features. Of the thirty-six features listed under "flirts" in the American Film Institute catalog for the 1910s, only five or six depicted female flirts as amusing and basically harmless, while dozens of films in the "vamp" genre preached that sexual aggression and promiscuity led directly to dissipation. Even innocent flirting could open the door to disaster. Lois Weber's *The Flirt* (1916), for example, concerned a habitual and thoughtless flirt who met her match in a con man, and the result was ruin and murder. See Patricia King Hansen, ed., *The American Film Institute Catalog of Motion Pictures Produced in the United States: Feature Films, 1911–1920* (Berkeley: University of California Press, 1988), 34–35, 46, 286, 425, 619, 773, 827 (hereafter cited as Hansen, *AFI Catalog . . . 1911–1920*).

3. Kalton C. Lahue, *Continued Next Week: A History of the Moving Picture Serial* (Norman: University of Oklahoma Press, 1964), 76.

4. Lois W. Banner, *American Beauty* (New York: Knopf, 1983), 87; Lois W. Banner, *Women in Modern America: A Brief History*, 2nd ed. (New York: Harcourt Brace Jovanovich, 1984), 21–22.

5. Kathy Peiss, *Cheap Amusements: Working Women and Leisure in Turn-of-the-Century New York* (Philadelphia: Temple University Press, 1986).

6. Ben Singer, *Melodrama and Modernity: Early Sensational Cinema and Its Contexts* (New York: Columbia University Press, 2001), 246–49.

7. Bruce A. McConachie, *Melodramatic Formations: American Theatre and Society, 1820–1870* (Iowa City: University of Iowa Press, 1992), 206, 214, 217, 220–21; Faye E. Dudden, *Women in the American Theatre: Actresses and Audiences, 1790–1870* (New Haven, CT: Yale University Press, 1994), 31.

8. Singer, *Melodrama and Modernity*, 150–52.

9. Lewis Jacobs, *The Rise of the American Film: A Critical History* (New York: Columbia University Press, 1968), 70; *The Girl from Montana* (Selig, 1907), viewed at MPBRSD-LC.

10. *The Lonedale Operator*, viewed through Em Gee Film Library, Reseda, CA, a film rental company; Eileen Bowser, *Biograph Bulletins, 1908–1912* (New York: Octagon, 1973), 240, 284, 390.

11. *MPW*, June 5, 1909, 672; Gene Gauntier, "Blazing the Trail," clipping from

*Woman's Home Companion*, in MOMA file, 54 (see note 41 of chap. 1 above); Eileen Bowser, *The Transformation of Cinema, 1907–1915* (Berkeley: University of California Press, 1994), 206.

12. *How She Triumphed: An Argument in Favor of Physical Culture* (1911) and *The Diving Girl* (1911), in Bowser, *Biograph Bulletins*, 294, 327.

13. *Ruth Roland, Kalem Girl* (Kalem, 1912), viewed at BFI.

14. Attendance figures in Peiss, *Cheap Amusements*, 148; "Play to the Ladies," *Nickelodeon*, Feb. 1909, 33–34.

15. Gauntier, "Blazing the Trail," TS, 46–55.

16. "Edison-McClure," *MPW*, June 29, 1912, 1212.

17. "Editor's Study," *Harper's Magazine*, Aug. 1915, 477–78; Louis Reeves Harrison, "What Happened to Mary," *MPW*, July 5, 1913, 26; Stephen L. Hansen, "Serials and the *Perils of Pauline*," in *Magill's Survey of Cinema—Silent Films*, ed. Frank N. Magill (Englewood Cliffs, NJ: Salem Press, 1982), 1:95–100; William Stedman, *The Serials: Suspense and Drama by Installment* (Norman: University of Oklahoma Press, 1977), 4–7.

18. Quoted in Anthony Slide, *Early American Cinema* (New York: A. S. Barnes, 1970), 158.

19. Quoted in Linda Mehr, "Down Off the Pedestal: Some Modern Heroines in Popular Culture, 1890–1917" (PhD diss., University of California at Los Angeles, 1974), 185.

20. Lahue, *Continued Next Week*, 7–8; Kalton C. Lahue, *Ladies in Distress* (New York: A. S. Barnes, 1971), 319–20. *The Adventures of Kathlyn* was novelized in 1914 and released under the same title as a ten-reel feature film in 1916. See Hansen, *AFI Catalog . . . 1911–1920*, 6.

21. "Popularity of the Serial Is on the Increase," and "Serials Supreme in Middle West," *MPN*, Jan. 2, 1915, 72; Alfred A. Cohn, "Harvesting the Serial," *Photoplay*, Feb. 1917, 22; Kalton C. Lahue, *Bound and Gagged: The Story of the Silent Serials* (New York: A. S. Barnes, 1968), 89; Stedman, *The Serials*, 14–15.

22. Singer, *Melodrama and Modernity*, 216.

23. Lahue, *Continued Next Week*, 36.

24. See synopses of *Perils of Pauline* episodes 1, 6, 9, 13, 15, 20, in *MPW* (1914), repr. in *Blackhawk Bulletin*, B–253, July 1974, 64–66, AMPAS.

25. I. G. Edmunds, *Big U: Universal in the Silent Days* (South Brunswick, NJ: A. S. Barnes, 1977), 37–38; Slide, *Early American Cinema*, 166.

26. Miriam Hansen, *Babel and Babylon: Spectatorship in American Silent Film* (Cambridge, MA: Harvard University Press, 1991), 118–19.

27. Mabel Condon, "The Statuesque Mutual Girl," *Photoplay*, March 1914, 51; Kuhn's cameo was significant; while investment firms were still cautious, Kuhn put his personal funds in Mutual, providing a critical endorsement for the movies and an indication of the serial's improving reputation; see Janet Wasko, *Movies and Money: Financing the American Film Industry* (Norwood, NJ: Ablex, 1982); Slide,

*Early American Cinema*, 159–60; for reference to "consumer fantasies" see Elizabeth Ewen, "City Lights: Immigrant Women and the Rise of the Movies," *Signs* 5, no. 3 (spring 1980): S45–S65."

28. "Creator of 'J. Rufus' Will Write Reliance Serial," *MPN*, Dec. 5, 1914, 40–41; "The 'Runaway June' California Trip Contest," advertisement, *MPN*, March 27, 1915, 50; " 'June' Winners Will Represent Ideal Womanhood," *MPW*, May 1, 1915, 46; Alfred A. Cohn, "Harvesting the Serial," *Photoplay*, Feb. 1917, 25; Lahue, *Continued Next Week*, 33; *MPW*, March 27, 1915, 32.

29. *The Ventures of Marguerite*, episodes 1, 4, and 15, viewed at MPBRSD-LC; "Ruth Roland Goes to Balboa for Three Years," *MPN*, Dec. 19, 1914, 57; "Balboa Co. Offers Prize for Solution," *Long Beach Daily Telegraph*, Oct. 30, 1914; "Noted People Laude Pathé Balboa *Who Pays?* Series," *MPN*, May 8, 1915, 54; Lahue, *Continued Next Week*, 34–35; Slide, *Early American Cinema*, 165.

30. "A Fact Worth Noting," exploitation guide for exhibitors [1912?], William N. Selig Collection, AMPAS.

31. *The Lightning Raider*, episode 13, reel 2 of 2, viewed at MPBRSD-LC.

32. *What Happened to Mary* (Edison, 1912), *The Perils of Pauline* (Pathé, 1914), and *The Ventures of Marguerite* (Kalem, 1915), among others, revolved around lost inheritances and/or ineffectual guardians; *The Ventures of Marguerite*, episode 1, viewed at MOMA; Lary May, *Screening Out the Past: The Birth of Mass Culture and the Motion Picture Industry* (Chicago: University of Chicago Press, 1980), 108.

33. Lahue, *Continued Next Week*, 13–14, 17–19.

34. Richard Willis, "Kathlyn the Intrepid," *Photoplay*, April 1914, 42.

35. Alan Burden, "The Girl Who Keeps a Railroad," *Photoplay*, July 1915, 91.

36. "Ruth Roland Creates Stir In 'Who Pays,'" *MPN*, June 5, 1915, 57.

37. "Mabel Normand . . . ," *Photoplay*, July 11, 1914, 239.

38. Kalton C. Lahue, *World of Laughter: The Motion Picture Comedy Short* (Norman: University of Oklahoma Press, 1966), xii.

39. Kalton C. Lahue and Samuel Gill, *Clown Princes and Court Jesters* (New York: A. S. Barnes, 1970), 149–51, 369–70.

40. "Mabel Normand . . . ," *Photoplay*, July 11, 1914, 239.

41. Sidney Sutherland, "Madcap Mabel Normand," *Liberty*, Sept. 5, 1930, 60; Sept. 13, 1930, 18, Kemp Niver Collection, AMPAS. For example, in Vitagraph comedies c. 1910, Flora Finch starred as the "thin, sour, henpecking wife," in contrast to her husband, the "jovial, fat John Bunny"; see Harvey Douglass, "Movie Stars Got Start on Flatbush Lots," *Brooklyn Daily Eagle*, Feb. 17, 1933, M2.

42. Henry Jenkins, *What Made Pistachio Nuts? Early Sound Comedy and the Vaudeville Aesthetic* (New York: Columbia University Press, 1992), chaps. 2 and 9.

43. Quoted in Betty Harper Fussell, *Mabel* (New York: Ticknor and Fields, 1982), 57.

44. Mack Sennett, *King of Comedy* (Garden City, NY: Doubleday, 1954), 47–49, 136.

45. "Mabel Normand . . . ," *Photoplay*, July 11, 1914, 239; Randolph Bartlett, "Would You Ever Suspect It?" *Photoplay*, Aug. 1918, 45. *Mabel's Dramatic Career* (1913), *Caught in a Cabaret* (1914), and *Mabel, Fatty, and the Law* (1915) viewed at Film Study Center, MOMA; *Mabel Lost and Won* (1915) viewed at MPBRSD-LC; Sennett, *King of Comedy*, 49; for an extended analysis of the Gibson Girl see Banner, *American Beauty*, 154–74.

46. *Mabel's Married Life* (Keystone, 1914), viewed at MOMA.

47. Janet Staiger, "The Hollywood Mode of Production to 1930," in David Bordwell, Kristin Thompson, and Janet Staiger, *The Classical Hollywood Cinema: Film Style and Mode of Production to 1960* (New York: Columbia University Press, 1985), 121–27.

48. Lahue, *World of Laughter*, 77–78.

49. Charles Chaplin, *My Autobiography* (New York: Simon and Schuster, 1964), 153, 158–59.

50. Ibid., 159–62.

51. James R. Quirk, "The Girl on the Cover," *Photoplay*, Aug. 1915, 40.

52. MPW, Dec. 13, 1913, 1289; Lahue, *World of Laughter*, 75; Slide, *American Film Industry*, 184. Like most female filmmakers of the 1910s, Normand received little attention when it was announced that she was going to direct. *Photoplay* simply remarked that "this will undoubtably make Keystone more popular than ever" (*Photoplay*, April 1914, n.p.).

53. Julian Johnson, "The Shadow Stage," *Photoplay*, Sept. 1916, 119–20; Sennett, *King of Comedy*, 160; Kalton C. Lahue, *Mack Sennett's Keystone: The Man, the Myth, and the Comedies* (New York: A. S. Barnes, 1971), 114. Two of the surviving films codirected by Normand, *Caught in a Cabaret* and *Mabel's Married Life*, viewed at MOMA, appear to bear this out.

54. "Helen Holmes Mourns Death of Her Father," *New York Review*, 1916, n.p.; Helen Holmes with Terry Ramsaye, Director of Publicity, "Shooting the Thrills," typed MS, (Mutual Film Corporation, 1916), n.p., Helen Holmes clippings file, Billy Rose Theatre Collection, NYPL-PA; "Helen Holmes Dies at 58," *Los Angeles Examiner*, July 10, 1950, n.p.

55. Slide, *Early American Cinema*, 161–62; Anthony Slide, *The American Film Industry: A Historical Dictionary* (New York: Limelight Editions, 1990), 13–14, 313; Alan Burden, "The Girl Who Keeps a Railroad," *Photoplay*, July 1915, 90; *Photoplay*, Jan. 1916, 109; Feb. 1916, 95; *Variety*, Sept. 24, 1915, 18; Helen Holmes biography file, AMPAS; and Kalton C. Lahue, *Winners of the West: The Sagebrush Heroes of the Silent Screen* (New York: A. S. Barnes, 1970), 159.

56. Brian Duryea, "The Necessity of Thrills," *Green Book*, April 1916, 741–43, Helen Holmes clippings file, Billy Rose Theatre Collection, NYPL-PA; *Hazards of Helen*, episode 13, viewed at MPBRSD-LC.

57. Slide, *American Film Industry*, 13–14, 313; Burden, "The Girl Who Keeps a Railroad," *Photoplay*, Jan. 1916, 109; *Photoplay*, Feb. 1916, 95; *Variety*, Sept. 24, 1915, 18.

58. *The Girl and the Game*, episodes 1, 10, 13, 15, viewed at MPBRSD-LC; *Mutual News*, March 24, 1917, 1862; *MPW*, March 24, 1917, 1956; July 21, 1917, n.p.; James C. Jewell, "Helen Holmes and the Hazards of Helen," *Worlds of Yesterday*, Dec. 1979, n.p., Helen Holmes biography file, AMPAS.

59. Lahue, *Ladies in Distress*, 52.

60. *Dramatic Review*, Jan. 31, 1914, n.p.; "Clever Grace Cunard," *Photoplay*, April 1914, 33; "West Coast Studio Jottings," *Photoplay*, April 1914, 99; William M. Henry, "Her Grace and Frances I," *Photoplay*, April 1916, 28−29; "Talking to a Director," *Motion Picture Stories*, June 23, 1916, n.p.; Anthony Slide, *Early Women Directors* (New York: A. S. Barnes, 1977), 57; Edmonds, *Big U*, 44−45.

61. Edmonds, *Big U*, 37−38; credits for *Lucille Love* list Francis Ford as the producer and Grace Cunard as the writer, *Lucille Love, Girl of Mystery* (Universal, 1914), viewed at MPBRSD-LC.

62. Henry, "Her Grace and Frances I," *Photoplay*, April 1916, 29; *MPN*, May 1, 1915, 59; *MPN*, May 22, 1915, 62; *MPN*, June 26, 1915, 50, 136; Dudley L. McClure, "The Broken Coin," *Coins*, Feb. 1970, 30−31; clipping in *The Broken Coin* (1915) production file, AMPAS; Edmund, *Big U*, 60; Lahue, *Continued Next Week*, 29−30.

63. Lahue, *Ladies in Distress*, 54.

64. Edmonds, *Big U*, 38.

65. *The Purple Mask*, episodes 2, 5, 7, 8, 11−13, 16, viewed at MPBRSD-LC. Episode 13 concerns the "home for unfortunate girls"; Edmonds, *Big U*, 90; Lahue, *Continued Next Week*, 44.

66. Russell Merritt, "Nickelodeon Theaters, 1905−1914: Building an Audience for the Movies," in *The American Film Industry*, ed. Tino Balio, rev. ed. (Madison: University of Wisconsin Press, 1987), 100−101.

67. *Photoplay*, July 1916, 77; *Photoplay*, Dec. 1916, 64.

68. Capt. Leslie T. Peacocke, "Enter—The Free Lance Writer," *Photoplay*, March 1917, 95.

69. Edmonds, *Big U*, 143−44; Lahue, *Continued Next Week*, vi, 126; Lahue, *Winners of the West*, 162−63.

70. Richard Koszarski, *An Evening's Entertainment: The Age of the Silent Feature Picture, 1915−1928* (Berkeley: University of California Press, 1994), 164−65; Miriam Hansen, *Babel and Babylon*, 99.

71. [?] Sheldon to George Kleine, April 19, 1917, George Kleine Papers, MD-LC; Cohn, "Harvesting the Serial," 25; Stedman, *The Serials*, 37.

72. Lahue, *Continued Next Week*, 60.

73. *MPW*, July 5, 1919, 57.

74. S. L. Rothapfel, "Creation of Atmosphere as Necessary as Presentation of Striking Pictures," *MPW*, Aug. 14, 1920, 883; Koszarski, *An Evening's Entertainment*, 164−65; Miriam Hansen, *Babel and Babylon*, 99.

75. Slide, *Early American Cinema*, 140−41.

76. Indeed, the Drew comedies may have provided inspiration for *I Love Lucy*. The date "12/10/54" and the name "Desi Arnaz" appear in the upper right-hand corner of a one-page typed filmography of the Mr. and Mrs. Sidney Drew comedies of 1917. Sidney Drew biography file, AMPAS.

77. Randolph Bartlett and Kitty Kelly, "The Shadow Stage," *Photoplay*, Feb. 1918, 68; *Fox Trot Finesse* (Vitagraph, 1915), viewed at MPBRSD-LC; Frederick James Smith, "Seeking the Germ," *Photoplay*, Sept. 1917, 28; Jacobs, *Rise of the American Film*, 268; advertisement, MPW, April 21, 1917, n.p.

78. "When Two Hearts Are Won," MPW Sept. 2, 1911, 610; Albert E. Smith, "The Beginning," Albert E. Smith Collection, SC-UCLA; Smith, "Seeking the Germ," 28; Sidney Valentine, "o-h S-i-d-n-e-y!" *Photoplay*, May 1919, 30; Henry MacMahon, "Women Directors of Plays and Pictures," LHJ, Dec. 1920, 13; Lahue, *World of Laughter*, 23.

79. Indeed, even the great D. W. Griffith believed that "the mental age of the average audience is about nine years" (Lahue, *World of Laughter*, 19); "Here are 'Henry and Polly,'" *Photoplay*, Oct. 1918, 44–45; Sidney Drew, "Comedy Picture Production," MPW, July 21, 1917, 412–13; Smith, "Seeking the Germ," 29.

80. *A Florida Enchantment* (1914), viewed at MPBRSD-LC.

81. Advertisement, MPW, April 21, 1917, n.p.

82. Ellis Paxton Oberholtzer, *The Morals of the Movies* (Philadelphia: Penn Publishing, 1922), 71–72.

83. Koszarski, *An Evening's Entertainment*, 175.

84. Paul H. Davis, "Investing in the Movies," *Photoplay*, June 1916, 137–38.

85. Lahue, *Continued Next Week*, 38.

86. Staiger, "Hollywood Mode of Production to 1930," 134–36.

87. MPW, June 1916, 107; MPW, Oct. 1917, 116; Edmonds, *Big U*, 89, 96, 100, 125–28, 143; Lahue, *Continued Next Week*, 43–44; "Grace Cunard Marries . . . ," in Cal York, Plays and Players, *Photoplay*, April 1917, 122; "Fordart Productions," in Slide, *American Film Industry*, 135; Lahue, *Ladies in Distress*, 58–59; J. P. McGowan also worked at Universal for several years after his breakup with Helen Holmes; see MPW, July 5, 1919, 57.

88. *Variety*, Sept. 24, 1915, 18; *Photoplay*, Jan. 1916, 109; *Photoplay*, Feb. 1916, 95; Slide, *Early American Cinema*, 161–62; Burden, "The Girl Who Keeps a Railroad," 90; Lahue, *Winners of the West*, 160–61; Jewell, "Helen Holmes and the Hazards of Helen," n.p.; "Helen Holmes Serials in Demand," *Mutual News*, March 24, 1917, 1862; "The Railroad Raiders," MPW, March 24, 1917, 1956; for Hutchinson and Mutual see Slide, *American Film Industry*, 13–14, 313, 231–32, 374; Lahue, *World of Laughter*, 64.

89. *Photoplay*, March 1918, 83; *Photoplay*, Sept. 1918, 88.

90. For McGowan and Eddie Polo see MPW, July 5, 1919, 57; for S. L. K. see C. S. Sewell, Among the Independent Producers, MPW, July 5, 1919, 100; Lahue, *Winners of the West* 162.

91. *Wid's Daily,* July 2, 1919, 1; C. S. Sewell, Among the Independent Producers, MPW, July 12, 1919, 100; "The Fatal Fortune," MPW, Sept. 20, 1919, 1867; *MPW,* Aug. 2, 1919, 720; Lahue, *Continued Next Week,* 67–68.

92. "Helen Holmes Sues," *Variety,* July 3, 1920, n.p.; Lahue, *Continued Next Week,* 4–5, 22, 88; Lahue, *Winners of the West,* 162–63.

93. After acting in independent westerns, Gibson joined the Wild West show in the Ringling Brothers and Barnum & Bailey Circus, appearing in bit parts until 1961 (Lahue, *Continued Next Week,* 25–27). It was reported that Gibson was at Holmes's deathbed in 1950, after both had been long forgotten by the industry; see "Helen Holmes Dies at 58," *Los Angeles Examiner,* July 10, 1950, n.p., Helen Holmes biography file, AMPAS.

94. The Drews defected just in time apparently. Vitagrapher James Morrison claimed that when he left the studio two years later, the efficiency experts "had people straightening nails"; see Albert E. Smith, "The Beginning," Albert E. Smith Collection, SC-UCLA; Kevin Brownlow *The Parade's Gone By* (New York: Knopf, 1968), 20; Anthony Slide, *The Big V: A History of the Vitagraph Company* (Metuchen, NJ: Scarecrow Press, 1976), 58.

95. When they moved from Vitagraph to Metro, the Drews received a reported joint salary of $90,000. Their earnings for 1917 are unknown. *Photoplay,* March 1916, 120.

96. *Photoplay,* May 1918, 85; "Diamonds and Coal," *Photoplay,* June 1918, 76; *Photoplay,* Aug. 1918, 79; "Here Are 'Henry and Polly,'" *Photoplay,* Oct. 1918, 44–45; *Photoplay,* Jan. 1923, 95; Lahue, *World of Laughter,* 23; "Public Shows Deep Interest in Drews," MPW, March 8, 1919, 1373; *Variety,* April 11, 1919, n.p.; "Sidney Drew Dies," unidentified clipping, biography file, AMPAS; Slide, *Big V,* 100–101; Slide, *Early Women Directors,* 106; Kenneth W. Munden, ed., *The American Film Institute Catalog of Motion Pictures Produced in the United States: Feature Films, 1921–1930* (New York: R. R. Bowker, 1971), 149 (hereafter cited as Munden, AFI Catalog ... 1921–1930).

97. "Mrs. Drew's Comedies Are Booked by Capitol," unidentified source, 1920, Mr. and Mrs. Drew clipping file, Billy Rose Theatre Collection, NYPL-PA.

98. "Mrs. Drew Begins Work on Series of Two-Reelers," *Exhibitor's Trade Review,* Dec. 13, 1919, n.p.; "First of Drew Series Filmed," *MPN,* Jan. 17, 1920, n.p.; "Announces Plans for Drew-Street Comics," *MPN,* Jan. 24, 1920, n.p.; "'The Charming Mrs. Chase' Has Been Finished," unidentified clipping; "Pathé Comedies," unidentified clipping, April 17, 1920; "Pathé Says Mrs. Drew's Comedy Is a Big Hit," unidentified clipping; "Drew Comedies Going Big All Over," unidentified clipping; all clippings in Mr. and Mrs. Drew clipping file, Billy Rose Collection, NYPL-PA.

99. "Mrs. Sidney Drew Acts in Lake George Locale," MPW, Sept. 6, 1919, 1464; MPW, Sept. 27, 1919, 1963; Henry MacMahon, "Women Directors of Plays and Pictures," LHJ, Dec. 1920, 13; "Pathé Points to Gains in Short Subject Field

and Discloses Plans for Next Year," *MPW*, July 24, 1920, 451; Albert E. Smith to Ronald Reader, telegram, Nov. 4, 1920, Albert E. Smith Collection, SC-UCLA; "Cousin Kate," *MPW*, Jan. 29, 1921, 598; "Mrs. Lucille McVey Drew," *Variety*, Nov. 11, 1925, n.p.

100. "The De Havens with 'Smiling Bill' Parsons," *MPW*, Jan. 25, 1919, 516; advertisement for Capitol Comedies, *MPW*, July 12, 1919, 171; De Haven product claimed to be first-run material in "Wm. A. Seiter, Director," advertisement, *Wid's Year Book* (Hollywood: Wid's Film and Film Folks, 1919–20), n.p.; First National Film Company of America advertisement, *Wid's Year Book* (1919–20), n.p.; "The Year in Headlines," *Wid's Year Book* (Hollywood: Wid's Film and Film Folks, 1921–22), 95; "Personnel of Important Producing and Distributing Organizations," *Wid's Year Book* (1921–22), 393; "Statistical Survey Reveals That First National Production Units Annually Spend about $5,376,000," *MPW*, July 17, 1920, 339; De Haven Studio not listed in *Wid's Year Book* (Hollywood: Wid's Film and Film Folks, 1922–23); "Short Subject Releases," *Wid's Year Book* (Hollywood: Wid's Film and Film Folks, 1924), 79; "Twin Beds," in Hansen, *AFI Catalog . . . 1911–1920*, 957; Anthony Slide, "RKO Radio Pictures, Inc." in Slide, *American Film Industry*, 289.

101. Lahue, *World of Laughter*, 18, 56.

102. Ibid., 76. The cost of making Keystone comedies is difficult to assess. According to Sennett negative costs for each two-reeler came in between $25,000 and $30,000, and each earned between $75,000 and $85,000; see Sennett, *King of Comedy*, 114. According to Kalton C. Lahue the New York office of Triangle "begged Sennett to spend more money on his comedies," and Keystone doubled its average cost per negative by raising actors' salaries, still only reaching $18,000; see Kalton C. Lahue, *Dreams for Sale: The Rise and Fall of the Triangle Film Company* (New York: A. S. Barnes, 1971), 133. This seems confirmed by Arbuckle's equally important reason for leaving Keystone, his relatively paltry salary of $500 a week. At Comique he made $7,000 a week and 25 percent of the profits; see Tony Scott, "In the Silent Era, Balboa Studios Forged Showbiz History," *Daily Variety*, 5th Anniversary Issue (1990), 121–22.

103. Sutherland, "Madcap Mabel Normand," *Liberty*, Sept. 20, 1930, 46; *Photoplay*, June 1916, 104.

104. *Variety*, Sept. 24, 1915, 16; Lahue, *Dreams for Sale*, 75–76; Kalton C. Lahue, *Mack Sennett and the Keystone Legend* (New York: A. S. Barnes, 1971), 51–53.

105. *Variety*, Oct. 8, 1915, 18; *Photoplay*, Dec. 1915, 60; *Photoplay*, June 1916, 104.

106. Sennett, *King of Comedy*, 195, 199; *Photoplay*, May 1916, 145; *Photoplay*, June 1916, 104; *Photoplay*, July 1916, 99; *Photoplay*, Aug. 1916, 126; *Photoplay*, Sept. 1916, 94.

107. *Photoplay*, April 1917, 123–24; Sennett, *King of Comedy*, 201.

108. Winnick to Sennett, telegram, Nov. 14, 1917; John Waldron to Sennett, telegram, June 17, 1917; Sennett to Waldron, telegram, June 27, 1917; Sennett to

Waldron, telegram, June 29, 1917; Waldron to Sennett, telegram, July 3, 1917, all in Mack Sennett Collection, AMPAS; "Triangle Company Makes Virulent Charges against H. E. and R. E. Aitken," *MPW*, March 5, 1921, 69; for *Mickey* see Lahue, *Dreams for Sale*, 189; Sennett, *King of Comedy*, 200–201, 211; Fussell, *Mabel*, 60–61; Sennett was named lessor in a 1919 contract to let the studio "formerly known as Mabel Normand Studio" to William S. Hart, Sennett Collection, AMPAS.

109. "Glorifying the Bathing Beauty," *MPW*, March 26, 1927, 390, 413.

110. Sutherland, "Madcap Mabel Normand," *Liberty*, Sept. 20, 1930, 46–47; *Wid's Year Book* (1920–21), 29, 71; *Wid's Year Book* (1921–22), 4, 97; Fussell, *Mabel*, 103–11.

111. Lahue, *Continued Next Week*, 41–42.

112. Ibid., 59.

113. "Finds Demand for Serials Increasing," *MPW*, April 5, 1919, 74; "Pathé's Sales Force Wide Awake," *MPW*, July 12, 1919, 227–30; "Serials Are Speeding to the Fore," *MPW*, Aug. 16, 1919, 954; "What of the Serial?" *Film Daily Yearbook* (1919–20), n.p.; " 'Velvet Fingers', New George Seitz Marguerite Courtot Serial for Pathé Is Widening Field for Episode Plays," *MPW*, Nov. 27, 1920, n.p.; Lahue, *Bound and Gagged*, 35.

114. Advertisement, "Ruth of the Rockies," *MPW*, Aug. 14, 1920, 886–87. In the long run Laemmle's policy proved disastrous; see Thomas Schatz, *The Genius of the System: Hollywood Filmmaking in the Studio Era* (New York: Pantheon, 1988), 21; Lahue, *Bound and Gagged*, 79–80.

115. James Bell, "Is It Impossible? Marie'll Do It!" *Photoplay*, Nov. 1916, 49; Stedman, *The Serials*, 41–42; advertisement, "The Red Glove," *MPW*, Feb. 1, 1919, 560–61; "Universal's New Serial, 'The Red Glove,' Released," *MPW*, March 15, 1919, 1514; *MPW*, March 22, 1919, 1684.

116. Marc Wanamaker, "Balboa Amusement Corporation (1913–1928)," *Encyclopedia of the Movie Studios*, Local History Collection, Long Beach, California, Main Public Library, 233–34; Lahue, *Continued Next Week*, 35, 62; Stedman, *The Serials*, 36.

117. *Wid's Film Yearbook* (1919–20), n.p.

118. "Serials Set until 1921 by Brunet," and "Men of Brains Write Pathé Serials for Stars of Renown on Stage and Screen," *MPW*, July 12, 1919, 228–29; Lahue, *Continued Next Week*, 67; Stedman, *The Serials*, 44. The prerelease press in *MPW* stated that William Parke was to direct *The Adventures of Ruth*, but Lahue believes it was Marshall; see "Ruth Roland Company Starts on First Serial," *MPW*, Nov. 19, 1919, 378; Roland listed under contract to, or a user of, Brunton rental studios; see *Wid's Year Book* (1920–21, 1921–22), n.p.; advertisement for *Ruth of the Rockies*, *MPW*, Aug. 14, 1920, 886–87; review of *Ruth of the Rockies*, *MPW*, Aug. 28, 1920, 1216; advertisement for *The Avenging Arrow*, *MPW*, Aug. 16, 1921, 723–24.

119. Advertisement for *Rowdy Ann*, *MPW*, May 10, 1919, n.p.; Lahue and Gill, *Clown Princes and Court Jesters*, 372.

120. "Sherrill to Make Two-Reel Westerns," *MPW*, Feb. 1, 1919, 602; "Miss Texas Guinan Is a Native Daughter of Texas," *MPW*, March 22, 1919, 1694; "Nearly All Rights Sold on Guinan and Swain Pictures," *MPW*, July 19, 1919, 404; Lahue, *Winners of the West*, 129–32.

121. Larry Langman, *A Guide to Silent Westerns* (Westport, CT: Greenwood Press, 1992), 188, 401.

122. "'I Am the Woman' to Be First of Kremer's Texas Guinan Westerns," *MPW*, Feb. 5, 1921, 718; "Texas Guinan in Two-Reelers Produced by Her Own Company," *MPW*, Aug. 8, 1921, 610; Lahue, *Winners of the West*, 129–32; *MPW*, Jan. 29, 1921, 572.

123. *MPW*, Jan. 29, 1921, 572.

124. "Fine Arts Studios for Texas Guinan," *MPW*, Oct. 1, 1921, 557; "J. J. Goldberg Heads Company Producing Texas Guinan Films," *MPW*, Aug. 13, 1921, 174; "Texas Guinan Westerns Have Rapid Sale, Reports Company," *MPW*, Sept. 3, 1921, 85; "Texas Guinan Sues Reelcraft Pictures for $50,000 over Bulls-Eye Contract," *MPW*, Sept. 24, 1921, 402; Lahue, *Winners of the West*, 133–36; Abel Green, "Texas Guinan Helped Make B'way History during Volstead Era," *Variety Anniversary* (Jan. 4, 1956), 423; Dorothy Manners, "Queen of Prohibition," *Los Angeles Herald Examiner*, July 11, 1974; James Doherty, "Texas Guinan, Queen of Whoopee!" *Chicago Tribune*, March 4, 1951.

125. "Serials Are Backbone of Several Programs a Week at 60 Per Cent. of Picture Houses," *MPW*, April 15, 1921, 707.

126. "Revival of Motion Picture Patronage Is Rolling Eastward from West Coast," *MPW*, Sept. 3, 1921, n.p.; see also John Izod, *Hollywood and the Box Office, 1895–1986* (New York: Columbia University Press, 1988), 68.

127. House Committee on Education, *Proposed Federal Motion Picture Commission: Hearings on HR 4094 and HR 6233*, 69th Cong., 1st sess., 1926, 147, 150; Oberholtzer, *Morals of the Movies*, 77.

128. "Find Many 'Movies' Bad for Children," *NYT*, April 10, 1920, 16; "Condemn Degraded Drama," *NYT*, July 2, 1920, 19; "Chicago Forbids All Films Showing Criminals in Action," *NYT*, Jan. 5, 1921, 1; "Newark Bars Crime Films," *NYT*, Jan. 7, 1921, 3; "Movies Inspire Wreck," *NYT*, Jan. 23, 1921, 16; Rev. Charles N. Lathrop, *The Motion Picture Problem* (New York: Federal Council of the Churches of Christ in America, 1922), 12; Lahue, *Continued Next Week*, 91.

129. It was common during the nickelodeon era for upscale theaters to get away with much more potentially objectionable material than the cheaper theaters, and this pattern was being repeated; see Merritt, "Nickelodeon Theaters, 1905–1914," 101–2.

130. On Arguments for censorship in New York State see Lathrop, *The Motion Picture Problem*, 32. Anticensorship advocates argued that the motion picture would squelch incipient bolshevism; see "Editorial against Censorship by Woman Exhibitor," *MPW*, April 12, 1919, 205; "Motion Picture Will Become National

Stabilizer in This Country, Says Hays at National Press Club Dinner," *MPW*, March 11, 1922, 155.

131. Oberholtzer, *Morals of the Movies*, 55.

132. "New Orleans Women Initiate Campaign to Ban Serial as Harmful to Nervous Tots," *MPW*, Nov. 19, 1921, 286.

133. Oberholtzer, *Morals of the Movies*, 56–57, 60.

134. "Bay State to Move to Censor Movies," *NYT*, May 11, 1920, 17; "Chicago Forbids All Films Showing Criminals in Action," *NYT*, Jan. 5, 1921, 1; "Newark Bars Crime Films," *NYT*, Jan. 7, 1921, 3; "New Orleans Women Initiate Campaign to Ban Serial as Harmful to Nervous Tots," *MPW*, Nov. 19, 1921, 286; William Lord Wright, *Photoplay Writing* (New York: Falk Publishing, 1922), 107. Most of these incidents occurred in 1921, but in Battle Creek, Michigan the serial was banned from 1919 to 1921; see "Serials Now Showing in Cereal City, Pathé Man Wins Over the Authorities," *MPW*, Oct. 15, 1921, 757.

135. Oberholtzer, *Morals of the Movies*, 74, 77.

136. "Nationwide Censorship Agitation," *MPW*, March 26, 1921; Francis G. Couvares, "Hollywood, Main Street, and the Church: Trying to Censor the Movies before the Production Code," *American Quarterly* 44, no. 4 (Dec. 1992): 586, 589; Morris L. Ernst and Pare Lorentz, *Censored: The Private Life of the Movies* (New York: Jonathan Cape and Harrison Smith, 1930), 103–21; May, *Screening Out the Past*, 54–55, 82, 86, 92.

137. Ernst and Lorentz, *Censored*, 115, 117.

138. "Political Pot Boiling at Albany" and "Albany Sees Two Big Picture Fights," *MPW*, Jan. 25, 1919, 463; "Indiana Women Plan Legislative Bill to Censor Films and Prevent Sunday Shows," *MPW*, Jan. 29, 1919, 534; "Many against Nebraska Bill," *MPW*, March 22, 1919, 1615; "As to Censorship," *NYT*, April 18, 1920, 5; "Legislature Besieged by Women Who Hope to Hamper the Moving Picture Industry," *MPW*, Jan. 19, 1921, 534; Lathrop, *The Motion Picture Problem*, 27, 47–48, 51; House Committee on Education, *Proposed Federal Motion Picture Commission*, 147, 150; Ernst and Lorentz, *Censored*, 115, 117; "Bill to Bar Film Thrillers," *NYT*, Jan. 8, 1920, 21; "'Bad Grammar' of House Bill Draws a Veto from Wilson," *NYT*, June 3, 1920, 1.

139. "Where Do the Women Stand Today?" *MPW*, Nov. 27, 1920, 453 (emphasis in original).

140. "Noticed and Noted," *NYT*, June 20, 1920, 2; "Agree to Clean Up Moving Pictures," *NYT*, March 7, 1921, 7; "Producers Take Drastic Step to Assure 100 Per Cent Clean Screen Productions," *MPW*, March 19, 1921, 240–41.

141. "Brady Off on Trip to Get Co-Operation of Women's Organizations with Screen," *MPW*, Nov. 5, 1921, 35; Couvares, "Hollywood, Main Street, and the Church," 587–91.

142. "Miller Signs New York Censorship Bill; Reconsiders Intention of Naming Woman," *MPW*, May 29, 1921, 378; "New York State Censors Begin Working after Frantic Week of Preparation," *MPW*, Aug. 13, 1921, 687; "Miller Names

Second Female Politician to Personnel of New York Censor Board," *MPW*, Aug. 27, 1921, 883.

143. "New Hallroom Boys Comedy Said to be Censor-Proof," *MPW*, Sept. 3, 1921, 85.

144. Lahue and Gill, *Clown Princes and Court Jesters*, 372–76.

145. Wright, *Photoplay Writing*, 110.

146. "Serials Are Backbone of Several Programs a Week at 60 Per Cent. of Picture Houses," *MPW*, April 15, 1921, 707; "Child Appeal Best Advertisement," *MPW*, July 5, 1919, 55.

147. See "Pathé Sales Force Wide Awake," *MPW*, July 12, 1919, 227–30; and "Finds Demand for Serials Increasing," *MPW*, April 5, 1919, 74.

148. Advertisements for *Elmo, the Mighty*, *MPW*, May 3, 1919, 597; *The Cyclone Smith Stories*, *MPW*, May 5, 1919, n.p.; and *The Midnight Man*, *MPW*, May 10, 1919, 736.

149. "Eight Pathé Serials for Coming Year; Will Continue 'Two-at-a-Time' Policy," *MPW*, June 24, 1920, 451.

150. Epes W. Sargent, "Pathé Offers Something Entirely New in a Series of One Reel Plays," *MPW*, March 5, 1921, 73; "Pathé Re-Affirms Its Intentions to Only Distribute Pictures That Are Wholesome," *MPW*, March 26, 1921, 364.

151. "Universal Has New Policy for Its Serials; New Department Created," *MPW*, Sept. 24, 1921, 445; "Universal to Present Serials Based upon Historical Facts," *MPW*, Sept. 3, 1921, 80; "Universal's 'Winners of the West' Finds Favor as Historical Serial," *MPW*, Nov. 5, 1921, 81; Epes. W. Sargent, "Two Big Features Will Sell Audiences on 'The Adventures of Tarzan' Serial," *MPW*, Aug. 27, 1921, 887; "'Adventures of Tarzan' Said to Be Booking at Rapid Rate," *MPW*, Sept. 3, 1921, 89; "Woman Censor Likes Outdoor Thrills," *MPW*, Aug. 20, 1921, 788.

152. Advertisement, *MPW*, April 21, 1921, 724; Roland also made *Ruth of the Rockies* (1920), *White Eagle* (1922), *Timber Queen* (1922), *Haunted Valley* (1923), and *Ruth Range* (1923).

153. The *Avenging Arrow* was booked into ten Paramount first-run houses in Missouri; see *Morning Telegraph* clipping, Feb. 27, 1921, *Avenging Arrow* production file, AMPAS.

154. Roland tried again to become an independent producer in 1923–24, this time making features with the respected director Tod Browning, but for unknown reasons this effort failed quickly. Nevertheless, Roland retired a wealthy woman thanks to her investments in Los Angeles real estate; see "Hal Roach Enlarged Activities and Will Make Next Serial Starring Ruth Roland," *MPW*, March 26, 1921, 362; "Pathé Exchange Reports New Ruth Roland Serial for Release January 1," *MPW*, Nov. 19, 1921, 324; Cal York, Gossip East and West, *Photoplay*, May 1923, 88; Lahue, *Continued Next Week*, 104–5, 108–9, 119; Stedman, *The Serials*, 44–45.

155. Advertisement, *Go-Get-'em-Hutch*, *MPW*, March 25, 1922, 362–63; "Eight Pathé Serials for Coming Year; Will Continue 'Two-at-a-Time' Policy," *MPW*, June 24, 1920, 451; "Pathé Sales Force Wide Awake," *MPW*, July 12, 1919,

227–30; "Finds Demand for Serials Increasing," *MPW*, April 5, 1919, 74; advertisement, *Elmo the Mighty*, *MPW*, May 3, 1919, 597; advertisement, *The Cyclone Smith Stories*, *MPW*, May 5, 1919, n.p.; advertisement, *The Midnight Man*, *MPW*, May 10, 1919, 736; Lahue, *Continued Next Week*, 16–17, 46: Stedman, *The Serials*, 50–61; Lahue, *Bound and Gagged*, 24.

156. Koszarski, *An Evening's Entertainment*, 206.

157. *MPW*, July 21, 1917, 468; *Photoplay*, Sept. 1917, 110; *Photoplay*, Jan. 1923, 95; *Wid's Year Book* (1921–22), 4, 97, 99, 316; Sutherland, "Madcap Mabel Normand," *Liberty*, Sept. 27, 1930, 67; and *Liberty*, Oct. 11, 1930, 67; Marjorie Rosen, *Popcorn Venus: Women, Movies, and the American Dream* (New York: Avon, 1973), 89, 95–96; Scott Eyman, *Mary Pickford, America's Sweetheart* (New York: Donald I. Fine, 1990), 169; Linda Martin and Kerry Segrave, *Women in Comedy* (Secaucus, NJ: Citadel Press, 1986), 87–93; Sennett, *King of Comedy*, 221–54, 268–73; Brownlow, *The Parade's Gone By*, 41; Fussell, *Mabel*, 198. Sennett and Normand were working at Paramount when the murder took place, and according to a letter from actress Louise Brooks to author Kevin Brownlow sent in May of 1971, Paramount did nothing to stop the publicity implicating Normand in the Taylor murder because her films were not doing well, and it was an easy way to get rid of her; see Barry Paris, *Louise Brooks* (New York: Anchor, 1989), 166. In his novelized version of director King Vidor's 1967 investigation of the murder, author Sidney D. Kirkpatrick implies that Paramount did all it could to hide Taylor's homosexuality, along the way distorting the facts and allowing Normand to be considered a suspect; see Sidney Kirkpatrick, *A Cast of Killers* (New York: Dutton, 1986), esp. 195, 248.

158. Couvares, "Hollywood, Main Street, and the Church," 132.

159. "$10,000 Free for Fans of Movies; 120 Pictures; Who Is This?" *Chicago Examiner*, 1922, in *The William Hays Papers*, ed. Douglas Gomery (Frederick, MD: University Publications of America, 1988), microfilm.

160. Oberholtzer, *The Morals of the Movies*, 87.

161. Robert S. Lynd and Helen Merrell Lynd, *Middletown: A Study in Contemporary American Culture* (New York: Harcourt, Brace, and World, 1956), 266; Couvares, "Hollywood, Main Street, and the Church," 587–91.

### Chapter 5. "The Real Punches"

1. Sumiko Higashi, *Cecil B. DeMille and American Culture: The Silent Era* (Berkeley: University of California Press, 1994), 28

2. For reference to "heart-interest" see *MPW*, July 21, 1917, 376–77; for reference to masculine "punch" see L. H. Johnson, " 'A Lady General in the Motion Picture Army,' Lois Weber-Smalley, Virile Director," *Photoplay*, June 1915, 42.

3. See Higashi, *Cecil B. DeMille*, 1–22.

4. Ibid., 28. Shelley Stamp makes the same comparison between Weber and DeMille in "Lois Weber, Progressive Cinema, and the Fate of 'The Work-a-Day Girl' in *Shoes*," *Camera Obscura* 56, vol. 19, no. 2 (2004): 145.

5. Stamp, "Lois Weber," 147–48.

6. Ibid.

7. Higashi, *Cecil B. DeMille*, 94–100.

8. Lee Grieveson, *Policing Cinema: Movies and Censorship in Early-Twentieth-Century America* (Berkeley: University of California Press, 2004), 186, 188.

9. Ibid., 148, 189–90.

10. Richard Koszarski, *An Evening's Entertainment: The Age of the Silent Feature Picture, 1915–1928* (Berkeley: University of California Press, 1994), 199–201.

11. Grieveson, *Policing Cinema*, 200–201.

12. Ibid., 202.

13. Edward Weitzel, review of *The Hand That Rocks the Cradle*, MPW, June 2, 1917, 1458.

14. "The Films Reviewed," *New York Dramatic Mirror*, May 26, 1917, 28.

15. *Variety Weekly*, May 18, 1917, 26.

16. "Mrs. Sanger to Tour with Her Film," *NYT*, March 28, 1917, 11; Brownlow, *Behind the Mask of Innocence: Sex, Violence, Prejudice, Crime: Films of Social Conscience in the Silent Era* (Berkeley: University of California Press, 1990), 48.

17. "Bars Birth Control Film," *NYT*, May 7, 1917, 18; Kay Sloan, *The Loud Silents: Origins of the Social Problem Film* (Chicago: University of Illinois Press, 1988), 87–89; "Would Restrain Commissioner Bell," *NYT*, May 10, 1917, 11.

18. "Upholds Mrs. Sanger's Film," *NYT*, June 7, 1917, 10.

19. Sloan, *The Loud Silents*, 89.

20. Ibid.

21. Ibid., 86.

22. Cal York, Plays and Players, *Photoplay*, April 1917, 122.

23. Arthur Denison, "A Dream in Realization," *MPW*, July 21, 1917, 417–18.

24. "The Center of the Stage," *Los Angeles Times*, November 14, 1921, 18.

25. Anthony Slide, *Lois Weber: The Director Who Lost Her Way in History* (Westport, CT: Greenwood Press, 1996), 25–28.

26. "Turning Out Masterpieces," *Moving Picture Stories*, Jan. 12, 1917, 28–29.

27. Shelley Stamp, "Lois Weber and the Celebrity of Matronly Respectability," forthcoming in *Looking Past the Screen: Case Studies in American Film History and Method*, ed. Jon Lewis and Eric Smoodin (Raleigh, NC: Duke University Press).

28. Quoted in Slide, *Lois Weber*, 106.

29. Elizabeth Peltret, "On the Lot with Lois Weber," *Photoplay*, Oct. 1917, 91.

30. See Steven J. Ross, *Working-Class Hollywood: Silent Film and the Shaping of Class in America* (Princeton, NJ: Princeton University Press, 1998), 115–42.

31. Jolo, "The Doctor and the Woman," *Variety*, April 26, 1918, n.p.

32. Jolo, "For Husbands Only," *Variety*, Sept. 6, 1918, n.p.; Jolo, "Borrowed Clothes," *Variety*, Nov. 22, 1918, n.p.

33. "Lois Weber Signed to Produce Picture at Enormous Salary," unidentified source, Dec. 7, 1918, Lois Weber biography file, MOMA.

34. "Weber Directs Anita Stewart," unidentified source, no date, Lois Weber clippings file, MOMA.

35. *A Midnight Romance*, in Patricia King Hansen, ed., *The American Film Institute Catalog of Motion Pictures Produced in the United States: Feature Films, 1911–1920* (Berkeley: University of California Press, 1988), 609 (hereafter cited as Hansen, *AFI Catalog . . . 1911–1920*); review of *A Midnight Romance*, *NYT*, March 10, 1919, 9; review of *A Midnight Romance*, *Variety*, May 9, 1919, n.p.

36. *Mary Regan*, in Hansen, *AFI Catalog . . . 1911–1920*, 595; review of *Mary Regan*, *Variety*, May 9, 1919, n.p.

37. Aline Carter, "The Muse of the Reel," *Motion Picture Magazine*, March 1921, 62–105; Winifred Aydelotte, "The Little Red Schoolhouse Becomes a Theatre," *Motion Picture Magazine*, March 1934, 35; Richard Koszarski, "The Years Have Not Been Kind to Lois Weber," *Village Voice*, Nov. 10, 1975, 140.

38. Description of Windsor from Observer [author], review of *What's Worth While*, n.d., n.p., Claire Windsor Collection, SC-USC. The beautiful Claire Windsor was originally named Ola Cronk.

39. Observer [author], review of *What's Worth While*, n.d., n.p.

40. Review of *To Please One Woman*, *NYT*, Dec. 20, 1920, 11.

41. Koszarski, "The Years Have Not Been Kind to Lois Weber," 140.

42. "Plays Up His Cartoons to Boom New Titles," *MPW*, Jan. 8, 1921, 202.

43. Higashi, *Cecil B. DeMille*, 140; see also Ruth Vasey, "Beyond Sex and Violence: 'Industry Policy' and the Regulation of Hollywood Movies, 1922–1930," in *Controlling Hollywood: Censorship and Regulation in the Studio Era*, ed. Matthew Bernstein (New Brunswick, NJ: Rutgers University Press, 1999).

44. Higashi, *Cecil B. DeMille*, 144.

45. Ibid., 144–58

46. "Class and Has Great Characterizations but Rough in Spots" [review of *Old Wives for New*], *Wid's Daily*, May 20, 1918, n.p.

47. Kathy Peiss, *Cheap Amusements: Working Women And Leisure in Turn-of-the-Century New York* (Philadelphia: Temple University Press, 1986).

48. Adolph Zukor, "Zukor Outlines Coming Year's Productions," *MPN*, June 29, 1918, 3869, quoted in Higashi, *Cecil B. DeMille*, 160.

49. Higashi, *Cecil B. DeMille*, 160–61.

50. "Class and Has Great Characterizations but Rough in Spots" [review of *Old Wives for New*], *Wid's Daily*, May 20, 1918, n.p.

51. Higashi, *Cecil B. DeMille*, 160–61.

52. Ibid., 162.

53. "Give Director De Mille a Prominent Place in Your Advertising" [review of *Don't Change Your Husband*], *Wid's Daily*, Jan. 26, 1919, n.p.

54. *Too Wise Wives* (1921), viewed at MPBRSD-LC.

55. Review of *The Affairs of Anatol*, *NYT*, Sept. 12, 1921, 16.

56. *The Blot*, viewed at MPBRSD-LC; review of *The Blot*, *Variety*, Aug. 19,

1921, n.p.; review of *The Blot*, *MPN*, Aug. 27, 1921, n.p., from Claire Windsor Collection, SC-USC.

57. Review of *What Do Men Want?* *NYT*, Nov. 14, 1921, 18.

58. Fritzi Remont, "The Lady behind the Lens," *Motion Picture Magazine*, May 1918, 126.

59. Aydelotte, "The Little Red Schoolhouse," 35.

60. Maescher was a Los Angeles real estate developer, which probably explains the source of her startup funds. See Mary Kelly, "*Night Life in Hollywood* Marks New Feminine Influence in Picture Making," *MPW*, July 29, 1922, 339.

61. E. Leslie Gilliams, "Will Woman's Leadership Change the Movies?" *Illustrated World*, Feb. 1923, 860.

62. "Universal Film Director Marries Leading Woman," *MPW*, Nov. 29, 1913, 993.

63. "Uses Hubby's Name," *New York Morning Telegraph*, Aug. 14, 1921, n.p., Dorothy Davenport Reid biography file, AMPAS.

64. Eric Shaefer, *"Bold! Daring! Shocking! True!" A History of Exploitation Films, 1919–1959* (Durham, NC: Duke University Press, 1999), 223.

65. Ibid.

66. Cal York, Plays and Players, *Photoplay*, April 1923, 76.

67. Minutes, Meeting of the Board of Directors, Feb. 4, 1924, Thomas Ince Collection, MD-LC.

68. "Continuity sheet on 'Human Wreckage,'" n.d., production file, 1914–55, box 29, Thomas H. Ince Collection, MD-LC. This continuity script may well have been revised before the film was shot.

69. Brownlow, *Behind the Mask of Innocence*, 90–91.

70. This interpretation is based on my reading of a continuity script that may have been revised before the film was shot (see note 67 above); for other drug-related social problem films see Brownlow, *Behind the Mask of Innocence*, 117–19.

71. Review of *Human Wreckage*, *Variety*, July 4, 1923, 32, quoted in Brownlow, *Behind the Mask of Innocence*, 116.

72. "Mrs. Reid's 'Human Wreckage' a Great Success," *Photoplay*, Sept. 1923, 45; see also Shaefer, *"Bold! Daring! Shocking! True!"*

73. "A Remarkable Monument to Wally Reid's Memory," *Photoplay*, Sept. 1924, 74; Brownlow, *Behind the Mask of Innocence*, 173–74.

74. *The Red Kimono* (1926), viewed at FTA-UCLA.

75. Anthony Slide, "The Red Kimono," in *Magill's Survey of Cinema*, 2:906; Anthony Slide, *Early Women Directors* (New York: A. S. Barnes, 1977), 78–79.

76. Obituary of Reid, *Variety*, Oct. 19, 1977, n.p.

## Chapter 6. A "'Her-Own-Company' Epidemic"

1. *Photoplay*, May 1917, 121.

2. Benjamin B. Hampton, *A History of the American Film Industry from Its Beginnings to 1931* (New York: Dover, 1970), 146–49.

3. Richard Koszarski, *An Evening's Entertainment: The Age of the Silent Feature Picture, 1915–1928* (Berkeley: University of California Press, 1994), 68, 265.

4. Hampton, *History of the American Film Industry*, 149–55; Koszarski, *An Evening's Entertainment*, 265–66; Tino Balio, "Stars in Business: The Founding of United Artists," in *The American Film Industry*, ed. Tino Balio, rev. ed. (Madison: University of Wisconsin Press, 1985), 159; Scott Eyman, *Mary Pickford, America's Sweetheart* (New York: Donald I. Fine, 1990), 73, 86–87, 89.

5. Hampton, *History of the American Film Industry*, 140.

6. Ibid., 146–49, 165, 168, 179–92.

7. "Close-Ups," *Photoplay*, Nov. 1916, 64.

8. *United States of America before Federal Trade Commission*, 51; Koszarski, *An Evening's Entertainment*, 71–72; Hampton, *History of the American Film Industry*, 147–48.

9. Tino Balio, "Struggles for Control, 1908–1930," in Balio, *American Film Industry*, 115; Hampton, *History of the American Film Industry*, 174; John Izod, *Hollywood and the Box Office, 1895–1986* (New York: Columbia University Press, 1988), 47; David Bordwell, "The Classical Hollywood Style, 1917–1960," in *The Classical Hollywood Cinema: Film Style and Mode of Production to 1960*, by David Bordwell, Janet Staiger, and Kristin Thompson (New York: Columbia University Press, 1985), 14, 99; Paul H. Davis, "Investing in the Movies," *Photoplay*, Aug. 1916, 119–21.

10. Mae D. Heuttig, *Economic Control of the Motion Picture Industry* (Philadelphia: University of Pennsylvania Press, 1944), 27; Hampton, *History of the American Film Industry*, 167–68.

11. Frank Woods, "The Academy of Motion Picture Arts and Sciences," *Transactions of the Society for Motion Picture Engineers* 12, no. 33 (April 1928): 25–32; Hampton, *History of the American Film Industry*, 167. For popular directors and stars the costs could go much higher. The negative cost (which does not include printing, distribution, or advertising) of Lois Weber's *Hypocrites* (1914) was $18,000; *It's No Laughing Matter* (1915) was $9,000, and *Captain Courtesy* (1915) was $25,000; see Paramount script file, Special Collections, AMPAS.

12. Janet Staiger, "The Hollywood Mode of Production to 1930," in *The Classical Hollywood Cinema: Film Style and Mode of Production to 1960*, by David Bordwell, Janet Staiger, and Kristin Thompson (New York: Columbia University Press, 1985), 124, 136–37; Charles Musser, "Pre-Classical American Cinema: Its Changing Modes of Production," in *Silent Film*, ed. Richard Abel (New Brunswick, NJ: Rutgers University Press, 1996), 85–108.

13. Staiger, "Hollywood Mode of Production to 1930," 135–36.

14. According to Blanche Sweet, "Mary would say I want so and so and so and so. [Zukor] would say no and she'd go away . . . Then he'd call her back and say, 'All right. Come on back' "; see Blanche Sweet, interview by Raymond Daum, June 1, 1981, 60, CUOHC.

15. Cal York, Plays and Players, *Photoplay*, June 1916, 108; Aug. 1916, 129; July 1916, 102; Nov. 1916, 101–3.

16. "Warren Stars Independent Distributing Firm to Have Thirty Features Yearly," *MPW*, April 23, 1921, 815; "Hodkinson to Have Own Exchange System after November 1," *MPW*, Sept. 10, 1921, 159; Hampton, *History of the American Film Industry*, 187; Slide, *American Film Industry*, 334, 389, 397.

17. Alfred A. Cohn, "The Middlemen of the Movies," *Photoplay*, March 1917, 79–80.

18. In an interesting anecdote related by Kevin Lewis regarding the value of the star, Selznick, while still at World, rushed to find out what caused a commotion in the office and expressed relief to find that it was only the sinking of the *Lusitania*: "Oh, is that all. I was afraid Clara Kimball Young was breaking her contract" (Kevin Lewis, "A World across from Broadway: The Shuberts and the Movies," *Film History* 1 [1987]: 39–45); review of *The Little Minister*, *MPW*, Oct. 12, 1912, 132; "Clara Kimball Young," *Photoplay*, April 1914, n.p.; George Blaisdell, "Clara Kimball Young, Artist," *MPW*, Oct. 3, 1914, 41–42; Monte M. Katterjohn, "Clara K. Young," *Photoplay*, Oct. 1914, 75–78; Hampton, *History of the American Film Industry*, 134.

19. "Clara Kimball Young to Have Own Company," unidentified clipping, Feb. 1, 1916, Florence Lawrence Collection, LACMNH; Lewis, "A World across from Broadway."

20. "Moving Picture World News Reel for June," *MPW*, July 21, 1917, 440.

21. "Close-Ups," *Photoplay*, Dec. 1916, 63.

22. Lewis, "A World across from Broadway," 45; Slide, *American Film Industry*, 195; Heuttig, *Economic Control of the Motion Picture Industry*, 32. The Kitty Gordon Film Corp. released only one feature, *Vera, the Medium*, in 1917; see Patricia King Hansen, ed., *The American Film Institute Catalog of Motion Pictures Produced in the United States: Feature Films, 1911–1920* (Berkeley: University of California Press, 1988), 986 (hereafter cited as Hansen, *AFI Catalog . . . 1911–1920*).

23. "Close-Ups," *Photoplay*, Dec. 1916, 64.

24. Cal York, Plays and Players, *Photoplay*, Dec. 1916, 90; "Norma Talmadge, Inc.," *Photoplay*, Feb. 1917, 84–85; "Norma Talmadge Company Buys Additional Building," *MPW*, Jan. 18, 1919, 314; "Norma Talmadge Looking for Plays Which Will Offer Strong, Dramatic Emotional Roles, While Constance Wants to Play in Subtle Comedies," *MPW*, Dec. 25, 1920, 1025; Hampton, *History of the American Film Industry*, 135–37; Tino Balio, *United Artists: The Company Built by the Stars* (Madison: University of Wisconsin Press, 1976), 52, 81–82.

25. Neal Gabler, *An Empire of Their Own: How the Jews Invented Hollywood* (New York: Anchor/Doubleday, 1988), 84–92.

26. Stewart had six months left on her contract, and Vitagraph sued for $250,000 in damages, but the out-of-court agreement amounted to $106,713.28; see "Price Waterhouse and Co., Reports and Accounts of Vitagraph," March 31,

1919, Albert E. Smith Collection, SC-UCLA. Interestingly, Stewart became a star in 1914 in *A Million Bid*, written by Mrs. Sidney Drew, which opened Vitagraph's first picture palace at 44th Street and Broadway in New York; see F. B. Fowler, "History of the Oldest Active Movie Pioneer," L.A. *Illustrated Daily News*, Feb. 11, 1924, n.p., Albert E. Smith Collection, SC-UCLA. In the summer of 1917 Stewart began feigning illness to get out of her Vitagraph contract; see *Photoplay*, Oct. 1916, 99; *Photoplay*, Sept. 1917, 117. For an account of her legal troubles see *Photoplay*, Dec. 1917, 93–94; Jan. 1918, 92; March 1918, 92; July 1918, 85; and Oct. 1918, 88; see also Dewitt Bodeen, "Anita Stewart," *Films in Review* 18, no. 3 (March 1968): 150–52. For distribution see *Wid's Year Book* (Hollywood: Wid's Film and Film Folks, 1918), n.p.; *Wid's Year Book* (Hollywood: Wid's Film and Film Folks, 1919–20), 41; *Wid's Year Book* (Hollywood: Wid's Film and Film Folks, 1921–22), 314; an Anita Stewart filmography is provided in Bodeen, "Anita Stewart," 156–61; for Mayer see Gabler, *An Empire of Their Own*, 90–92.

27. *Photoplay*, May 1917, 121.

28. Slide, *American Film Industry*, 397; Cal York, Plays and Players, *Photoplay*, June 1918, 83.

29. "Robertson-Cole to Furnish Films to Affiliated-Mutual," *Wid's Daily*, Nov. 21, 1918, 1.

30. *Wid's Year Book* (1921–22). The Great Northern Finance Corporation advertised that it would "lend money, discount or purchase negotiable paper, notes, drafts, acceptances, contracts, producers' and distributor's accounts and evidences of debt generally. We will finance and participate in financing motion picture enterprises, including production, distribution, and exhibition" (52).

31. Hansen, *AFI Catalog . . . 1911–1920*, 192, 514, 519, 697, 912; Cal York, Plays and Players, *Photoplay*, March 1917, 86; Elizabeth Peltret, "Bessie Barriscale's Nemesis," *Photoplay*, March 1918, 39–40; "Superlative Company Launched" [referring to Lois Meredith], *MPW*, June 16, 1917, 1793; Cal York, Plays and Players, *Photoplay*, July 1918, 82; Oct. 1918, 91; Year in Headlines, *Wid's Year Book* (1918), n.p.; Slide, *American Film Industry*, 397; Cal York, Plays and Players, *Photoplay*, June 1918, 83; P. A. Parsons, "Pathé Sales Force Wide Awake," *MPW*, July 12, 1919, 227; *Wid's Year Book* (1919–20), n.p.; *Wid's Year Book* (1920–21), 69, 83, 85, 397; *Wid's Year Book* (1921–22), 89, 97, 103, 123, 316; Hansen, *AFI Catalog . . . 1911–1920*, 182; Cal York, Gossip East and West, *Photoplay*, Dec. 1923, 80.

32. Hayakawa formed his company in April 1920, Mix formed his in June 1920, Kerrigan formed his in Dec. 1920, and Pickford formed his in July of 1921; see *Wid's Year Book* (1920–21), 87, 93, 97, 979.

33. *Federal Trade Commission v. Famous Players Lasky*, Brief for the Commission, pt. 1, 46; Hampton, *History of the American Film Industry*, 188.

34. Cal York, Plays and Players, *Photoplay*, Dec. 1917, 110; April 1918, 79.

35. Labor organizations, for example, were able to make their own films during

this period; see Steven J. Ross, *Working-Class Hollywood: Silent Film and the Shaping of Class in America* (Princeton, NJ: Princeton University Press, 1998), chap. 6; Hampton, *History of the American Film Industry*, 211; Lewis Jacobs, *The Rise of the American Film: A Critical History* (New York: Columbia University Press, 1968), 293.

36. The Mandarin Film Company built its own studio, based on "Chinese ideas," and released one film, *The Curse of the Quon Gwon*, partially filmed in China; see Hansen, *AFI Catalog . . . 1911–1920*, 140, 204, citing "Marion E. Wong, Chinese Film Producer," *MPW*, July 7, 1917, 63.

37. "A Successful Woman Sales Manager," *MPW*, [1913?], n.p.; *America Preparing* was released by the Kemble Film Corp. Unfortunately, there is little information regarding what happened to the short-lived Egan Film Corp.; "America Preparing," in Hansen, *AFI Catalog . . . 1911–1920*, 20; advertisement, "Motion Picture Enterprises," *Film Year Book* (1922–23), 216.

38. Louella O. Parsons, "Propaganda!" *Photoplay*, Sept. 1918, 110; Hansen, *AFI Catalog . . . 1911–1920*, 511; *Wid's Year Book* (1921–22), 394.

39. "Stage Women Make Twelve Pictures," *MPW*, Jan. 18, 1919, 326; "Films Aid Soldiers and Sailors," *MPW*, July 12, 1919, 223–25; "Cleaning Up with the Stage Stars," *MPW*, July 12, 1919, 266.

40. The Helen Keller Film Corporation was not set up by George Kleine, who financed the film, but just who originated the company is unknown. The film was copyrighted by Edwin Leibfreed, who brought suit against George Kleine, alleging that he was never paid his salary as treasurer of the corporation—$500 a week, or $13,250; see George Kleine to Secretary of State, Albany, March 19, 1924; Dept. of State to Kleine, Aug. 20, 1924; clipping from *Film Daily*, May 27, 1926, 3; State of New York Dept. of Taxation and Finance to Kleine, Oct. 4, 1927; Henry Melville to Kleine, March 5, 1928; Kleine to Melville, April 5, 1928; Melville to Kleine, April 6, 1928; Kleine to Melville, April 18, 1928; for records of Kleine's financial support of the Keller Film Corp. see Kleine to [?] Hinkley, Nov. 1, 1919; Kleine to Chicago office, Aug. 17, 1921; all documents in George Kleine Collection, MD-LC.

41. "Miss Keller's Own Work," *NYT*, Aug. 24, 1919, 4.

42. Interestingly, Keller declined to make a personal appearance at the premiere of her film in New York when she learned that there was an actor's strike against the Shuberts, who owned the Lyric Theater where the premiere was being held; see The Screen, *NYT*, Aug. 19, 1919, 10.

43. Quote from letter from Henry Melville, Kleine's attorney, to Harry G. Edwards, attorney for Edwin Leibfreed, April 19, 1928; for accounting and college rentals see file "College Reports, 1922–25," Nov. 4, 1919 and April 1, 1926, George Kleine Collection, MD-LC.

44. "Doraldina, Actress and Dancer, Enters Independent Picture Producing Field," *MPW*, Aug. 9, 1919, 852.

45. "Doraldina Prepares for Her First Metro," *MPW*, July 24, 1920, 502; "Doraldina in *The Woman Untamed*," *MPW*, Aug. 14, 1920, 901.

46. "Doraldina Is to Begin Making Pictures on Her Own Account," *MPW*, April 23, 1921, 866.

47. Hansen, *AFI Catalog . . . 1911–1920*, 242, citing *Variety*, June 16, 1916, 5; Cal York, Plays and Players, *Photoplay*, Aug. 1916, 126; Charles Samuels and Louise Samuels, *Once upon a Stage: The Merry World of Vaudeville* (New York: Dodd, Mead, 1974), 52, 54, 56, 57, 58–59; Robert W. Snyder, *Voice of the City: Vaudeville and Popular Culture in New York* (New York: Oxford University Press, 1989), 149–50.

48. Cal York, Plays and Players, *Photoplay*, Aug. 1916, 126.

49. *The Wild Girl* (1917) was also self-promotional but considerably tailored for screen entertainment. Tanguay played an heiress raised as a gypsy boy to hide her identity. Thus Tanguay's character allowed her to display the "wildness" that made her a vaudeville attraction without transgressing gender boundaries. The film concludes with a gender-appropriate "happy ending" (marriage) after her true sexual identity is revealed. The film was directed by Howard Estabrook and written by George M. Rosener; *The Wild Girl* (1917), viewed at FTA-UCLA; see also Hansen, *AFI Catalog . . . 1911–1920*, 1027–38.

50. Nell Shipman, *The Silent Screen and My Talking Heart: An Autobiography*, 2nd ed. (Boise, ID: Boise State University, 1988), 43–44, 50, 64–65, 215.

51. Kay Armatage, "Nell Shipman: A Case of Heroic Femininity," in *Feminisms in the Cinema*, ed. Laura Pietropaolo and Ada Testaferri (Bloomington: Indiana University Press, 1995), 129.

52. Kay Armatage, *The Girl from God's Country: Nell Shipman and the Silent Cinema* (Toronto: University of Toronto Press, 2003), 161.

53. Ibid., 172–74.

54. Shipman, *Silent Screen*, 103–5; release date for *The Girl from God's Country* from *Film Daily Yearbook* (1925), ed. Joseph Dannenberg (Hollywood: Film Daily, 1925), n.p.; "Girl from God's Country," *Variety*, Nov. 18, 1921, n.p.; Armatage, "Nell Shipman," 131.

55. Shipman, *Silent Screen*, 114–17, 124–25; Armatage, *Girl from God's Country*, 222.

56. The films were *Little Dramas of the Big Places: Trail of the North Wind* (1923) and *Little Dramas of the Big Places: The Light on Lookout* (1923); see Shipman, *Silent Screen*, 137, 146–59, 208–9, 218.

57. *NYT*, Feb. 4, 1916, 9.

58. *Photoplay*, Oct. 1914, 78.

59. "Film Star Sues Selznick," *NYT*, May 24, 1917, 11; "Clara Kimball Young Sues Selznick," *MPW*, June 9, 1917, 1580; "Miss Young Makes Answer to Selznick," *MPW*, July 21, 1917, 432.

60. "Selznick Would Enjoin Miss Young," *MPW*, June 16, 1917, 1761.

61. "New C. K. Young Company Announced," *MPW*, July 7, 1917, 66.

62. "Clara Kimball Young Discusses Picture Art," *MPW*, July 21, 1917, 461.

63. "Selznick Wins Suit from Equity," *MPW*, Dec. 11, 1920, 707.

64. *MPW*, Jan. 1, 1921, 45.

65. "Clara Kimball Young's Triumphant Tour," MPW, July 23, 1921, 396.

66. Henry R. Davis, "Clara Kimball Young," Films in Review 12, no. 7 (Aug.-Sept. 1961): 421–22.

67. "Clara Kimball Young Has Relapse," NYT, April 19, 1924, 13; "Clara Kimball Young loses $75,000 in Jewels," NYT, April 20, 1925, 21.

68. Hampton, History of the American Film Industry, 174–76.

69. "First National Exhibitor's Circuit, Incorporated," Harvard Business Reports, ed. Charles I. Gragg, comp. Graduate School of Business Administration, George F. Baker Foundation, Harvard University, vol. 8 (New York: McGraw-Hill, 1930), 14–18.

70. Balio, "Stars in Business," 162; Eyman, Mary Pickford, 118, 120–21; "Mrs. Pickford Heads Two Companies," MPW, Feb. 1, 1919, n.p.

71. "First National Gets Norma Talmadge," Wid's Daily, Dec. 10, 1918, 1.

72. Hampton, History of the American Film Industry, 189–90.

73. A. H. Geibler, MPW, Feb. 1, 1919, 619, quoted in Balio, United Artists, 3.

74. Charles Chaplin, My Autobiography (New York: Simon and Schuster, 1964), 222, quoted in Balio, United Artists, 12.

75. "Motion Picture Stars' Reasons to Combine," Motion Picture Classic, April 1919, 17.

76. "Associated First National to Handle A. P. Retains Identity, with Lichtman," MPW, Sept. 17, 1921, 260; Wid's Year Book (1921–22), 4.

77. Sumiko Higashi, Cecil B. DeMille and American Culture: The Silent Era (Berkeley: University of California Press, 1994), 19–20.

78. Hampton, History of the American Film Industry, 194.

79. Ibid., 216–18.

80. "Abolish Film Star System," NYT, Oct. 31, 1920, 22.

81. Hampton, History of the American Film Industry, 195. Weber's 1921 contract with the unknown Claire Windsor stipulated that Weber would "endeavor to cause her to become known as a star in motion picture work"; see contract between Ola Cronk [Claire Windsor] and Lois Weber Productions, Jan. 12, 1921, Claire Windsor Collection, SC-USC.

82. Wid's Year Book (1920–21), 31, 81; Dewitt Bodeen, "Betty Compson," Films in Review 17, no. 7 (Aug.-Sept. 1966): 402.

83. "Paramount Famous Lasky Corporation," in Gragg, Harvard Business Reports, 182–200; Koszarski, An Evening's Entertainment, 110.

84. Hampton, History of the American Film Industry, 240. Janet Wasko, Movies and Money: Financing the American Film Industry (Norwood, NJ: Ablex, 1982), 18–19; Koszarski, An Evening's Entertainment, 75.

85. "First National Exhibitor's Circuit, Incorporated," 21.

86. Ibid., 17–20, 22–25.

87. Balio, "Struggles for Control, 1908–1930," 121.

88. Douglas Gomery, Shared Pleasures: A History of Movie Presentation in the United States (Madison: University of Wisconsin Press, 1992), 38.

89. Cal York, Plays and Players, *Photoplay*, Feb. 1923, 74; Oct. 1923, 80; unsigned draft of contract between Corinne Griffith Productions, Inc., and Associated First National Pictures, Inc., July 9, 1923; unsigned draft of contract between Corinne Griffith and Edward Small and Charles R. Rogers, July 9, 1923, WBA; Cal York, Gossip East and West, *Photoplay*, Sept. 1923, 68.

90. "Western Representative" [?] to Alfred Wright, Oct. 6, 1923, WBA.

91. Corinne Griffith to Messrs. Edward Small and Charles R. Rogers, Oct. 16, 1923, WBA.

92. Griffith to Associated First National Pictures, Inc., Nov. 9, 1923; Griffith to Small and Rogers, Nov. 9, 1923, WBA.

93. Contract between Corinne Griffith and Corinne Griffith Productions, Oct. 20, 1924; Corinne Griffith to First National, July 14, 1925, WBA.

94. *Photoplay*, April 1923, 68; Bodeen, "Anita Stewart," 153–55; Peter Harry Brown and Pamela Ann Brown, *The MGM Girls: Behind the Velvet Curtain* (New York: St. Martin's, 1983), 5, 7.

95. "Famous Players Revolutionizes Policy of Production, Forming Stock Company," *MPW*, Nov. 13, 1920, 173.

96. "Star Series Production Plan Abandoned by Selznick, Who Explains Capitulation," *MPW*, July 1, 1922, 23; Jacobs, *Rise of the American Film*, 282.

97. Hampton, *History of the American Film Industry*, 213.

98. "Public Unalterably Opposes Censorship, Star System Passing, Ince Survey Shows," *MPW*, March 4, 1922, 37.

99. "Paramount School to Help Stars and Players to Perfect Their Work," *MPW*, July 1, 1922, 21; Jesse L. Lasky, *I Blow My Own Horn* (Garden City, NY: Doubleday, 1957), 192–94.

100. Koszarski, *An Evening's Entertainment*, 21–23.

101. Gomery, *Shared Pleasures*, 55.

102. Ibid., 51–58.

103. Ibid., 38; Wasko, *Movies and Money*, 21; Thomas Schatz, *The Genius of the System: Hollywood Filmmaking in the Studio Era* (New York: Pantheon, 1988), 29–30.

104. Schatz, *Genius of the System*, 11, 176–77.

105. Douglas Gomery, *The Hollywood Studio System* (New York: St. Martin's, 1986), 13.

106. Independents observed this occurring as early as 1921; see C. S. Lewis, "In the Independent Field," *MPW*, July 9, 1921, 213; "Propositions of Financiers Worrying Independent Producers on West Coast," *MPW*, April 29, 1922, 945; "Elimination of Distributor Impends," *MPW*, May 20, 1922, 242; "Producers Finance Corporation Plans to Encourage Production among State Rights Men by Establishment of Credit," *MPW*, July 29, 1922, 339.

107. Schatz, *Genius of the System*, 21.

108. "Goldwyn Tells Why Producers Demand Advance Deposits and Acquire Theaters," *MPW*, Aug. 21, 1920, 980; F. W. Heathcote, "How Producing Compa-

nies May Receive Adequate Financial Support from Banks," *MPW*, Oct. 15, 1921, 751–52; "Taking the Flicker Out of Movie Finances," *Literary Digest*, Sept. 13, 1924, 85.

109. "Bankers Join in War on Wild-Cat Stock; to Keep Tabs on Fly-By-Night Promoters," *MPW*, Nov. 19, 1921, 273–74. Speculation and fraudulent companies were loudly derided in the late 1910s and 1920s as the film industry attempted to attract Wall Street financing; see "Speculating in Movie Making," *New York Daily Telegram*, Dec. 12, 1915, 8; "Big Loss in 'Wildcat' Films," *NYT*, Dec. 6, 1920, 27.

110. "Trust-Busting the Movies," *Literary Digest*, Dec. 12, 1925, 15.

111. "A Banking Endorsement of the Industry," *Film Daily Yearbook* (1925), 55, 57.

112. Schatz, *Genius of the System*, 21.

113. Balio, *United Artists*, 49–50, 64–65.

114. Ibid., 58–60, 68.

115. "Nazimova in Film of War Brides Play," *NYT*, Dec. 4, 1916, 23; Dewitt Bodeen, "Nazimova," *Films in Review* 23, no. 10 (Dec. 1972): 577–88; Alexander Kirkland, "The Woman from Yalta," *Theatre Arts*, Dec. 1949, 48. Nazimova did not direct, but she did fight for her own vision with her directors; see Anthony Slide, *Early Women Directors* (New York: A. S. Barnes, 1977), 111–12.

116. Patricia White, "Nazimova's Veils: *Salome* at the Intersection of Film History," in *A Feminist Reader in Early Cinema*, ed. Jennifer M. Bean and Diane Negra (Durham, NC: Duke University Press, 2002), 60–87.

117. Cal York, Plays and Players, *Photoplay*, Jan. 1923, 95.

118. Quoted in Frances Marion, *Off with Their Heads!* (New York: Macmillan, 1972), 88–89.

119. *Salome*, viewed at MPBRSD-LC. The *NYT* reviewer praised *Salome*; see The Screen, Jan. 1, 1923, 18. *Variety* warned that only Nazimova "devotees" and "a few that like higher art" will enjoy it, and "then its box office value will end"; see "Salome," Jan. 5, 1923, n.p.; Nazimova to Nina Lewton (sister), Sept. 11, 1923, Alla Nazimova Collection, MD-LC. When the film was re-released in 1967, Bosley Crowther, a critic for the *New York Times*, called it "one of the silent movies' more notorious Tiffany lamps, relic of a style of artsy acting and a fancy decor that blazes as present-day camp" (*NYT*, Feb. 15, 1967, 42).

120. Gloria Swanson, *Swanson on Swanson: An Autobiography* (New York: Pocket Books, 1980), 276.

121. Balio, *United Artists*, 58.

122. Swanson, *Swanson on Swanson*, 294–98, 303, 309–21.

123. Balio, *United Artists*, 83.

124. Swanson, *Swanson on Swanson*, 361, 383–87; Balio, *United Artists*, 83–84.

125. Halsey, Stuart, and Co., "The Motion Picture Industry as a Basis for Bond Financing," May 27, 1927, repr. in *The American Film Industry*, ed. Tino Balio, rev. ed. (Madison: University of Wisconsin Press, 1985), 204.

126. Schatz, *Genius of the System*, 42–44.

127. Ruth Wing, ed., *The Blue Book of the Screen* (Hollywood: Blue Book of the Screen, 1923), 360–61.

128. Cathy Klaprat, "The Star as Market Strategy: Bette Davis in Another Light," in *The American Film Industry*, ed. Tino Balio, rev. ed. (Madison: University of Wisconsin Press, 1985), 351, 375; "Universal Puts Morality Clause into All Contracts to Protect Company and Prevent Offenses," *MPW*, Oct. 1, 1921, 526.

129. "A Banking Endorsement of the Industry," *Film Daily Yearbook* (1925), 55.

## Chapter 7. "Doing a 'Man's Work'"

1. Sydney Valentine, "The Girl Producer," *Photoplay*, July 1923, 55, 110; Joseph Danning, ed., *Film Year Book* (Hollywood: Film Daily, 1924), 63.

2. Lewis Jacobs, *The Rise of the American Film: A Critical History* (New York: Columbia University Press, 1968), 296–97.

3. Untitled review of an unidentified Lois Weber film, *MPN*, Aug. 27, 1921, n.p., Claire Windsor Collection, SC-USC; review of Ida May Park's *The Butterfly Man*, *Variety*, May 21, 1920, n.p. For a discussion of director Dorothy Arzner and her "butch style" see Judith Mayne, *Directed by Dorothy Arzner* (Bloomington: Indiana University Press, 1994); for an early description of filmmaking as "man's work" see Frances Denton, "Lights! Camera! Quiet! Ready! Shoot!" *Photoplay*, Feb. 1918, 48–50.

4. Feature films directed by women from 1922 through 1928 (with distributor): *1922*: Nell Shipman, *The Girl from God's Country* (F. B. Warren); May Tully, *Our Mutual Friend*, *The Old Oaken Bucket* (Wid Gunning, Inc.); Marion Fairfax, *The Lying Truth* (American Releasing Corp.); Lois Weber, *What Do Men Want?* (Film Booking Offices of America/Robertson-Cole); Mrs. and Mrs. George Randolph Chester, *The Son of Wallingford* (Vitagraph); Ruth Bryan Owen, *Once upon a Time* (states' rights [?]). *1923*: Julia Crawford Ivers, *The White Flower* (Paramount); Lois Weber, *A Chapter in Her Life* (Universal-Jewel); Grace Haskins, *Just like a Woman* (W. W. Hodkinson). *1924*: Jane Murfin, *Flapper Wives* (Selznick); Lillian Ducey, *Enemies of Children* (Mammoth Pictures). *1925*: No feature films directed by women. *1926*: May Tully, *This Old Gang of Mine* (states' rights [?]); Lois Weber, *The Marriage Clause* (Universal-Jewel). *1927*: Dorothy Arzner, *Fashions for Women* (Paramount), *Ten Modern Commandments* (Paramount), *Get Your Man* (Paramount); Lois Weber, *Sensation Seekers* (Universal-Jewel), *Angel of Broadway* (DeMille Pictures/Pathé). *1928*: Dorothy Arzner, *Manhattan Cocktail* (Paramount). After 1928, with only two exceptions, Dorothy Arzner was the only woman to work as a Hollywood director until the 1970s; information from *Film Year Book* (Hollywood: Film Daily, 1922–28).

5. Wendy Holliday, "Hollywood's Modern Women: Screenwriting, Work Culture, and Feminism, 1910–1940" (PhD diss., New York University, 1995), 174–76.

6. "Photoplayers Now in Its Third Year of Success," *MPN*, April 3, 1915, 171.

7. K. Owen, "The Club, James!" *Photoplay*, Feb. 1917, 67.

8. Anthony Rotundo, *American Manhood: Transformations in Masculinity from the Revolution to the Modern Era* (New York: Basic Books, 1993), 199–203, 250.

9. Lillian R. Gale, "Motion Picture Director's Association," in *Film Year Book, 1920–21*, ed. Joseph Dannenberg (New York: Wid's Films and Film Folks, [1921?]), 211.

10. "Directors' Association Honors Lois Weber," unidentified source, Nov. 21, 1916, Lois Weber clippings file, MOMA.

11. Motion Picture Directors Association, *Souvenir: Annual Ball, Feb. 17, 1923, Alexandria Hotel, Los Angeles*, AMPAS.

12. Ibid., 69–70.

13. "Studio Efficiency: Scientific Management as Applied to the Lubin Western Branch by Wilbert Melville," *MPW*, Aug. 9, 1913, 624.

14. Janet Staiger, "Blueprints for Feature Films: Hollywood's Continuity Scripts," in *The American Film Industry*, ed. Tino Balio, rev. ed. (Madison: University of Wisconsin Press, 1985), 189–91.

15. Ibid., 189.

16. Lizzie Francke, *Script Girls: Women Screenwriters in Hollywood* (London: British Film Industry, 1994), 6. See also Cari Beauchamp and Mary Anita Loos, *Anita Loos Rediscovered: Film Treatments and Fiction* (Berkeley: University of California Press, 2003).

17. Francke, *Script Girls*, 18.

18. Clara Beranger, "Feminine Sphere in the Field of Movies," *MPW*, Aug. 2, 1919, 662.

19. Staiger, "Blueprints for Feature Films," 190.

20. Louis Reeves Harrison, "Directorial Censorship," *MPW*, April 12, 1913, n.p.; Louis Reeves Harrison, "A Wondrous Training School," *MPW*, Aug. 16, 1913, 720; "Studio Efficiency: Scientific Management as Applied to the Lubin Western Branch by Wilbert Melville," *MPW*, Aug. 9, 1913, 624; Captain Leslie T. Peacocke, "Studio Conditions as I Know Them," *Photoplay*, June 1917, 127–30.

21. Captain Leslie T. Peacocke, "The Scenario Writer and the Director," *Photoplay*, May 1917, 112; Jacobs, *Rise of the American Film*, 219.

22. From *Moving Picture Stories*, n.d., n.p., quoted in Anthony Slide, *Early Women Directors* (New York: A. S. Barnes, 1977), 104.

23. Nell Shipman, *The Silent Screen and My Talking Heart: An Autobiography*, 2nd ed. (Boise, ID: Boise State University, 1988), 43–44; Peter Morris, "The Taming of the Few: Nell Shipman in the Context of Her Times," in Shipman, *Silent Screen*, 215.

24. Alice Martin, "From 'Wop' Parts to Bossing the Job," *Photoplay*, Oct. 1916, 96–97.

25. Slide, *Early Women Directors*, 54–55.

26. Ibid., 105–8; "Creator of 'J. Rufus' Will Write Reliance Serial," *MPW*, Dec. 5, 1914, 25–26; "George Randolph Chester Forms Company; Plans to Make Two Big Films Yearly," *MPW*, Nov. 19, 1921, 290.

27. Obituary, *Los Angeles Times*, May 9, 1930, Ivers biography file, AMPAS;

*Majesty of the Law*, reel 4 of 5, viewed at MPBRSD-LC; "The Shadow Stage," review of *Call of the Cumberlands*, *Photoplay*, April 1916, 104; Patricia King Hansen, ed., *The American Film Industry Catalog of Motion Pictures Produced in the United States: Feature Films, 1911–1920* (Berkeley: University of California Press, 1988), 117–18, 567, 863; review of *The White Flower*, *Variety*, March 8, 1923, n.p.; Slide, *Early Women Directors*, 110.

28. Marion codirected one more film, *The Song of Love* (1923), after its original director became ill, but that was the end of her directing career. Dewitt Bodeen, "Frances Marion," *Films in Review* 20, no. 2 (Feb. 1969): 75–87; MPW, April 5, 1919, 75; The Screen, *NYT*, Jan. 19, 1920, 16; review of *The Love Light*, *Variety*, Jan. 14, 1921, n.p.; "Frances Marion," in *American Screenwriters*, ed. Robert E. Morsberger, Stephen O. Lesser, and Randall Clark (Detroit: Gale Research, 1984), 232.

29. E. Leslie Gilliams, "Will Woman's Leadership Change the Movies?" *Illustrated World*, Feb. 1923, 38, 860, 956; Slide, *Early Women Directors*, 114–15.

30. See *Photoplay*, Dec. 1915, 159; William H. Henry, "Cleo the Craftswoman, *Photoplay*, Jan. 1916, 110. The two five-reel features were *A Soul Enslaved* and *Her Bitter Cup*; see "Cleo Madison in *Her Bitter Cup*," *Moving Picture Weekly*, April 16, 1916, 24–25, 32.

31. "Such a Little Director," *Moving Picture Weekly*, March 24, 1917, 19; "One of the Universal Faithful—Lule Warrenton, Character Leads," *Universal Weekly*, Feb. 7, 1914, 5; "Mrs. Warrenton Starts Children's Photoplays," MPW, Feb. 17, 1917, 1030; Cal York, Plays and Players, *Photoplay*, Nov. 1917, 81; Slide, *Early Women Directors*, 56–58.

32. "Where Work Is Play and Play Is Work, Universal City, California, the Only Incorporated Moving Picture Town in the World," *Universal Weekly*, Dec. 27, 1913, 4–5.

33. Richard Koszarski, *An Evening's Entertainment: The Age of the Silent Feature Picture, 1915–1928* (Berkeley: University of California Press, 1994), 87.

34. "Such a Little Director," *Moving Picture Weekly*, March 24, 1917, 19; Cal York, Plays and Players, *Photoplay*, Feb. 1918, 118.

35. Cal York, Plays and Players, *Photoplay*, Jan. 1917, 83; Cal York, Plays and Players, *Photoplay*, Oct. 1917, 114.

36. "Mrs. Warrenton Starts Children's Photoplays," MPW, Feb. 17, 1917, 1030; Cal York, Plays and Players, *Photoplay*, Nov. 1917, 81; Slide, *Early Women Directors*, 56–57.

37. Slide, *Early Women Directors*, 60; *Film Comment*, Nov. 1972, n.p., in Ida May Park biography file, AMPAS; *Wid's Year Books* for 1918, 1919–20, 1922, n.p.; review of *The Butterfly Man*, *Variety*, May 21, 1920, n.p.

38. Wilson directed one more film, *Insinuation* (1922), but it was not released widely, if at all; see *Wid's Year Book* (1921–22), 323; Slide, *Early Women Directors*, 62–66; *Insinuation* does not appear in *Wid's Year Book*, the industry's official chronicle of the year's productions.

39. "Vera McCord's 'The Good-Bad Wife' Arousing Big Interest among Buyers," MPW, Aug. 28, 1920, 1168; "Vera McCord Productions Asks Receiver for Walgreene in Suit over Receipts," MPW, Sept. 24, 1921, 402; review of The Good-Bad Wife, Variety, Jan. 21, 1921, n.p.

40. "Marion Fairfax, Dramatist, Is with Lasky," MPW, June 12, 1915, 64; "Initial Marion Fairfax Picture Is Completed on the West Coast," MPW, June 11, 1921, 634; "Preview of Marion Fairfax Film," New York Morning Telegraph, July 10, 1921, n.p., in Marion Fairfax biography file, AMPAS; ad for "Marion Fairfax Productions," Wid's Year Book (1921–22), 40.

41. "The Son of Wallingford a Winner," Exhibitor's Herald, Oct. 8, 1921, 45; "Creator of 'J. Rufus' Will Write Reliance Serial," MPW, Dec. 5, 1914, 25–26; "George Randolph Chester Forms Company; Plans to Make Two Big Films Yearly," MPW, Nov. 19, 1921, 290.

42. "A Society Girl in Filmdom's Whirl," Photo-Play Journal, Oct. 1919, n.p.; advertisement, Wid's Year Book (1919–20), n.p.; Wid's Year Book (1920–21), 85, 91, 314, 394; Hansen, AFI Catalog . . . 1911–1920, 730; MPW, Jan. 1, 1921, 92; "Brought into Focus," NYT, Feb. 27, 1921, 2; " 'The Sky Pilot' Prologue Conceived by First National and Strand Scores Hit," MPW, May 7, 1921, 81; review of The Sky Pilot, Variety, April 22, 1921, n.p.

43. By July Unsell had accepted employment as head of the Robertson-Cole scenario department. "Eve Unsell Creates Scenario Service Bureau: Will Also Dispose of Film Rights to Books," MPW, Jan. 15, 1921, 288; advertisement, "Eve Unsell Photoplay Staff, Inc.," MPW, Jan. 15, 1921, 254. For the recession see Wid's Year Book (1921–22), 99, 105, 107; "Eve Unsell Will Sail for England," MPW, July 26, 1921, 515; "Eve Unsell Heads Scenario Department of Robertson-Cole on the West Coast," MPW, July 23, 1921, 412.

44. David Puttnam, with Neil Watson, Movies and Money (New York: Knopf, 1998), 94.

45. Wasko, Movies and Money, 11–12.

46. According to Angel Kwolek-Folland the banking industry's approach to women was firmly grounded in nineteenth-century gender ideals; see Angel Kwolek-Folland, Engendering Business: Men and Women in the Corporate Office, 1870–1930 (Baltimore: Johns Hopkins University Press, 1994), 171–72.

47. Denton, "Lights! Camera! Quiet! Ready! Shoot!" 48–50.

48. Henry MacMahon, "Women Directors of Plays and Pictures," LHJ, Dec. 1920, 140; Remodeling Her Husband (1920) was not well reviewed, but it made money; see Lillian Gish, with Ann Pinchot, The Movies, Mr. Griffith, and Me (Englewood Cliffs, NJ: Prentice-Hall, 1969), 222–37; Anthony Slide, "Remodeling Her Husband," in Magill's Survey of Cinema, ed. Frank N. Magill (Englewood Cliffs, NJ: Salem Press, 1982), 1:918–20.

49. Emma Carus, "The Feminine Stage Producer," Green Book, Nov. 1911, Robinson Locke Collection, Scrapbook, ser. 2, NYPL-PA.

50. Louis Reeves Harrison, "Studio Saunterings," *MPW*, June 15, 1912, 1007–10.

51. Unidentified clipping, with photos, 152, Kevin Brownlow private collection.

52. L. H. Johnson, "'A Lady General in the Motion Picture Army,' Lois Weber-Smalley, Virile Director," *Photoplay*, June 1915, 42.

53. "Miss Robins, of Accounting Department, Goes Flying," *Universal Weekly*, Sept. 6, 1913, 5; "Ella Hall Auto Enthusiast," *MPW*, April 16, 1916, 16; "Marie Walcamp Tackles Motorcycle," *Moving Picture Weekly*, June 3, 1916, 14.

54. "Ida Schnall Organizes Fair Sex Baseball Team at Universal City," *Moving Picture Weekly*, Dec. 4, 1915, 13; for reference to male team see "Manager Pops Question While His Club Plays Ball," *Universal Weekly*, May 16, 1914, 17.

55. Mlle. Chic, "The Dual Personality of Cleo Madison," *Moving Picture Weekly*, July 1, 1916, 24; Denton, "Lights! Camera! Quiet! Ready! Shoot!" 48.

56. "Where Work Is Play," 4–5.

57. I. G. Edmonds, *Big U: Universal in the Silent Days* (South Brunswick, NJ: A. S. Barnes, 1977), 46; "Where Work Is Play," 4–5.

58. "Where Work Is Play," 4–5.

59. "Lady Cops Captivate," *Universal Weekly*, March 14, 1914, 17.

60. Susan A. Glenn, *Female Spectacle: The Theatrical Roots of Modern Feminism* (Cambridge, MA: Harvard University Press, 2000), 151.

61. William H. Henry, "Cleo the Craftswoman," *Photoplay*, Jan. 1916, 109–10.

62. Mlle. Chic, "The Dual Personality of Cleo Madison," 25.

63. Ibid., 24.

64. Denton, "Lights! Camera! Quiet! Ready! Shoot!" 48–50.

65. "Such a Little Director," *Moving Picture Weekly*, March 24, 1917, 19.

66. Wig-Wag at the Movies, "At the Gene Gauntier Studio," *New York Star*, May 9, 1914, 20, in Gene Gauntier scrapbook, p. 113, NYPL-PA.

67. *Motion Picture Magazine*, March 1921, quoted in Slide, *Early Women Directors*, 47.

68. Johnson, "Lady General," 42.

69. Review of *The Female of the Species*, "Manufacturers Advance Notes," *MPW*, Dec. 27, 1913, 1557.

70. "Anna Pavlova Finally Decides to Appear on Screen," *MPW*, June 9, 1915, 52.

71. Bertha H. Smith, "A Perpetual Leading Lady," *Sunset, the Pacific Monthly*, March 1914, 635; Frances Marion, *Off with Their Heads!* (New York: Macmillan, 1972), 12.

72. Fritzi Remont, "The Lady behind the Lens," *Motion Picture Magazine*, May 1918, 59–61, 126; "we're all in business" quote from Arthur Denison, "A Dream in Realization," *MPW*, July 21, 1927, 417–18.

73. Denton, "Lights! Camera! Quiet! Ready! Shoot!" 48–50.

74. "Such a Little Director," *Moving Picture Weekly*, March 24, 1917, 19; "One of the Universal Faithful," 5; "Mrs. Warrenton Starts Children's Photoplays," 1030; Cal York, Plays and Players, *Photoplay*, Nov. 1917, 81; Slide, *Early Women Directors*, 56–58.

75. *Photoplay*, Sept. 1920, quoted in Anthony Slide, "Remodeling Her Husband," in *Magill's Survey of Cinema*, ed. Frank N. Magill (Englewood Cliffs, NJ: Salem Press, 1982), 1:920.

76. Puttnam and Watson, *Movies and Money*, 76–79.

77. Thomas Schatz, *The Genius of the System: Hollywood Filmmaking in the Studio Era* (New York: Pantheon, 1988), 23–25.

78. Ibid., 16–25; MPW, Jan. 25, 1920, 6.

79. Slide, *Early Women Directors*, 103.

80. Benjamin B. Hampton, *A History of the American Film Industry from Its Beginnings to 1931* (New York: Dover, 1970), 248–51; *Wid's Year Book* (1921–22), 99, 105, 107.

81. Puttnam and Watson, *Movies and Money*, 97, 101.

82. Wasko, *Movies and Money*, 14; Tino Balio, *United Artists: The Company Built by the Stars* (Madison: University of Wisconsin Press, 1976), 36.

83. Helen Bullitt Lowry, "Wall Street's Heel on the Prodigal Movies," *NYT*, July 24, 1921, 6, 36.

84. Janet Staiger, "The Hollywood Mode of Production to 1930," in *The Classical Hollywood Cinema: Film Style and Mode of Production to 1960*, by David Bordwell, Janet Staiger, and Kristin Thompson (New York: Columbia University Press, 1985), 135–36.

85. Puttnam and Watson, *Movies and Money*, 97.

86. "Efficiency," *Wid's Daily*, March 7, 1921, 1.

87. Jacobs, *Rise of the American Film Industry*, 295; Hampton, *History of the American Film Industry*, 244; Wasko, *Movies and Money*, 295; Balio, *United Artists*, 36.

88. Lowry, "Wall Street's Heel on the Prodigal Movies," 6.

89. "1907–1927, Too Much Efficiency," MPW, March 26, 1927, 277.

90. Schatz, *Genius of the System*, 31–36, 39, 44–45.

91. Ibid., 46.

92. Jacobs, *Rise of the American Film Industry*, 296–97.

93. Kristin Thompson, "The Formulation of the Classical Style, 1909–1928," in *The Classical Hollywood Cinema: Film Style and Mode of Production to 1960*, by David Bordwell, Janet Staiger, and Kristin Thompson (New York: Columbia University Press, 1985), 205–6; Schatz, *Genius of the System*, 46–47; Helen Starr, "Putting It Together," *Photoplay*, July 1918, 52–54.

94. Staiger, "Hollywood Mode of Production to 1930," 152.

95. The Motion Picture Art Directors Association first appeared in *Wid's Year Book* (1920–21), 215; there were forty-two names, none of them female; in *Wid's Year Book* (1921–22), 282, there were forty-four names, again, no women; for development of art direction, see Staiger, "Hollywood Mode of Production to 1930," 147–48.

96. Brownlow, *The Parade's Gone By*, 79.

97. Ibid.

98. Charles S. Dunning, "The Gate Women Don't Crash," *Liberty*, May 14, 1927, 31, 33.

99. See William C. DeMille, *Hollywood Saga* (New York: Dutton, 1939).

100. Dunning, "The Gate Women Don't Crash," 31, 33; Jane Murfin, "Sex and the Screen," in *The Truth about the Movies, by the Stars*, ed. Laurence Hughes (Hollywood: Hollywood Publishers, 1924), 459–60.

101. Dunning, "The Gate Women Don't Crash," 31, 33.

102. Brownlow, *The Parade's Gone By*, 15.

103. Swanson, *Swanson on Swanson: An Autobiography* (New York: Pocket Books, 1980), 103.

104. Photograph of DeMille's office, [1920], Cecil B. DeMille Collection, AMPAS.

105. Marion, *Off with Their Heads!* 48.

106. "Cecil B. DeMille Carries On," *Motion Picture Director*, Jan. 1926, 34–35.

107. Richard Koszarski, *The Man You Love to Hate: Erich von Stroheim and Hollywood* (Oxford: Oxford University Press, 1983), 166; review of *Blind Husbands*, *Variety*, Dec. 12, 1919, n.p.

108. Koszarski, *An Evening's Entertainment*, 234–36; Laurence Hughes, ed., *The Truth about the Movies, by the Stars* (Hollywood: Hollywood Publishers, 1924), n.p.; Jim Tully, "The Man You Love to Hate," *Motion Picture Classic*, May 1924, 20.

109. Slide, *Early Women Directors*, 62–66.

110. Brownlow, *The Parade's Gone By*, 71.

111. Dunning, "The Gate Women Don't Crash," 31, 33.

112. Ibid., 33.

113. Hughes, *Truth about the Movies*, n.p.

114. Puttnam and Watson, *Movies and Money*, 95.

115. Ivan St. Johns, "Glyn and Glynne," *Photoplay*, May 1924, 53, 129.

116. Cal York, Plays and Players, *Photoplay*, May 1923, 87; "Our Foremost Woman Director," *Photoplay*, July 1924, 74.

117. Nora B. Geibler, "Rubbernecking in Filmland," *MPW*, June 18, 1921, 719.

118. Koszarski, *An Evening's Entertainment*, 240–41.

119. Kathleen Lipke, "Most Responsible Job Ever Held by a Woman," *Los Angeles Times*, June 3, 1923, 13.

120. Koszarski, *The Man You Loved to Hate*, 144–45.

121. *Ben Hur* continued to be plagued by sundry and epic disasters; see Brownlow, *The Parade's Gone By*, 380–409.

122. Koszarski, *An Evening's Entertainment*, 241.

123. Thompson, "Formulation of the Classical Style," 152; Kristin Thompson, "Film Style and Technology to 1930," in *The Classical Hollywood Cinema: Film Style and Mode of Production to 1960*, by David Bordwell, Janet Staiger, and Kristin Thompson (New York: Columbia University Press, 1985), 278.

124. Margaret Booth, interview by Rudy Behlmer, in *An Oral History with Mar-*

*garet Booth* (Beverly Hills, CA: AMPAS, 1991); Francis Humphrey, "The Creative Woman in Motion Picture Production" (master's thesis, University of Southern California, 1970), 123–25; Barrie Pattison, "Thirty Leading Film Editors," in *International Film Guide*, ed. Peter Cowie (New York: A. S. Barnes, 1973), 67; Dick Cantwell, "Anna Spiegel's Story in Series on Old-Timers," *Metro-Goldwyn-Mayer Studio Club News*, Dec. 24, 1938, 12, editors general file, AMPAS; for references to male cutters see Captain Leslie T. Peacocke, "Studio Conditions as I Know Them," *Photoplay*, June 1917, 127–30; and Starr, "Putting It Together," 52.

125. Humphrey, "Creative Woman," 111.

126. "Film Editor [Lawrence] Retires; Put in 50 Years," *Los Angeles Herald-Examiner*, Aug. 9, 1962, A-17; see also Anne Bauchens biography file, AMPAS.

127. Douglas Gomery, "Margaret Booth," in *International Dictionary of Films and Filmmakers*, ed. Christopher Lyon and Susan Doll (New York: Putnam, 1985), 4:53.

128. Humphrey, "Creative Woman," 114–15, 125–26, 130.

129. Ibid., 110–12, 114–17, 123–27.

130. Susan Coultrap-McQuinn, *Doing Literary Business: American Writers in the Nineteenth Century* (Chapel Hill: University of North Carolina Press, 1990), 48.

131. Wendy Gamber, *The Female Economy: The Millinery and Dressmaking Trades, 1860–1930* (Urbana: University of Illinois Press, 1997), 158–59.

132. Kathy Peiss, *Hope in a Jar: The Making of America's Beauty Culture* (New York: Henry Holt, 1998), 71, 100–101.

133. Catherine Oglesby, *Fashion Careers American Style* (New York: Funk and Wagnalls, 1935), 126, 127, quoted in Peiss, *Hope in a Jar*, 107.

134. Jane R. Plitt, *Martha Matilda Harper and the American Dream: How One Woman Changed the Face of Modern Business* (Syracuse: Syracuse University Press, 2000), 135.

135. Gamber, *The Female Economy*, 193.

## Epilogue. Getting Away with It

1. Charles S. Dunning, "The Gate Women Don't Crash," *Liberty*, May 14, 1927, 33.

2. Myrtle Gebhart, "Business Women in Film Studios," *Business Woman*, Dec. 1923, 26, 68; Lois Hutchinson, "A Stenographer's Chance in Pictures," *Photoplay*, March 1923, 42–43, 107; Helen Carlisle, "They're Not Afraid to Fight," *Motion Picture Magazine*, Feb. 1924, 36–37, 86; Reina Wiles Dunn, "Off-Stage Heroines of the Movies," *Independent Woman*, July 1934, 202–3, 214.

3. Mlle. Chic, "The Dual Personality of Cleo Madison," *Moving Picture Weekly*, July 1, 1916, 24; Lois Hutchinson, "A Stenographer's Chance in Pictures," *Photoplay*, March 1923, 42–43, 107.

4. Constance Talmadge, "What Opportunities Are There for a Girl Who Is Willing to Work to Stardom," in *The Truth about the Movies, by the Stars*, ed. Laurence Hughes (Hollywood: Hollywood Publishers, 1924), 85, 87, 89.

5. Robert H. Wiebe, *The Search for Order, 1877–1920* (New York: Hill and Wang, 1967).

6. Judith Mayne, *Directed by Dorothy Arzner* (Bloomington: Indiana University Press, 1994), 1, 35–79, 81; the two exceptions were Lois Weber's *White Heat* (1934) and Wanda Tuchock's *Finishing School* (1933); Tuchock was primarily a scenarist, and this is the only film she directed; see Gerald Peary, "Sanka, Pink Ladies, and Virginia Slims," *Women and Film* 1, nos. 5–6 (1974): 84.

7. Quoted in Mayne, *Directed by Dorothy Arzner*, 13–17, 23–26.

8. Ibid., 13–17, 23–27, 33, 48.

9. Ibid., 33–35.

10. Ibid, 35–54.

11. Amazingly, Arzner completed *Fashions for Women* in less than two weeks; see Mayne, *Directed by Dorothy Arzner*, 33–54.

12. Ibid., 32, 39, 48, 53–54.

13. Ibid., 64.

14. Women began to reappear as directors and producers in the late 1960s and 1970s, but with few exceptions their films did not gain wide release; see J. Pyros, "Notes on Women Directors," *Take One* 3, no. 2 (Nov.–Dec. 1970): 7–9; Richard Henshaw, "Women Directors: 150 Filmographies," *Film Comment* 8, no. 4 (Nov.–Dec. 1972): 33–45; Peary, "Sanka, Pink Ladies, and Virginia Slims," 82–84; Nancy Dowd, "The Woman Director through the Years," *Action* 8, no. 4 (July–Aug. 1973): 15–18; Barbara Koenig Quart, *Women Directors: The Emergence of a New Cinema* (New York: Praeger, 1988); Ally Acker, *Reel Women: Pioneers of the Cinema, 1896 to the Present* (New York: Continuum, 1991), 29–44, 78–91, 137–40; for reference to the new wave of actor-producers during and after World War II and a brief description of the decay of the studio system see Douglas Gomery, *The Hollywood Studio System* (New York: St. Martin's, 1986), 9–10; for a fuller description of the decline of the studio era see Thomas Schatz, *The Genius of the System: Hollywood Filmmaking in the Studio Era* (New York: Pantheon, 1988), 411–62; for female directors in contemporary Hollywood and the percentage of films directed by women see Sharon Bernstein, "A Change in Direction? Women Directors Make Slight Inroads in Film, TV," *Los Angeles Times*, March 12, 1990, H17, H26.

# Essay on Sources

When I first conceived of this study, there was little on the subject of early female filmmakers. Now both empirical and theoretical scholarship is exploding. Interested scholars can begin their journey with an online visit to the Women Film Pioneers Project at Duke University (www.duke.edu/web/film/wfp). In addition to providing online information, this project is creating a book series, including a sourcebook to "jumpstart" research on women in the silent film industry, and a monograph series on individual filmmakers and films.

## Primary Sources

The bad news is that studio records, much like the films themselves, have mostly disappeared, leaving the discourse of trade journals, fan magazines, newspapers, the occasional memoir, and extant films to provide the bulk of primary research. As rich as these sources are, they were written with various motivations. In other words, publicity writers often lie, and so do autobiographers, but not always. The good news is that several primary sources are online, and films that once required a visit to the Library of Congress or the Museum of Modern Art can be added to one's private collection with a brief internet search and a credit card. This is a tremendous advance for the researcher with limited travel funds, but researchers should keep in mind that it is also useful to see the marginalia on the primary documents, the advertising between articles of *Moving Picture World*, and to sit in a darkened archive experiencing the physicality of the reels of film.

The single most important source of primary information came from the pages of the industry's trade journals, particularly *Moving Picture World*. Published from 1907 to 1927, *Moving Picture World* aimed to inform exhibitors about the industry in a balanced manner. It covered both licensed and independent film manufacturers, and it provided information on release dates of various companies, as well as film reviews, interviews, business news, gossip, and plenty of advertising. Other important early trade journals include *Photoplay*, *Motion Picture News*, *Exhibitors' Herald*,

*Motion Picture Magazine, Moving Picture News, Motion Picture Stories,* and *Wid's Year Book.* Mostly unindexed, these magazines require time and dedication from the researcher, but they are incredibly rich and thus indispensable for early film research. Scholars are undoubtedly spending less time in front of the microfilm reader, however, since the 1997 publication of *Filmmakers in the Moving Picture World: An Index of Articles, 1907–1927,* edited by Annette M. D'Agostino (Jefferson, NC: McFarland).

Most of the primary research for this study took place at the Margaret Herrick Library of the American Academy of Motion Picture Arts and Sciences in Los Angeles. The Herrick Library houses not only an extensive collection of trade magazines on microfilm but a comprehensive library of film-related books, including rare copies, as well as oral histories and special collections. The latter is a scholar's treasure trove of stills, publicity materials, production files, and clippings files listed by individual, film, and subject. The Cinema-Television Library at the University of Southern California offers its own film collections and special collections, including the Warner Bros. Archives; and the University of California at Los Angeles houses an extensive oral history collection, as well as special collections, including the Dorothy Arzner papers and the Albert E. Smith (Vitagraph) papers. The university's Film and Television Archive houses several rare films relevant to this study, including *The Love Light* (1921, directed by Frances Marion). The Florence Lawrence and Florence Turner papers are housed at the Seaver Center for Western Research at the Los Angeles Natural History Museum in Los Angeles, and the Film Department at the Museum of Modern Art in New York City offers films, stills, and a special collections department. The librarians at the National Film and Television Archive of the British Film Institute in London were particularly helpful in arranging the viewing of several Florence Turner films, as well as Alice Guy's *La vie du Christ* (1906), *Ruth Roland, Kalem Girl* (1912), *Hazards of Helen: Girl at Lone Point* (1914?), and Lois Weber's *Suspense* (1912); the institute also provided several stills for this study. The bulk of my film viewing took place at the Motion Picture, Broadcast, and Recorded Sound Division of the Library of Congress. Boasting the largest collection of silent films in the United States, the Library of Congress is especially rich in early films thanks to the Paper Print Collection (1894–1912). Films once copyrighted by providing each sequential frame on paper were made into 16mm films and now offer clearer viewing than the often-damaged post-1912 films. In addition, the Library of Congress's Manuscript Division houses the very useful George Kleine Collection and the Alla Nazimova papers.

One of the late joys of this project was my discovery of the New York Public Library for the Performing Arts at Lincoln Center. Between the Billy Rose Theatre Collection and the Locke Collection, voluminous scrapbooks seem to cover virtually every person who ever set foot on the stage in American history. The library also offers clippings and still files on individuals, productions, and companies, as well as information on dance and recorded sound. The nearby main library houses the collection of the National Board of Censorship/National Board of Review.

Finally, several primary sources are in print. The books most useful to this study include *The Memoirs of Alice Guy Blaché*, ed. Anthony Slide, trans. Roberta Blaché and Simone Blaché (Metuchen, NJ: Scarecrow Press, 1986); David S. Hulfish, *Cyclopedia of Motion-Picture Work* (Chicago: School of Correspondence, 1911); Ellis P. Oberholtzer, *The Morals of the Movies* (Philadelphia: Penn Publishing, 1922); Gloria Swanson, *Swanson on Swanson: An Autobiography* (New York: Pocket Books, 1980); Fred F. Balshofer and Arthur C. Miller, *One Reel a Week* (Berkeley: University of California Press, 1967); Charles Chaplin, *My Autobiography* (New York: Simon and Schuster, 1964); Albert E. Smith, *Two Reels and a Crank* (Garden City, NY: Doubleday, 1952); Anita Loos, *A Girl like I* (New York: Viking, 1966); and Nell Shipman, *The Silent Screen and My Talking Heart: An Autobiography* (Boise, ID: Boise State University, 1988).

## Secondary Sources

This study began when I encountered an intriguing secondary source: Anthony Slide's *Early Women Directors* (New York: A. S. Barnes, 1977). Slim and profusely illustrated, Slide's account of women filmmakers was suggestive rather than conclusive. Slide and Jeffrey Goodman then produced a forty-five-minute documentary entitled *The Silent Feminists: America's First Women Directors* (1993). Any scholar wishing to pursue silent film history should become familiar with Slide's numerous, detailed, and accurate narrative histories of the silent era. Some of the titles used here include *Aspects of American Film History Prior to 1920* (Metuchen, NJ: Scarecrow Press, 1978); *The Big V: A History of the Vitagraph Company*, with Alan Gevinson (Metuchen, NJ: Scarecrow Press, 1987); *Early American Cinema*, with Paul O'Dell (New York: A. S. Barnes, 1970); and *The Griffith Actresses* (South Brunswick: A. S. Barnes, 1973). Slide's *The American Film Industry: A Historical Dictionary* (1986; repr., New York: Limelight, 1990), proved an extremely useful desk reference. In addition, three classic overviews of this period written by participants and eyewitnesses were worthwhile: Lewis Jacobs's *The Rise of the American Film: A Critical History* (1939; repr., New York: Columbia University Press, 1968); Benjamin B. Hampton's *A History of the American Film Industry from Its Beginnings to 1931* (New York: Covici, Friede, 1931; repr., New York: Dover, 1970); and Terry Ramsaye's *A Million and One Nights: A History of the Motion Picture* (New York: Simon and Schuster, 1926).

A comprehensive film history series from the University of California Press appeared in the 1990s, entitled *History of the American Cinema*, edited by Charles Harpole. The first three volumes provide an indispensable reference guide to the silent era: volume 1, Charles Musser, *The Emergence of Cinema: The American Screen to 1907*; volume 2, Eileen Bowser, *The Transformation of Cinema, 1907–1915*; and volume 3, Richard Koszarski, *An Evening's Entertainment: The Age of the Silent Feature Picture, 1915–1928*. One of the newer collections of essays on silent film history is *American Cinema's Transitional Era: Audiences, Institutions, Practices*, ed. Charlie Keil and Shelley Stamp. In particular, I found the expression of a "transitional" period between the rise of the nickelodeon and the rise of the classical film form in 1917 helpful in thinking about my own periodization.

With regard to business histories of the film industry, one of the books I found useful initially was a collection of essays edited by Tino Balio entitled *The American Film Industry*, rev. ed. (Madison: University of Wisconsin Press, 1985). Many of the articles in this volume, such as Robert C. Allen's "The Movies in Vaudeville," are classics, and several essays are actually primary sources, such as "The Motion Picture Industry as a Basis for Bond Financing," authored by Halsey, Stuart, and Co. in 1927. David Puttnam's *Movies and Money* (New York: Knopf, 1998), summarizes the business history and financial history of the film industry, detailing the importance of investment bankers, but has no citations. Another initially indispensable reference was *The Classical Hollywood Cinema: Film Style and Mode of Production to 1960* (New York: Columbia University Press, 1985), authored jointly by David Bordwell, Janet Staiger, and Kristin Thompson. The section authored by Janet Staiger entitled "The Hollywood Mode of Production to 1930," which describes the shifts in the mode of production from the cameraman system to the director-unit system and then to the central-producer system, proved critical to my understanding of studio work during the silent era. *The Classical Hollywood Cinema* also offers very useful appendices that trace the chronology of the major silent film companies as they came and went, merged and split, and transmogrified into different companies with different alliances. Recently, Charles Musser challenged Staiger's periodization in "Pre-Classical American Cinema: Its Changing Modes of Production," in *Silent Film*, ed. Richard Abel (New Brunswick: NJ: Rutgers University Press, 1996). Abel's volume is part of the excellent Rutgers Depth of Field series of edited volumes, which includes *The Studio System*, ed. Janet Staiger (1995), and *Controlling Hollywood: Censorship and Regulation in the Studio Era*, ed. Matthew Bernstein (1999).

Censorship is a primary theme in my analysis, and I found an older volume to provide the best material for this particular study: Ruth A. Inglis's *Freedom of the Movies* (Chicago; University of Chicago Press, 1947; repr., New York: Da Capo, 1974). However, it must be read in tandem with more recent work. Lee Grieveson's *Policing Cinema: Movies and Censorship in Early-Twentieth-Century America* (Berkeley: University of California Press, 2004) is insightful and was instrumental in offering a periodization that corresponded with my own primary research; *Movie Censorship and American Culture*, ed. Francis G. Couvares (Washington, DC: Smithsonian Institution Press, 1996), offers contemporary essays that I found very useful, particularly his own. Many of the films made by women were "social problem films," and three volumes provided particularly helpful contextualization: Kay Sloan, *The*

*Loud Silents: Origins of the Social Problem Film* (Chicago: University of Illinois Press, 1988); Steven J. Ross, *Working-Class Hollywood: Silent Film and the Shaping of Class in America* (Princeton, NJ: Princeton University Press, 1998); and Kevin Brownlow, *Behind the Mask of Innocence: Sex, Violence, Prejudice, Crime: Films of Social Conscience in the Silent Era* (Berkeley: University of California Press, 1990).

Feminist film scholarship is at a particularly cogent moment, poised between the theoretical studies of the past and a new appreciation for history. The place to start is *A Feminist Reader in Early Cinema*, ed. Jennifer M. Bean and Diane Negra (Durham, NC: Duke University Press, 2002). I used several of the essays from this pathbreaking collection. Shelley Stamp's work is among the best for this subject and time period. Her book *Movie-Struck Girls: Women and Motion Picture Culture after the Nickelodeon* (Princeton, NJ: Princeton University Press, 2000) was particularly insightful regarding "playing to the ladies" and understanding the white slavery cycle, and her essays on Lois Weber in *Camera Obscura* and *Looking Past the Screen: Case Studies in American Film History and Method*, ed. Jon Lewis and Eric Smoodin (Durham, NC: Duke University Press, forthcoming) indicate that her forthcoming biography of Weber will be comprehensive. Kathryn Fuller-Seeley's work in women in exhibition and her 1996 book *At the Picture Show: Small-Town Audiences and the Creation of Movie Fan Culture* (Charlottesville: University Press of Virginia) also enhanced my understanding of the relationship between the audience and the industry, as did Miriam Hansen's *Babel and Babylon: Spectatorship in American Silent Film* (Cambridge, MA: Harvard University Press, 1991). Ben Singer's work on serial heroines in *Melodrama and Modernity: Early Sensational Cinema and Its Contexts* (New York: Columbia University Press, 2001) is particularly smart and historically well-informed.

Several biographical encyclopedias include entries on early women filmmakers. Ally Acker's *Reel Women: Pioneers of the Cinema, 1896 to the Present Day*, was published by Chrysalis Books in 1991. A more recent encyclopedia is *The St. James Women Filmmakers Encyclopedia: Women on the Other Side of the Camera*, ed. Amy L. Unterburger (Boston: Visible Ink, 1999). These encyclopedias are useful to those interested in an overview, but most of my information on individual female filmmakers came from primary sources or biographical essays published in older journals, such as *Films in Review*. I did use four biographies rather extensively, however. First, I am indebted to Alison McMahan for *Alice Guy Blaché: Lost Visionary of the Cinema* (New York: Continuum, 2002). During a ten-year period McMahan tracked down 111 extant films by this earliest of female filmmakers and gained access to private collections of remaining family members. Only a devoted multilingual biographer could fill in the holes left by Guy's cryptic memoir, mentions in trade journals, and scattered films, and McMahan's opus solved several mysteries regarding Guy Blaché's life and work. Second, I am indebted to Judith Mayne for her 1994 biography, *Directed by Dorothy Arzner* (Bloomington: Indiana University Press). The very last woman in my study, Arzner is in some ways the most difficult, and Mayne provides just the type of gender analysis required to explain Arzner's exceptionalism. Third, I was grateful for Kay Armatage's *The Girl from God's Country: Nell Shipman and the Silent Cinema* (Toronto: University of Toronto Press, 2003), as Shipman, like Guy Blaché, left a cryptically written autobiography that is more intriguing than comprehensible. Finally, I benefited from Anthony Slide's empirical *Lois Weber: The Director Who Lost Her Way in History* (Westport, CT: Greenwood Press, 1996). As I write, film scholar Shelley Stamp is completing her biography on Lois Weber, which will integrate the perspectives of women's history, cultural history, and film studies. I would also recommend that interested readers pick up Cari Beauchamp's lively history *Without Lying Down: Frances Marion and the Powerful Women of Early Hollywood* (New York: Scribner's, 1997). This is an ideal piece to read in tandem with this study, as my work focuses on the structural changes of the industry, while Beauchamp's offers extensive information on the biographies and personal relationships of filmmakers named here.

At the onset of this project sociology offered the most detailed analyses of the gendering of work, and I found Barbara Drygulski Wright, Myra Marx Ferree, Gail O. Mellow, Linda H. Lewis, Mara-Luz Daza Sampler, Robert Asher, and Kathleen Claspell, eds., *Women, Work, and Technology* (Ann Arbor: University of Michigan Press, 1987) to be an extremely useful starting point. Ava Baron offered a landmark essay incorporating theory into the historical study of gender in the workplace in her introduction to *Work Engendered: Toward a New History of American Labor* (Ithaca, NY: Cornell University Press, 1991). Shortly thereafter Angel Kwolek-Folland's *Engendering Business: Men and Women in the Corporate Office, 1870–1930* (Baltimore: Johns Hopkins University Press, 1994) provided an analysis of women's entry into the masculine space of the office, which was followed by her synthesis *Incorporating Women: A History of Women and Business in the United States* (New York: Twayne, 1998). In this regard the work of Susan Coultrap-McQuinn, *Doing Literary Business: American Women Writers in the Nineteenth Century* (Chapel Hill: University of North Carolina Press, 1990), Wendy Gamber, *The Female Economy: The Millinery and Dressmaking Trades, 1860–1930* (Urbana: University of Illinois Press, 1997), and Kathy Peiss, *Hope in a Jar: The Making of America's Beauty Culture* (New York: Henry Holt, 1998), provided valuable context. For information on the closely allied business of theater I relied heavily on Alfred L. Bernheim, *Business of the Theatre* (New York: Actor's Equity Association, 1932), and Fay E. Dudden, *Women in the American Theatre: Actresses and Audiences, 1790–1870* (New Haven, CT: Yale University Press, 1994).

# Index

feature films, 68; costs of making, 118, 157; focus on stars, 157; independent studios for, 69–72; of literary classics, 88; Motion Picture Patents Company and, 68; movie stars and, 68; strong female characters in, 70; studio competition and, 156–57

feminism, in Guy Blaché's films, 50–51

fiction films. *See* narrative films

film, as technological product, 43–44, 49

Film Booking Offices of America, 122

Film D'Art, 59, 88

film development. *See* film processing

film distribution, 30; block booking, 146, 168; change from purchase to rental, 31–32; control by Motion Picture Patents Company, 46–47; control by studios, 36, 170; exhibition services, 31; by General Film Company, 46–47; independent distributors, 158; road-show method, 68–69; states' rights method, 69, 158, 174; by unlicensed distributors, 47. *See also* movie theaters

film editing: hostility to female editors, 201–2; as men's job, 21–22; women as cutters, 201–2; women's work in, 22, 200–201. *See also* film processing

film exchanges, 31–32

film exhibition: as family businesses, 31–32; of feature films, 68–69; female exhibitors idealized, 82; locales for, 9, 30, 32; mixed film programs, 68; road-shows, 68–69; social context for viewing, 9. *See also* audiences

film factories, 19

film industry: adaptation of theater methods, 27; centralization of, 171; cultural legitimacy as profitmaking strategy, 87; dominated by big business, 195–96, 204; efficiency movement, 182; gender ideology and, 7–8; gendering of, 85–86; gender revolution in, 29–30; innate immorality of, 131; investors' influence on, 7, 8, 134, 154, 174, 186, 193, 194; periods of growth, 5; producer-distributor-exhibitor companies' impact on, 173–74, 193; recession impact on, 193–94; self-censorship wanted for, 78; shift from theater to business paradigm,

202; vertical integration, 7, 134, 155, 170; World War I impact on, 192. *See also* central-producer system; star system; uplift movement

film industry, beginnings: Edison's monopoly attempts, 13; equipment emphasis, 15, 24–25; film factories, 19; filmmaking as manly adventure, 13; masculine engineering environment, 10–11, 15, 25; occupational sex-typing, 24–25; patent disputes, 19; sexual division of labor, 24–25. *See also* cameramen; film processing

filmmaking. *See* film industry; film production

film processing: cinematographer's role in, 19–20, 21; procedures, 19–20, 21–24; specialization in, 157; streamlining of, 157–58; women's roles in, 20, 21, 22–23. *See also* film editing; film industry beginnings

film production: actors' involvement in, 39–40; applying management principles to, 48; collaborative work culture, 39–40, 134, 157; compartmentalization of, 180–82, 195–96; continuity script, 182–83, 195–96; costs of, 118, 156–57, 193, 194; director's role in, 40, 61, 157; director units, 61; investment bankers' involvement in, 194–95; as manly adventure, 13; masculinization of, 180–81, 193, 203; postwar resurgence of women producers, 149–50; regimentation of, 195–96; sex-typing of crafts in, 196; transitions in, 60. *See also* central-producer system; independent studios; producers; star producers; studios; *names of specific producers and directors*

film projection, 33–34. *See also* nickelodeons

film quality, brand-name identity and, 48

film studios. *See* independent studios; studios

Film Supply Company of America, 75

Finch, Flora, 122

Fine Arts Film Corporation, 165

First National, 166–71, 172, 173, 200; Corinne Griffiths Productions and, 171–72

First National (DeHaven studio), 121

Hackett, James J., 67
Haggen, Mrs. S. M., 86
Hall, Ella, 180, 188
Hallroom Boys Photoplays, 129
Hampton, B. B., 13, 156
*Hands Up* (serial films), 124–25
*The Hand That Rocks the Cradle* (film), 139–40
Hansen, Juanita, 131
Harris, Mildred, 141
Harrison, Louis Reeves, 69, 85, 89, 187
Hart, William S., 168
Haskins, Grace, 179, 180, 192
Hatch, Mrs. Alonzo, 32
Hayakawa, Sessue, 160
Hayes, Will, 132, 150
Hays Office, 132, 149
*The Hazards of Helen* (serial films), 110, 113–14, 119, 120
Hearst, William Randolph, 172
*The Heart of a Painted Woman* (film), 75, 89
Helen Gardner Picture Players, 69
Helen Gibson Productions, 120–21
Helen Holmes Production Corporation, 120
Helen Keller Film Corporation, 161–62
Henry, William H., 189
Hepworth Manufacturing Company, 65
Herald Square Theater (New York), 57
Heron, Nan, 201, 205
Higashi, Sumiko, 147
*A Hitherto Unrelated Incident of the Girl Spy* (film), 105
Hodkinson, W. W., 156, 160
Holliday, Wendy, 41
Hollywood: innate immorality of, 131; masculinzation of, 143; scandals, 131, 150, 172
Holmes, Helen, 4, 110, 113–14, 115, 118, 120; independent company ownership, 119, 120
Holt, Jack, 180
*Home* (film), 142
Hoover, Herbert, 192
*Hop, The Devil's Brew* (film), 97
Horkheimer, H. M., 67
Hosmer, Mrs. Eli T., 130
Houdini, 131
*The House of Bondage* (film), 86

*The House with the Closed Shutters* (film), 50, 62
Howell, Alice, 122
*How She Triumphed* (film), 105
*Human Wreckage*, 150–51, 152
humor, middle class attitudes toward, 111
*Hush* (film), 164
Hutchinson, Charles, 131
Hutchinson, Samuel S., 119
Hyland, Peggy, 160
*Hypocrites* (film), 92–97, 136, 140, 187

*I Am the Woman* (film), 126
*Idle Wives* (film), 97
Ince, Elinor Kershaw, 150
Ince, Thomas, 122, 150, 151–52, 169, 172
Independent Moving Picture Company (IMP), 13, 40, 53; antitrust suit against Motion Picture Patents Company, 61–62
independent studios, 62, 63, 65–68, 154, 166, 194; defections from established studios and, 67–68, 69, 157–58; demise of, 203; for feature films, 69–72; female stars' vulnerability to manipulation, 74–76; financial investment in, 159, 160, 161, 166; headed by women, 62, 63, 65–66, 126, 154–55, 158–60, 159, 160, 175, 176; for highbrow comedy, 119; limiting of creative freedom by, 172; for lowbrow comedy, 122; male-female partnerships, 72–76, 203; of movie stars, 62, 63, 65–68, 158–59, 160; picture palaces and, 174; quasi-independent companies, 160, 170; for serial films, 124–26; stimulated by central-producer system, 158. *See also* studios; *names of independent studios*
Ingleton, E. Magnus, 183
Ingram, Rex, 200
*The Inside of the White Slavery Traffic* (film), 86–87
*In Swift Waters* (film), 64
International Alliance of Theatrical Stage Employees and Moving Picture Operators, 34
*In the Power of the Ku Klux Klan* (film), 71
*In the Watches of the Night* (film), 70
*In the Year 2000* (film), 51